An Atlas of Spectral Interferences in ICP Spectroscopy

An Atlas of Spectral Interferences in ICP Spectroscopy

M. L. Parsons
Alan Forster
and
Donn Anderson
Arizona State University
Tempe, Arizona

c.2

PLENUM PRESS • NEW YORK AND LONDON

Library of Congress Cataloging in Publication Data

Parons, Michael Loewen, 1940-
An atlas of spectral interferences in ICP spectroscopy.

Includes index.
1. Atomic spectra–Tables, etc. I. Forster, Alan, joint author. II. Anderson, Donn, joint author. III. Title.

QD96.A8P37 543′.08 79-24222
ISBN 0-306-40334-X

A Division of Plenum Publishing Corporation
227 West 17th Street, New York, N.Y. 10011

Printed in the United States of America

Acknowledgments

We wish to express our gratitude to the following: Professor F. M. Phelps III, of Central Michigan University, for making the MIT Tables available; the Arizona State University Department of Chemistry, for support in acquiring the invaluable resource of the MIT Tables on computer tape; and the Arizona State University Computer Center, for computing time.

M.L.P.
A.F.
D.A.

Contents

Introduction

With the advent of the highly energetic inductively coupled plasma (ICP) excitation source for atomic emission analysis, several old and some new and quite different spectral problems are being encountered. This is in part due to the lack of thermodynamic equilibrium in the ICP. In particular, since the ICP generates abnormally high ionization, many more transitions from ionic species are observed. In fact, for a large number of elements, the most intense spectra originate from ions rather than atoms. This results in two new problems for the analytical spectroscopist. First, new analysis lines must often be utilized. Second, spectral interferences arising from ionic spectra become much more important.

These problems may seem rather unimportant until the magnitude of total spectral emission from an ICP is considered. In the MIT Tables[1] about 80,000 (mostly atomic) transitions are tabulated as originating from the DC arc and about 50,000 (mostly ionic) transitions from spark excitation. The ICP emits both arc and spark spectra, and there will probably be even more ionic transitions observed than are listed in either the DC arc or the spark tables. We estimate that at least 120,000 transitions are excitable in the ICP.

Because of the different and expanded spectra from ICP emission, the current tabulations have been put forth. Unfortunately, there is no experimental atlas of ICP transitions. Further, there are no "standard" experimental methods in ICP analyses. The only published information to date consists of "prominent" lines by Fassel.[2] We have, therefore, had to rely on data compiled from other sources, namely arc, spark, and other types of discharges. In fact, these tables consist of data from all sources available. *Because of this, intensity data cannot be considered reliable.* Any transition should be considered potentially intense in the ICP and warrants investigation.

This work was made possible mainly through the acquisition of the MIT Wavelength Tables on computer tape from Professor F. M. Phelps III of Central Michigan University. We have updated the tables with the addition of argon lines from NBS compilations[3] and signal-to-background data from Fassel's compilation of prominent ICP lines.[2] Furthermore, we have extended the tables into the UV to 1850 angstroms using both Fassel[2] and the NRL compilations.[4] A search program based on Boolean strategies similar to the one developed in our laboratory for the lunar data base[5] has been developed to generate the main body of this tabulation.

The book is divided into four tables. Table I is a list of most commonly reported atomic and ionic lines used for elemental analysis. Table II is subdivided: for each analytical line denoted by an asterisk in Table I, the first part of Table II consists of all transitions within ± 1 angstrom of the analytical line; the second part consists of all prominent lines within ± 5 angstroms of the analytical line. Table III is a compilation of lines below 2000 angstroms. Table IV contains all the argon transitions, for quick reference. A concise explanation of its format and use immediately precedes each table.

References

1. G.R. Harrison, *Massachusetts Institute of Technology Wavelength Tables* (John Wiley & Sons, New York, 1969).
2. R.K. Winge, V.J. Peterson, and V.A. Fassel, *Applied Spectroscopy,* **33,** 3, pp. 206–219 (1979).
3. National Bureau of Standards, *Atomic Transition Probabilities,* NSRDS-NBS, **22,** II, pp. 187–210 (1969).
4. R.L. Kelley and L.J. Palumbo, *Atomic and Ionic Emission Lines below 2000 Angstroms—Hydrogen through Krypton,* NRL Report 7599 (Naval Research Laboratory, Washington, D.C., 1973).
5. D.L. Anderson, C.B. Moore, M.L. Parsons, and D.D. Pratt, *Computers & Geosciences,* **4,** pp. 162–172 (1978).

Explanation of Table I

Table I represents an evaluation of experimental ICP analysis lines reported in the literature. A conscientious effort has been made to tabulate most of the analysis lines found in the current literature. It is hoped that all of the genuinely useful lines have been included.

The table is arranged alphabetically according to the atomic symbols in column one. A "I" or "II" following the symbol designates a transition originating from an atomic or ionic species, respectively. An asterisk (*) preceding the atomic symbol indicates an expansion about that line in Table II, while the symbol "#" signifies that the line appears in Table III.

Column two contains the wavelength of the analytical transition in angstroms. In most cases, six or seven significant figures are included, depending upon the current knowledge of wavelengths measured in air.

Limits of detection reported in the literature are included in the next two columns. A value found in the third column indicates that the line was reported by Fassel[25] in a self-consistent study of numerous elements with a single experimental ICP apparatus. A value in column four represents the best reported limit of detection in the literature other than Fassel's. If a limit of detection is reported in column four, the reference for that value is given in column five. The units for both limit of detection columns are nanograms per milliliter (ng/ml), or, for convenience, parts per billion (ppb).

References

1. P.W.J.M. Boumans and R. Barnes, *ICP Information Newsletter,* **3,** 11, pp. 445–448 (1978).
2. R.H. Scott, A. Strasheim, and A.R. Oakes, *ICP Information Newsletter,* **3,** 11, pp. 448–457 (1978).
3. S. Greenfield, H.McD. McGeachin, and P.B. Smith, *Talanta,* **23,** 1, pp. 1–14 (1976).
4. P.W.J.M. Boumans and F.J. DeBoer, *ICP Information Newsletter,* **1,** 5, pp. 100–103 (1975).
5. F. Breck and R.L. Crawford, *ICP Information Newsletter,* **1,** 1, pp. 9–11 (1975).
6. R.H. Crawford, *ICP Information Newsletter,* **3,** 3, p. 116 (1977).
7. M.B. Denton and D.L. Windson, *ICP Information Newsletter,* **1,** 11, pp. 263–264 (1976).
8. R. Barnes, *ICP Information Newsletter,* **1,** 8, pp. 164–167 (1976).
9. A.E. Watson, G.M. Russell, and G. Balaes, *ICP Information Newsletter,* **2,** 7, pp. 205–220 (1976).
10. J.A.C. Broekaert, F. Leis, and K. Laqua, *ICP Information Newsletter,* **3,** 9, pp. 381–387 (1978).

11. P.W.J.M. Boumans, L.C. Bastings, F.J. DeBoer, and L.W.J. Van Lollenburg, *ICP Information Newsletter,* **3,** 5, pp. 201–207 (1977).
12. P.W.J.M. Boumans, A.W. Witmer, F.J. DeBoer, and M. Bosueld, *ICP Information Newsletter,* **3,** 6, pp. 213–218 (1977).
13. C. Allemand and G. Benzie, *ICP Information Newsletter,* **1,** 12, pp. 273–277 (1976).
14. A. Ward, *ICP Information Newsletter,* **3,** 10, pp. 434–435 (1978).
15. S. Greenfield and H.McD. McGeachin, *ICP Information Newsletter,* **2,** 6, pp. 167–173 (1976).
16. L.A. Stelmack, *ICP Information Newsletter,* **1,** 11, pp. 253–257 (1976).
17. G.R. Garbarino and H.E. Taylor, *ICP Information Newsletter* **3,** 10, pp. 423–424 (1978).
18. P.W.J.M. Boumans, F.J. DeBoer, J. Dahmen, H. Holzel, and A. Meier, *ICP Information Newsletter,* **1,** 5, pp. 111–113 (1975).
19. P.W.J.M. Boumans and F.J. DeBoer, *Spectrochimica Acta,* **27B,** 9, pp. 391–414 (1972).
20. R.H. Scott and M.L. Kokot, *ICP Information Newsletter,* **1,** 1, pp. 22–23 (1975).
21. G. Horlick, S. Merali, and R.C.L. Ng, *ICP Information Newsletter,* **3,** 11, pp. 460–461 (1978).
22. A.E. Watson and G.M. Russel, *ICP Information Newsletter,* **3,** 7, pp. 273–284 (1977).
23. R.K. Winge and V.A. Fassel, *ICP Information Newsletter,* **3,** 10, pp. 420–421 (1978).
24. R.H. Scott, A. Strasheim, and M.L. Kokot, *ICP Information Newsletter,* **1,** 12, pp. 282–283 (1976).
25. R.K. Winge, V.J. Peterson, and V.A. Fassel, *Applied Spectroscopy,* **33,** 3, pp. 206–219 (1979).

Table I
Analytical Lines

	Element	Wavelength (Angstrom)	Detection Limit (25)	(ng/ml)	Ref
*	AG I	3280.683	7	2	1
*	AG I	3382.891	13		
*	AG II	2437.791	120		
	AG II	2246.41	130		
	AG II	2413.188	200		
*	AL I	3961.527	28	0.2	1
*	AL I	3082.155	45	0.4	1
*	AL I	3944.032	47	20	2
*	AL I	3092.713	23		
*	AL I	3092.842	23		
*	AL I	2373.132	30		
#	AS I	1936.96	53	2	1
*	AS I	2288.12	83	6	1
#	AS I	1971.97	76		
	AS I	2349.84	142	90	3
	AS I	2003.34	120		
#	AS I	1890.42	136		
*	AU I	2675.95	31	0.9	1
*	AU I	2427.95	17		
#	AU I	1978.19	38		
*	AU II	2082.09	42		
	AU I	2012.00	55		
	AU II	2110.68	63		
*	B I	2497.733	4.8	0.1	1
*	B I	2496.778	5.7	5	3
*	B I	2089.59	10		
*	B I	2088.93	12		
*	BA II	4554.042	1.3	0.01	1
*	BA I	5535.551		2	4
	BA II	4934.086	2.3		
	BA II	4130.664	32		
*	BA II	2335.269	4.0	6	5
	BA II	2304.235	4.1		
*	BE I	2348.610	0.31	0.003	1

Table I
Analytical Lines
(con't)

	Element	Wavelength (Angstrom)	Detection (25)	Limit (ng/ml)	Ref
*	BE II	3130.416	0.27		
*	BE II	3131.072	0.73	0.3	6
	BE I	2494.733	3.8		
	BE I	2650.470	4.7		
*	BI I	2897.975	330	10	1
*	BI I	2230.608	34		
	BI I	3067.716	75		
	BI I	2228.251	83		
	BI I	2061.70	85		
#	BI II	1902.41	300		
*	BR II	4704.86			7
#	C I	1930.91	44		
	C I	1933.62	8800	100	1
	C I	2478.573	176		
*	CA II	3933.666	0.19	0.0001	1
*	CA II	3968.468	0.5	0.002	3
*	CA I	3644.410		0.5	8
	CA II	3179.332	10	50	9
	CA I	4226.728	10		
*	CD II	2265.017	3.4	0.07	1
*	CD I	2288.018	2.7	0.2	1
*	CD II	2144.382	2.5		
	CD I	3261.057	330	3	3
	CD I	3610.510	230		
*	CE II	4186.599	52	0.4	1
*	CE II	4012.388	75	30	10
*	CE II	4137.646	48		
	CE II	4133.800	50		
	CE II	3931.088	60		
	CE II	3942.746	68		
*	CL II	4132.48			7

Table I
Analytical Lines
(con't)

Element	Wavelength (Angstrom)	Detection Limit (25)	(ng/ml)	Ref
* CO II	2388.915	6	0.1	1
* CO II	2389.565		0.4	11
* CO I	2378.622	9.7	0.8	11
CO I	3453.505		3	3
CO I	3502.279		7	11,12
CO I	3405.120		7	11
CO II	2286.156	7		
CO II	2378.622	9.7		
CO II	2307.857	9.7		
CO II	2363.787	11		
CO II	2311.604	13		
CO II	2383.46	14		
CO II	2314.054	16		
CO II	2353.42	17		
CO II	2386.363	21		
CO II	2344.260	21		
CO II	2314.98	23		
CO II	2347.39	23		
* CR II	2677.159	7.1	0.08	1
* CR II	3578.687	2[illegible]	1	11
* CR II	2055.52	6.1		
CR II	2061.49	7.1		
CR II	2835.633	7.1	10	9
CR II	2843.252	8.6		
CR II	2065.42	9.7		
* CR I	4254.346		20	2
CS II	4526.725	43000		
* CU I	3273.962	9.7	0.06	1
* CU I	3247.54	5.4	1.9	6
* CU II	2246.995	7.7		
CU I	2199.583	9.7		
CU II	2135.976	12		
* DY II	3531.712	10	4	1
* DY II	4000.454	35	9	10
DY II	3645.416	23		
DY II	3407.80	27		
DY II	3536.024	30		
DY II	3944.692	31		

Table I
Analytical Lines
(con't)

	Element	Wavelength (Angstrom)	Detection (25)	Limit (ng/ml)	Ref
*	ER II	4007.967		1	1
	ER II	3692.652	18	10	10
*	ER II	3372.750	10		
	ER II	3499.104	17		
	ER II	3230.585	18		
	ER II	3264.781	18		
*	EU II	3819.66	2.7	0.06	1
*	EU II	4205.046	4.3	3	10
	EU II	4129.737	4.3		
	EU II	3930.48	5.7		
	EU II	3907.110	7.7		
*	FE II	2599.396	6.2	0.09	1
*	FE I	3719.935		0.3	3
*	FE II	2611.872	12	0.5	1
*	FE II	2382.039	4.6		
	FE II	2395.625	5.1		
	FE II	2631.051		10	13
	FE II	2343.492	10		
	FE II	2404.882	11		
	FE II	2598.369	12		
*	GA I	4172.056	66	0.6	1
*	GA I	2943.637	46		
	GA I	2874.244	78		
	GA I	4032.982	111		
	GA I	2500.173	187		
*	GD II	3422.466	14	0.4	1
*	GD II	3646.196	30	10	10
	GD II	3362.244	20		
	GD II	3350.482	21		
	GD II	3358.628	21		
	GD II	3100.508	23		
*	GE I	2651.575	83	0.5	1
	GE I	2651.178	48	4	19
*	GE I	2094.23	40		
#	GE I	1998.24	88	50	14
	GE I	2068.65	60		

Table I
Analytical Lines
(con't)

Element	Wavelength (Angstrom)	Detection (25)	Limit (ng/ml)	Ref
GE I	2198.701	63		
* H I	6562.849			15
H I	4861.327			15
* HF II	3399.795		10	1
* HF II	2773.360	15		
HF II	2738.760	16		
HF II	2641.410	18		
HF II	2322.47	18		
HF II	2638.712	18		
HG I	1849.6		1	1
* HG I	2536.519	61	1	3
# HG II	1942.27	25		
* HO II	3456.00	5.7	3	1
* HO II	3398.98	13		
I I	2061.6		10	1
I	1830			16
* IN I	4511.323	187	30	1
* IN II	2306.062	63		
IN I	3256.090	120		
IN I	3039.356	150		
* IR II	2242.68	27		
* IR II	2126.81	30		
* IR I	2052.22	61		
IR II	2152.68	68		
IR I	3221.281		70	1
* K I	7664.907		30	1
* K I	4047.201	43000		17
* LA II	4086.714	10	0.1	1

Table I
Analytical Lines
(con't)

Element	Wavelength (Angstrom)	Detection (25)	Limit (ng/ml)	Ref
LA II	3988.52	11	6	10
LA II	3794.773	10		
LA II	3337.488	10		
LA II	4123.228	10		
LA II	2379.42		10	12
LA I	5211.870		80	12
* LI I	6707.844		0.02	1
* LI I	4602.863	857		
LI I	6103.642			18
* LU II	2615.42	0.10		
* LU II	2911.39	0.62		
LU II	2195.54	0.83		
* LU I	4518.57		8	1
LU II	3472.48		10	3
* MG II	2795.53	0.15	0.003	1
* MG II	2802.695	0.30	0.01	3
* MG I	2852.129	1.6	0.2	3
* MN II	2576.104	1.4	0.01	1
* MN I	4030.755	44	0.6	3
* MN II	2593.729	1.6		
MN II	2605.688	2.1		
MN II	2933.063	13		
MN I	2801.064	21		
* MO I	3798.252		0.2	19
* MO I	3864.110		0.3	1
* MO II	2020.30	7.9		
* MO II	2038.44	12	34	5
MO II	2816.154	14		
MO II	2775.400	25		
* NA I	5889.953	29	0.1	1
* NA I	5895.923	69		
* NB II	3094.183	36	0.2	1

Table I
Analytical Lines
(con't)

Element	Wavelength (Angstrom)	Detection (25)	Limit (ng/ml)	Ref
* NB II	3163.402	40		
* NB I	3130.786	50		
NB II	2697.06	69		
* ND II	4012.250	50	0.3	1
* ND II	4061.085	96	10	14
ND II	4177.321	136	40	10
ND II	4303.575	75		
* NI I	3524.541	45	0.2	1
* NI I	3414.765	48	1	1
* NI II	2316.037	15	2.7	6
NI I	3515.054		10	20
NI II	2216.473	10		
NI I	2320.03	15		
NI II	2394.516	38	300	9
* OS II	2255.847	0.36		
* OS II	2282.26	0.63		
# OS	1899.00	1.2		
OS II	2336.80	1.2		
* OS I	2909.061		6	1
* P I	2535.65	272	15	1
* P I	2136.199	76		21
* P I	2149.108	76		21
P I	2152.950	882		21
P I	2554.93			7
* PB II	2203.505	42	1	1
* PB I	2833.069	142	2	1
* PB I	4057.820	272	2	3
PB I	2169.99	90		
* PD I	3634.695	54	2	3
* PD I	3609.548	85	2	1
* PD II	2488.921	103	2	1
PD I	3404.580	44		
PD II	2296.513	68		

Table I
Analytical Lines
(con't)

Element	Wavelength (Angstrom)	Detection Limit (25)	(ng/ml)	Ref
* PR II	4225.327	42	10	1
* PR II	3908.431	37	30	3
PR II	4179.422	41	30	14
* PR II	4143.136	37		
* PT II	2144.231	30		
* PT I	2659.454	81	0.9	1
* PT II	2036.463	55		
PT I	2049.373	71		
RB I	4201.851	38000		
# RE II	1973.13	6		
* RE II	2214.261	6		
* RE II	2275.25	6		
RE II	2092.41	93	25	1
* RH I	3434.893	60	30	14
* RH II	2334.77	44		
RH II	2490.770	57		
RH II	2520.533	76		22
* RU II	2402.717	30		
* RU II	2456.568	30		
RU II	2678.758	36		
* RU I	3798.052		60	1
RU II	2692.065	90		
# S I	1820.3		30	1
S I	2168.9			7
* SB I	2175.81	44	15	1
* SB I	2068.38	32	10	6
SB I	2311.469	61		
SB I	2528.535	107		
SB I	2598.062	107	200	3
* SC II	3613.836	15	0.4	1

Table I
Analytical Lines
(con't)

Element	Wavelength (Angstrom)	Detection (25)	Limit (ng/ml)	Ref
* SC II	3572.523	20		
SC II	3630.740	21		
SC II	3642.785	27		
SC II	4246.829	27		
# SE I	1960.26	75	1	1
* SE I	2039.851	115		
SE I	2062.786	300		
* SI I	2516.123	12	2	1
* SI I	2881.595	27	10	14
* SI I	2124.12	16		
* SM II	3592.595	43	1	11
* SM II	3568.258	76	2	11
SM II	3885.284	83	2	11
SM II	3661.351		2	11
SM II	3739.199		2	11
SM II	4424.342	54	10	14
SM II	3928.272		10	10
* SN I	3034.121		3	1
# SN II	1899.89	25	6	1
SN I	2839.989	111	10	1
SN I	2354.845	96		
SN I	2429.495	96		
SN I	3175.05	214	500	9
* SR II	4077.714	0.42	0.002	3
* SR II	4215.524	0.77	0.2	14
SR II	4607.331	68	0.4	4
SR II	2165.96	8.3		
* TA II	2965.133	60	5	1
* TA II	2262.30	25		
TA II	2400.63	28	50	1
TA	3012.527		70	3
* TB II	3509.17	23	0.1	1

Table I
Analytical Lines
(con't)

Element	Wavelength (Angstrom)	Detection Limit (25)	(ng/ml)	Ref
* TB II	3848.73	55		
* TE I	2385.78	176	15	1
* TE I	2142.81	41	20	14
TE I	2259.02	176		
* TH II	4019.137	83	3	1
* TH II	2837.299	65		
TH II	2832.319	71		
TH II	2747.161	83		
* TI II	3349.406	3.8	0.03	1
* TI II	3349.035	7.5	0.2	3
* TI II	3361.213	5.3		
TI II	3234.516	5.4		
TI II	3372.800	6.7		
TI II	3347.11		12	5
# TL II	1908.64	40		
* TL I	3775.72	230	75	1
TL I	2767.87	120		
TL I	3519.24	200		
TL II	1988.6			23
* TM II	3462.20	8.1	0.15	1
* TM II	3131.26	5.2		
TM II	3848.018	9.7		
TM II	3761.333	11	10	10
TM II	3425.08	10		
* U II	3859.580	250	1.5	1
* U II	3670.072	300		
U II	3782.841		300	24
* V	3710.3		0.06	4
* V II	3093.108	5.0	0.06	1
* V II	3110.706	10	0.06	1
* V I	4379.238		0.2	19
V II	3102.299	6.4		

Table I
Analytical Lines
(con't)

Element	Wavelength (Angstrom)	Detection (25)	Limit (ng/ml)	Ref
V II	2924.025	7.5		
V II	2908.817	8.8		
V II	2893.320	10		
V II	2687.956	10		
V II	2924.644	16		
* W II	2764.266		0.8	1
* W I	4008.753		3	4
* W II	2079.106	30		
W II	2248.750	44		
* Y II	3710.290	3.5	0.04	1
* Y II	4177.552	11	7.4	5
* YB II	3694.203	3.0	0.02	1
* YB II	3289.37	1.8		
* ZN I	2138.56	1.8	0.1	1
* ZN II	2025.51	4.0		
ZN II	2061.91	5.9		
* ZR II	3438.230	7.1	0.06	1
* ZR II	3391.975	7.7		
ZR II	2571.391	9.7		
ZR II	3496.210	10		

Explanation of Table II

The analytical line to be expanded is listed in the heading at the top of each page and is labeled with an asterisk (*) preceding the appropriate element symbol in column one. For each line, Table IIa lists all lines found within ±1 angstrom and Table IIb all prominent lines found within ±5 angstroms.

In both parts of the table, column one contains an atomic symbol or molecular formula. The state of ionization or indication of a molecular compound is found to the right of the symbol. A "I" indicates transitions originating from an atom, a "II" represents those from a singly ionized atom, a "3" denotes doubly ionized atoms, a "C" represents molecular emission, and a "B" indicates a band spectrum. Column two contains the air wavelength of the element denoted in column one as it appears in our expanded MIT Tables. (Due to the computer generation method of Tables IIa and IIb, seven digits are always printed although only five or six may be significant.)

Column three contains relative intensity data from DC arc emission, while column four contains relative intensity data from high-voltage spark emission. These come from the original MIT Tables only.[1] Column five contains relative intensity data from one of two plasma sources. If no reference appears to the right of an entry in column five, the entry comes from the discharge tube column in the MIT Tables. If the initials VAF appear to the right of an entry, it comes from the previously cited works by Fassel.[2] The intensities in all three columns are reported in a linear range to two significant figures at best.

Caution should be exercised in using the intensity data. The values in the arc and spark columns have no real relationship to ICP emission intensity as they were measured from DC arc and high-voltage spark discharges rather than an ICP. It is recommended that they be used as rough guides only, as it is quite possible that the arc data will be overstated and the spark data understated with respect to the ICP. Furthermore, the data from Fassel[2] are in terms of signal-to-background intensity and not in relative intensity terms. Thus, the plasma intensity column cannot be related to the arc and spark columns. Unfortunately, relative intensity data are not yet readily available for the ICP. The data presented here are the best currently available.

Table IIa contains all lines found in the expanded MIT Tables within ±1 angstrom of the analytical line of interest. The ranges of ±0.25, ±0.50, and ±0.75 angstrom are designated by double, triple, and quadruple dashes in the left margin.

In Table IIb, lines meeting the proper criteria are included within ±5 angstroms of the analytical line of interest. In all cases, lines originating in the previously mentioned reference[2] or in NRL[3] and NBS[4] compilations are included. Normal MIT lines have been selected by the values of their arc or spark intensity. If a "(1)" is present in the table heading, any line having an arc intensity greater than or equal to 10% of the analytical line's arc intensity or a spark intensity greater than or equal to 5% of the analytical line's spark intensity will be included. A "(2)" denotes limits of 50% and 25% for arc and spark, respectively, while a "(3)" represents 100% and 50% limits, respectively. This algorithm was adopted in order to maintain a reasonable length for most of the tables. A "(3)" in the heading most probably indicates a worse spectral interference problem than a "(1)."

References

1. G.R. Harrison, *Massachusetts Institute of Technology Wavelength Tables* (John Wiley & Sons, New York, 1969).
2. R.K. Winge, V.J. Peterson, and V.A. Fassel, *Applied Spectroscopy,* **33,** 3, pp. 206–219, (1979).
3. R.L. Kelley and L.J. Palumbo, *Atomic and Ionic Emission Lines below 2000 Angstroms—Hydrogen through Krypton,* NRL Report 7599 (Naval Research Laboratory, Washington, D.C., 1973).
4. National Bureau of Standards, *Atomic Transition Probabilities,* NSRDS–NBS, **22,** II, pp. 187–210 (1969).

Table IIa. AG I 3280.683 A

	Element	Wavelength (Angstrom)	Intensity Arc	Spark	Plasma
	FE	3279.739	6	3	
	CU I	3279.816	25	32	
	NB	3279.826	4		
	CE I	3279.842	32	5	
	CE II	3279.842	32	5	
	V	3279.845	20	8	
	PB II	3279.850		10	
	GD	3279.923	2		
---	NB	3279.974		5	
	HF II	3279.980	25	25	
	YB	3279.980	5		
	TI II	3279.995	10	40	
	U	3280.004	8	8	
	DY	3280.100	63	5	
---	P	3280.200			32
	ND	3280.202	2	4	
	SM	3280.218	4		
	ER	3280.220	13	3	
	FE	3280.261	160	160	
	TB	3280.280	32	16	
	MO	3280.318	5	5	
	TH	3280.374	10	13	
	TI I	3280.391	3		
	U	3280.399	5		
--	XE	3280.480			4
	CE II	3280.485	16		
	LU	3280.500	10		
	RH I	3280.550	32	10	
	CL	3280.600			8
	U	3280.608		2	
	CE	3280.668	6		
	PD	3280.678		2	
	CO	3280.681	2		
	*AG I	3280.683	2000	1000	38 VAF
	IN	3280.692		3	
	TH	3280.736	2	2	
	ZR II	3280.748	3	2	
	MN	3280.756	63	32	
	SM	3280.842	20	6	
	TA	3280.872	3	2	

Table IIa. AG I 3280.683 A
(con't)

	Element	Wavelength (Angstrom)	Intensity Arc	Spark	Plasma
	MO	3280.877		25	
	Y II	3280.913	8	13	
--	OS	3280.915	5		
	TH	3281.034	2	5	
	MO	3281.068	25	3	
	CE	3281.095	20		
	U	3281.115	2		
	V	3281.115	3	50	
---	HO	3281.170		4	
	XE II	3281.260			8
	TH	3281.283	4	5	
	FE II	3281.300	16	100	
	MO	3281.345	5		
	TB	3281.400	50	16	
----	TH	3281.415	8	10	
	CA	3281.480		4	
	ND	3281.487	4	6	
	RB II	3281.490			20
	BA I	3281.500	25		
	U	3281.549	5		
	CE	3281.587	3		
	CO I	3281.588	6		
	GD	3281.611	5	5	
	MO	3281.617		20	
	SI	3281.640		3	
	U	3281.652	2		

Table IIb.(1) AG I 3280.683 A

Element	Wavelength (Angstrom)	Intensity Arc	Spark	Plasma
YB	3275.810	13	100	
V II	3276.124	50	200	19 VAF
PB	3276.190		63	
FE I	3276.468	100	50	
TI II	3276.774	13	63	
FE II	3277.346	40	200	

Table IIb.(1) AG I 3280.683 A
(con't)

Element	Wavelength (Angstrom)	Intensity Arc	Spark	Plasma
TI II	3278.290	25	100	
FE	3278.734	100	63	
MO	3278.881		50	
TI II	3278.922	40	160	
ZR II	3279.265	50	50	21 VAF
FE	3280.261	160	160	
*AG I	3280.683	2000	1000	38 VAF
V	3281.115	3	50	
FE II	3281.300	16	100	
TI II	3282.329	32	160	
ZN I	3282.333	500	320	1 VAF
V	3282.533	13	79	
FE	3282.892	79	79	
NB	3283.463	2	100	
SN II	3283.510		100	
FE I	3284.588	200	130	

Table IIa.   AG I   3382.891 A

	Element	Wavelength (Angstrom)	Intensity Arc	Spark	Plasma
	U	3381.950	4	6	
	AU II	3382.000	6	8	
	ER	3382.067	16		
	CR	3382.080	32		
	ND	3382.093	4		
	W	3382.096	8	6	
----	BI	3382.280	2		
	MO	3382.292	10	6	
	U	3382.304	3		
	CE	3382.308	8		
	TI I	3382.312	32	6	
---	NB	3382.407	5	40	
	SM II	3382.407	100	40	
	FE I	3382.409	50	10	
	EU	3382.412	2		
	MO	3382.484	16	16	
	CE	3382.512	8		
	V	3382.530		130	
	PD II	3382.566		2	
	W	3382.606	10	13	
--	PR	3382.660	10	2	
	U	3382.675	4	4	
	O I	3382.680			13
	CR II	3382.683	32	200	
	CE	3382.698	8		
	TB	3382.800	16	8	
	Y	3382.830	3	3	
	CE	3382.888	8		
	*AG I	3382.891	1000	630	23 VAF
	ER	3382.892	20		
	ZR	3382.900	3		
	Y	3383.050	3	3	
	W	3383.120		20	
	TH	3383.120	5	10	
--	SB	3383.137	40	50	
	CE	3383.279	8		
---	PR	3383.376	16	3	
	CE	3383.394	13		
	U	3383.402	6	3	
	ND	3383.449	2		

Table IIa. AG I 3382.891 A
(con't)

Element	Wavelength (Angstrom)	Intensity Arc	Spark	Plasma
-- MO	3383.554	5		
CE	3383.687	20	2	
FE I	3383.698	100	63	
PR	3383.730	13	3	
TI II	3383.761	63	320	
ER	3383.780	13		
TB	3383.780	8		
NB	3383.800	16	5	
PT	3383.820		8	
I II	3383.857			3

Table IIb.(1) AG I 3382.891 A

Element	Wavelength (Angstrom)	Intensity Arc	Spark	Plasma
CR II	3378.337	25	160	
FE	3378.685	160	79	
FE I	3379.022	79	50	
NB	3379.300		100	
CR II	3379.371	6	100	
TA	3379.515	20	50	
CR I	3379.825	16	100	
CR II	3379.825	16	100	
FE	3380.111	200	25	
TI II	3380.280	25	160	
NI I	3380.574	630	100	
PD I	3380.674	160	2	
SR II	3380.711	160	200	9 VAF
NI I	3380.885	200	13	
LA II	3380.910	200	100	17 VAF
NB	3380.938		200	
CO I	3381.498	100		
NB	3382.407	5	40	
SM II	3382.407	100	40	
V	3382.530		130	
CR II	3382.683	32	200	
*AG I	3382.891	1000	630	23 VAF

Table IIb.(1) AG I 3382.891 A
(con't)

Element	Wavelength (Angstrom)	Intensity Arc	Spark	Plasma
SB	3383.137	40	50	
FE I	3383.698	100	63	
TI II	3383.761	63	320	
FE I	3383.980	200	100	
MO	3383.981		32	
V I	3384.599	63	40	
DY II	3385.027	25	8	9 VAF
ER II	3385.087	25	16	9 VAF
RU	3385.144	63	32	
CO I	3385.224	250	16	
NB	3386.244	5	100	
OS I	3387.836	100	16	
TI	3387.837	63	130	
ZR II	3387.872	100	100	

Table IIa. AG II 2437.791 A

	Element	Wavelength (Angstrom)	Intensity Arc	Spark	Plasma
	W	2436.835	6		
	SN	2436.884		2	
	RU	2436.923	25		
	DY	2436.930	3		
	B	2436.950		2	
	CO II	2436.979	2	20	
---	HO	2437.070		10	
	TA	2437.070	20		
	PR	2437.080		4	
	BH B	2437.100	250		
	CS	2437.100			2
	LA II	2437.130		10	
	FE II	2437.154	2	2	
	NB	2437.166	5		
	AS I	2437.230	25	2	
	FE II	2437.267			
---	W	2437.339		6	
	NB	2437.415	4	100	
	MN	2437.424		40	
	W	2437.484		10	
--	TH	2437.560	5	4	
	I	2437.610			20
	FE II	2437.666		3	
	U	2437.669	4	4	
	NB	2437.721		5	
	MO	2437.736	10	79	
	CL	2437.740			3
	PD	2437.760		3	
	*AG II	2437.791	63	500	3 VAF
	PD II	2437.797		16	
	NI II	2437.888	40	200	
	MN	2437.914		25	
	PD II	2437.930		16	
	W	2437.961	6		
--	LA II	2437.999		20	
	V	2438.044		16	
	FE I	2438.181	32	4	
	MN	2438.222		16	
--	I II	2438.280			5
	TI I	2438.310	10		

Table IIa. AG II 2437.791 A
(con't)

Element	Wavelength (Angstrom)	Intensity Arc	Spark	Plasma
MO	2438.343		16	
W	2438.353		6	
LA II	2438.390		10	
CO I	2438.411	2	3	
RE	2438.473	25		
---- CR	2438.475		32	
W	2438.556		3	
MO	2438.570		10	
PR	2438.640		25	
TA	2438.641	40	4	
IN	2438.750		3	
AR	2438.780			2
SI I	2438.782	32	20	
RH	2438.786	2	100	

Table IIb.(2) AG II 2437.791 A

Element	Wavelength (Angstrom)	Intensity Arc	Spark	Plasma
RU	2432.927	63		
TI I	2433.223	32		
RE	2433.291	32		
BI	2433.446	32		
HF II	2433.564	63	320	
NI I	2434.415	40		
RU	2434.882	50		
BH C	2435.000	32		
SI I	2435.159	160	79	36 VAF
V I	2435.521	32	32	
W	2435.962	32	10	
CO	2436.657	50	25	
NI	2436.672	32	20	
PT I	2436.689	320	20	
BH B	2437.100	250		
*AG II	2437.791	63	500	3 VAF
NI II	2437.888	40	200	
FE I	2438.181	32	4	

Table IIb.(2) AG II 2437.791 A
(con't)

Element	Wavelength (Angstrom)	Intensity Arc	Spark	Plasma
TA	2438.641	40	4	
SI I	2438.782	32	20	
PT I	2440.057	790	100	
TI I	2440.985	32		
RE	2441.480	40		
CU I	2441.637	200	100	
FE	2442.567	63	10	

Table IIa. AL I 3961.527 A

	Element	Wavelength (Angstrom)	Intensity Arc	Spark	Plasma
	SB II	3960.550		20	
	IR I	3960.555	4		
	RE	3960.584	16		
	PR	3960.599	50	25	
	TB	3960.695	5		
	EU	3960.733	3		
	CR	3960.763	40	8	
----	U	3960.804	5	8	
	W	3960.871		10	
	FE	3960.895	3	3	
	CE II	3960.914	40	8	
	TH	3960.947	2		
	NB	3960.984	8	5	
	CO I	3960.995	63	10	
	OS	3961.016	130	20	
---	RE I	3961.033	32		
	U	3961.064	6		
	EU	3961.140	50		
	FE I	3961.145	25	6	
	W	3961.178		10	
	ER	3961.210	6		
--	PR	3961.284	10	5	
	AG	3961.300	16	5	
	CE	3961.386	2		
	MO	3961.503	5	500	
	U	3961.515	8		
	*AL I	3961.527	3200	2000	10 VAF
	YB	3961.544	32		
	S I	3961.550			10
	ZR	3961.587	500	8	
	NB	3961.620		10	
	U	3961.660		3	
	CE	3961.661	6	2	
--	W	3961.760	5	5	
	SM	3961.808	25	10	
	TB	3961.970	3		
	U	3961.984	6	2	
	MO	3961.988	3		
---	EU	3961.990	10		
	LA	3962.030	2	3	

Table IIa. AL I 3961.527 A
(con't)

Element	Wavelength (Angstrom)	Intensity Arc	Spark	Plasma
IN II	3962.041			25
CE	3962.086	16	4	
GD	3962.105	25	25	
NI I	3962.120	10		
SM	3962.136	5	3	
NB	3962.153	3		
IN II	3962.159			20
U	3962.160	5		
NB	3962.163		10	
ZR	3962.167	2		
PR	3962.186	10	3	
CR	3962.187	8		
ND	3962.216	32	20	
SM II	3962.244	10	5	
U	3962.271	4	2	
-- IN II	3962.275			25
CA	3962.280	3	2	
W	3962.332	10	10	
KR II	3962.340			10
TH	3962.346	8	8	
FE I	3962.353	6		
IN II	3962.418			5
IR	3962.443	6		
PR	3962.445	63	50	
U	3962.470		4	
RE I	3962.476	100		

Table IIb.(1) AL I 3961.527 A

Element	Wavelength (Angstrom)	Intensity Arc	Spark	Plasma
FE I	3956.681	160	160	
GD	3957.681	320	200	
ZR II	3958.218	500	160	
PD I	3958.642	500	200	
MO	3961.503	5	500	
*AL I	3961.527	3200	2000	10 VAF

Table IIb.(1) AL I 3961.527 A
(con't)

Element	Wavelength (Angstrom)	Intensity Arc	Spark	Plasma
ZR	3961.587	500	8	
ND II	3963.114	32	25	17 VAF
OS I	3963.628	500	50	
CR I	3963.690	320	320	

Table IIa.   AL I   3082.155 A

	Element	Wavelength (Angstrom)	Intensity Arc	Spark	Plasma
	MO	3081.155	25		
	U	3081.188	4	2	
	W II	3081.210		3	
	FE	3081.225	2	2	
	CE	3081.245	3		
	V	3081.253	5	50	
	MN	3081.330	79	25	
	W	3081.375	6	3	
	ER	3081.380	3		
	RU	3081.384	4	50	
	U	3081.388	2		
---	LA II	3081.420	4	6	
	O	3081.460			5
	LU	3081.470	79	8	
	CA I	3081.547	2		
	SC I	3081.548		2	
	FE	3081.550	2	2	
	TI II	3081.575		40	
	CD II	3081.580		2	
	Y II	3081.600		2	
	MO	3081.650		20	
---	FE	3081.656	2	2	
	TH	3081.662	10	5	
	I	3081.664			100
	U	3081.666	6	4	
	NB	3081.769		16	
	FE	3081.840	2	2	
	TA	3081.850	50	5	
	W	3081.864	10	6	
--	MO	3081.950	25		
	S	3081.980			16
	CE	3081.984	5		
	GD II	3082.000	100	63	9 VAF
	V I	3082.012	16	2	
	U	3082.020	8	8	
	TH	3082.031	13	8	
	MN	3082.052	50		
	ER	3082.080	13	3	
	V I	3082.111	79	2	
	*AL I	3082.155	790	790	7 VAF

Table IIa. AL I 3082.155 A
(con't)

	Element	Wavelength (Angstrom)	Intensity Arc	Spark	Plasma
	Y II	3082.167	4	10	
	TH	3082.176	10	10	
	MO	3082.220	4	40	
	SM	3082.261		2	
	RB I	3082.270	10		
	RN	3082.300			10
	CE	3082.304	20	2	
--	TB	3082.360	3	8	
	RE I	3082.432	100		
	TA	3082.447	16		
	DY	3082.515	16	5	
	V	3082.523	3	32	
	SC II	3082.560	2	2	
	TM	3082.570	10		
	U	3082.587	6	4	
	TI I	3082.615	6		
	CO	3082.618	160	50	
---	XE II	3082.620			13
	CD I	3082.680	32		
	MN	3082.703	13	13	
	RE	3082.771	2		
	CO I	3082.844	32		
	NB	3082.857	3	5	
----	I	3082.875			25
	AR II	3082.990			5
	U	3083.022	5	4	
	FE II	3083.027		2	
	TH	3083.050	13	6	
	CE	3083.067	5		
	RU I	3083.148	50	3	
	FE	3083.150			

Table IIb.(1) AL I 3082.155 A

Element	Wavelength (Angstrom)	Intensity Arc	Spark	Plasma
FE II	3077.168		320	
TA	3077.245	160	50	
CR	3077.247		40	
OS	3077.437	79	8	
W	3077.519	2	40	
LU II	3077.600	100	200	17 VAF
MO	3077.661	20	130	
OS	3077.720	100	32	
CR I	3077.831	25	130	
FE	3078.018	100	79	
OS	3078.113	130	16	
OS	3078.383	130	16	
FE I	3078.434	79	50	
TI II	3078.645	63	500	37 VAF
TB	3078.860	32	79	
CO I	3079.398	79	2	
MN	3079.627	130	40	
NB	3080.350	8	100	
HF II	3080.657	32	100	
NI I	3080.755	200	63	
CD I	3080.827	160	100	
V	3081.253	5	50	
MN	3081.330	79	25	
RU	3081.384	4	50	
LU	3081.470	79	8	
TI II	3081.575		40	
GD II	3082.000	100	63	9 VAF
V I	3082.111	79	2	
*AL I	3082.155	790	790	7 VAF
MO	3082.220	4	40	
RE I	3082.432	100		
CO	3082.618	160	50	
V	3083.214	2	50	
FE I	3083.742	500	500	
RH I	3083.965	160	2	
CD	3084.866	10	40	
MO	3085.615	130	25	
CO I	3086.396	79	2	
V	3086.503	2	40	
RU	3086.520	32	40	

Table IIb.(1) AL I 3082.155 A
(con't)

Element	Wavelength (Angstrom)	Intensity Arc	Spark	Plasma
CO I	3086.777	200		
Y II	3086.858	13	50	
TM	3087.020	32	63	
NI II	3087.077		160	
U	3087.112	6	100000	

Table IIa. AL I 3944.032 A

	Element	Wavelength (Angstrom)	Intensity Arc	Spark	Plasma
	MO	3943.044	10	6	
	HF	3943.060	3		
	EU	3943.085	50	16	
	MO	3943.086	10	10	
	CE II	3943.138	13	3	
	ER	3943.191	10		
	IR	3943.205	6		
	SM II	3943.235	40	25	
	GD	3943.246	40	40	
--	FE I	3943.348	40	8	
	PR	3943.371	5	3	
	AS II	3943.470		3	
	CE	3943.497	6		
	U	3943.498	6	10	
	MO	3943.507	5	4	
--	XE II	3943.570			10
	FE	3943.594	2		
	EU	3943.602	3		
	CR I	3943.614	20	4	
	SM	3943.621	8	3	
	GD	3943.631	20	20	
	TB	3943.660	8		
	V I	3943.664	50	20	
	NB	3943.666	20	50	
	TH	3943.694	10	10	
	PR	3943.753	10	8	
--	PB	3943.800		5	
	U	3943.820	32	5	
	CE II	3943.888	40	16	
	EU	3944.010	8		
	ER	3944.014	2		
	RE	3944.016	16		
	RH	3944.019	5	4	
	*AL I	3944.032	2000	1000	6 VAF
	IR	3944.058	2		
	SE	3944.070			20
	CE II	3944.093	8		
	NI I	3944.126	5		
	U	3944.130	8	16	
	PR	3944.137	10	5	

Table IIa. AL I 3944.032 A
(con't)

	Element	Wavelength (Angstrom)	Intensity Arc	Spark	Plasma
	LA II	3944.150		2	
	RU I	3944.190	10	4	
	TB	3944.200	6		
	TH	3944.254	3		
--	AR	3944.270			50
	GA	3944.290		2	
	F II	3944.330			20
	RE	3944.349	16		
	IR I	3944.375	20	4	
	ND	3944.421	20		
---	ER	3944.423	13		
	SB II	3944.580		8	
	EU	3944.592	5		
	PR	3944.617	8	3	
	U	3944.621	8	2	
	Y	3944.687	3		
	DY II	3944.692	320	160	10 VAF
	RE	3944.727	32		
	SM	3944.744	5	5	
----	FE I	3944.752	4		
	RU	3944.785	4		
	W	3944.798	6	6	
	CE	3944.835	5		
	FE I	3944.896	16	8	
	PR	3944.899	32	13	
	CE	3944.924	6		
	CO I	3944.950	5	2	

Table IIb.(1) AL I 3944.032 A

Element	Wavelength (Angstrom)	Intensity Arc	Spark	Plasma
TB	3939.600	200	200	
FE I	3940.882	160	79	
MO	3941.478	5	160	
CR I	3941.490	200	63	
ND II	3941.512	63	32	19 VAF
CO I	3941.732	200		
FE I	3942.443	100	63	
CE II	3942.746	50	20	4 VAF
MN	3942.855	79	79	
NB	3943.666	20	50	
*AL I	3944.032	2000	1000	6 VAF
DY II	3944.692	320	160	10 VAF
CO I	3945.328	200	16	
GD	3945.544	200	160	
RU	3945.572	50	100	
PR	3947.633	130	63	
FE I	3948.107	130	50	
SM II	3948.108	50	50	
AS II	3948.740		50	
FE	3948.779	160	100	

Table IIa.   AL I   3092.713 A

	Element	Wavelength (Angstrom)	Intensity Arc	Spark	Plasma
	HF II	3091.790	8	3	
	MO	3091.840	10		
	RU	3091.873	50	5	
	CE	3091.918	20		
	ER	3091.930	5		
----	U	3091.976	2	2	
	CE	3092.035	2		
	TA	3092.057	16		
	GD	3092.064	3	2	
	MO	3092.074	32	100	
	CE	3092.197	3		
---	CL II	3092.220			50
	ZR	3092.240	3		
	HF II	3092.245	20	20	
	W	3092.285	6	3	
	TE	3092.320			16
	CD II	3092.393	10	16	
	FE	3092.399	4	5	
	IR	3092.401	5		
	XE II	3092.410			10
	SC	3092.420	3		
	TA	3092.444	50	16	
--	SC II	3092.519	3	4	
	YB	3092.560		32	
	MO	3092.700	20	2	
	Y	3092.712	8		
	*AL I	3092.713	1000	1000	13 VAF
	U	3092.720	5	3	
	V	3092.720	100	50	
	CE	3092.724	4		
	NA II	3092.729	50		200
	FE	3092.778	50	32	
	CE	3092.818	4		
	AL I	3092.842	50	20	13 VAF
	NB	3092.886		5	
	NE II	3092.910			4
	ND	3092.915	8	6	
--	MO	3092.920		10	
	MG I	3092.991	130	20	
	TA	3092.994	20		

Table IIa. AL I 3092.713 A
(con't)

Element	Wavelength (Angstrom)	Intensity Arc	Spark	Plasma
RN	3093.000			20
CL	3093.000			4
U	3093.012	20	20	
TH	3093.055	13	6	
DY	3093.108	16	5	
V II	3093.108	100	400	60 VAF
-- TM	3093.120	32	63	
CE	3093.243	6		
SI	3093.280		6	
AU II	3093.300		5	
ZR I	3093.321	3		
CE II	3093.339	13		
FE	3093.357	63	40	
U	3093.370	4	3	
AR II	3093.410			50
YB	3093.440		2	
RU	3093.447	4		
-- CA	3093.460	2	3	
RH I	3093.480	2	3	
CR	3093.488		100	
W	3093.512	13	10	
BI I	3093.580	10	8	
OS	3093.587	130	16	
CE	3093.614	20		
RE I	3093.652	63		
MO	3093.680	10		

Table IIb.(1) AL I 3092.713 A

Element	Wavelength (Angstrom)	Intensity Arc	Spark	Plasma
TA	3087.760	5	63	
TI II	3088.025	63	500	39 VAF
BH B	3088.600	100		
TI II	3089.401	13	100	
CO I	3089.595	100		
TI II	3090.051		100	

Table IIb.(1) AL I 3092.713 A
(con't)

Element	Wavelength (Angstrom)	Intensity Arc	Spark	Plasma
OS	3090.085	100	16	
OS	3090.296	100	13	
FE I	3091.578	320	200	
TL II	3091.660		50	
MO	3092.074	32	100	
*AL I	3092.713	1000	1000	13 VAF
V	3092.720	100	50	
AL I	3092.842	50	20	13 VAF
MG I	3092.991	130	20	
V II	3093.108	100	400	60 VAF
TM	3093.120	32	63	
CR	3093.488		100	
OS	3093.587	130	16	
RU	3093.901	32	100	
CU I	3093.992	160	50	
NB II	3094.183	100	1000	83 VAF
V	3094.199	20	130	
RU	3094.558	6	50	
MO	3094.664	160	25	
CR	3095.859	130	3	
CR	3096.130		100	
RU	3096.568	63	63	
MG I	3096.899	160	25	
TM	3096.970	16	50	
NI I	3097.118	200	50	
NB	3097.122	3	100	
TI II	3097.186	20	160	
EU	3097.456	200	5	

Table IIa. AL I 3092.842 A

	Element	Wavelength (Angstrom)	Intensity Arc	Spark	Plasma
	RU	3091.873	50	5	
	CE	3091.918	20		
	ER	3091.930	5		
	U	3091.976	2	2	
	CE	3092.035	2		
	TA	3092.057	16		
	GD	3092.064	3	2	
	MO	3092.074	32	100	
----	CE	3092.197	3		
	CL II	3092.220			50
	ZR	3092.240	3		
	HF II	3092.245	20	20	
	W	3092.285	6	3	
	TE	3092.320			16
---	CD II	3092.393	10	16	
	FE	3092.399	4	5	
	IR	3092.401	5		
	XE II	3092.410			10
	SC	3092.420	3		
	TA	3092.444	50	16	
	SC II	3092.519	3	4	
	YB	3092.560		32	
--	MO	3092.700	20	2	
	Y	3092.712	8		
	AL I	3092.713	1000	1000	13 VAF
	U	3092.720	5	3	
	V	3092.720	100	50	
	CE	3092.724	4		
	NA II	3092.729	50		200
	FE	3092.778	50	32	
	CE	3092.818	4		
	*AL I	3092.842	50	20	13 VAF
	NB	3092.886		5	
	NE II	3092.910			4
	ND	3092.915	8	6	
	MO	3092.920		10	
	MG I	3092.991	130	20	
	TA	3092.994	20		
	RN	3093.000			20
	CL	3093.000			4

Table IIa. AL I 3092.842 A
(con't)

	Element	Wavelength (Angstrom)	Intensity Arc	Spark	Plasma
	U	3093.012	20	20	
--	TH	3093.055	13	6	
	DY	3093.108	16	5	
	V II	3093.108	100	400	60 VAF
	TM	3093.120	32	63	
	CE	3093.243	6		
	SI	3093.280		6	
	AU II	3093.300		5	
	ZR I	3093.321	3		
---	CE II	3093.339	13		
	FE	3093.357	63	40	
	U	3093.370	4	3	
	AR II	3093.410			50
	YB	3093.440		2	
	RU	3093.447	4		
	CA	3093.460	2	3	
	RH I	3093.480	2	3	
	CR	3093.488		100	
	W	3093.512	13	10	
	BI I	3093.580	10	8	
----	OS	3093.587	130	16	
	CE	3093.614	20		
	RE I	3093.652	63		
	MO	3093.680	10		
	CD II	3093.740	3	2	
	ER	3093.750	6		
	Y II	3093.770	10	16	
	V I	3093.792	32		
	FE I	3093.806	50	40	
	TI I	3093.813	5		
	DY	3093.820	10		

Table IIb.(3) AL I 3092.842 A

Element	Wavelength (Angstrom)	Intensity Arc	Spark	Plasma
NB	3087.859	2	10	
CR	3087.884		40	
TI II	3088.025	63	500	39 VAF
TH	3088.031	10	10	
IR I	3088.038	50	2	
OS	3088.266	63	13	
TH	3088.473	20	20	
BH B	3088.600	100		
RE	3088.767	63		
U	3088.987	20	16	
YB	3089.092	3	13	
PB I	3089.093		32	
RU I	3089.145	63	13	
FE II	3089.388		10	
TI II	3089.401	13	100	
TB	3089.580	32	16	
CO I	3089.595	100		
V	3089.628		10	
TH	3089.631	10	10	
RU I	3089.801	63	5	
TI II	3090.051		100	
OS	3090.085	100	16	
TH	3090.102	16	10	
FE	3090.209	32	16	
RU	3090.229	50	6	
CO I	3090.254	79		
OS	3090.296	100	13	
OS	3090.492	79	16	
MG I	3091.077	79	10	
OS	3091.248	40	16	
U	3091.250	16	13	
FE I	3091.578	320	200	
TL II	3091.610		25	
TL II	3091.660		50	
RU	3091.873	50	5	
MO	3092.074	32	100	
HF II	3092.245	20	20	
CD II	3092.393	10	16	
TA	3092.444	50	16	
YB	3092.560		32	

Table IIb.(3) AL I 3092.842 A
(con't)

Element	Wavelength (Angstrom)	Intensity Arc	Spark	Plasma
AL I	3092.713	1000	1000	13 VAF
V	3092.720	100	50	
NA II	3092.729	50		200
FE	3092.778	50	32	
*AL I	3092.842	50	20	13 VAF
MO	3092.920		10	
MG I	3092.991	130	20	
U	3093.012	20	20	
V II	3093.108	100	400	60 VAF
TM	3093.120	32	63	
FE	3093.357	63	40	
CR	3093.488		100	
W	3093.512	13	10	
OS	3093.587	130	16	
RE I	3093.652	63		
Y II	3093.770	10	16	
FE I	3093.806	50	40	
TA	3093.869	50	16	
YB	3093.879	3	16	
FE	3093.883	40	32	
RU	3093.901	32	100	
CR	3093.947		25	
CU I	3093.992	160	50	
IR I	3094.009	20	10	
W	3094.032	8	10	
NB II	3094.183	100	1000	83 VAF
V	3094.199	20	130	
RU	3094.393	50	3	
RU	3094.558	6	50	
MO	3094.664	160	25	
U	3094.830	10	10	
FE I	3094.900	32	16	
CR	3094.927		16	
U	3095.041	13	13	
U	3095.231	10	10	
TA	3095.393	63	20	
CR	3095.488		13	
CO I	3095.718	63	2	
U	3095.748	10	10	
CR	3095.859	130	3	

Table IIb.(3) AL I 3092.842 A
(con't)

Element	Wavelength (Angstrom)	Intensity Arc	Spark	Plasma
W	3095.873	2	20	
CR	3096.130		100	
FE II	3096.296	2	32	
CO I	3096.405	63	3	
TI II	3096.424	3	32	
RU	3096.568	63	63	
CO I	3096.700	63	3	
FE	3096.836	32	20	
MG I	3096.899	160	25	
TM	3096.970	16	50	
MN	3097.063	79	40	
NI I	3097.118	200	50	
NB	3097.122	3	100	
TI II	3097.186	20	160	
EU	3097.456	200	5	
RU	3097.605	2	20	
MO	3097.689	10	32	
FE	3097.815	20	10	
IR I	3097.815	13	10	

Table IIa. AL I 2373.132 A

	Element	Wavelength (Angstrom)	Intensity Arc	Spark	Plasma
	PD II	2372.154	5	50	
	V	2372.175		100	
	I	2372.230			13
	NB	2372.230		8	
	TI I	2372.250	3		
	MO	2372.270	25	6	
	HF II	2372.330	3	4	
	ZR	2372.330	6		
	CE	2372.370		50	
----	CO	2372.510		2	
	ZR I	2372.570	8		
	W	2372.580	3	13	
	V	2372.583		16	
	FE II	2372.627		3	
---	OS	2372.650	20	5	
	SE	2372.710			25
	NB	2372.725	2	50	
	IR I	2372.774	100	40	12 VAF
	W	2372.790	5		
	TA	2372.800	3	32	
	CO I	2372.831	16	2	
	TE	2372.860			10
--	CR	2372.890	40	3	
	IN II	2372.900			8
	EU	2372.910	2		
	OS	2372.910	25	3	
	ZR II	2372.930	16	8	
	U	2372.940	2		
	MO	2372.960		32	
	PD	2372.980		10	
	IN II	2373.040			10
	YB	2373.060		5	
	V	2373.060		200	
	NB	2373.075	5		
	CO	2373.090	4	6	
	BA I	2373.108	13	2	
	OS	2373.120		10	
	*AL I	2373.132	100	32	10 VAF
	AU II	2373.140		5	
	ND	2373.170		6	

Table IIa. AL I 2373.132 A
(con't)

	Element	Wavelength (Angstrom)	Intensity Arc	Spark	Plasma
	IR	2373.172	5	10	
	RB	2373.220			63
	PR	2373.250		8	
	RE	2373.290	6	3	
	PT	2373.320		10	
	AL I	2373.362	200	100	10 VAF
--	MN II	2373.370	4	50	
	CO	2373.385	20	2	
	CS	2373.400			2
	W	2373.430	4	4	
	RE	2373.520	6	16	
	OS	2373.620	4	5	
	SB	2373.623	79	25	
---	FE I	2373.624	2		
	KR II	2373.680			4
	AG II	2373.700		2	
	CR I	2373.730	63		
	FE II	2373.730	6	16	
	ER	2373.820	10		
	DY	2373.830	2		
	TH	2373.846	3	3	
---	CO I	2373.862	10		
	YB	2373.890	2	8	
	PD II	2373.920		2	
	MO	2373.930		5	
	TA	2373.940		32	
	RU	2373.952		4	
	NB	2373.960	2	5	
	HG I	2374.020	10		
	U	2374.070	2		

Table IIb.(2) AL I 2373.132 A

Element	Wavelength (Angstrom)	Intensity Arc	Spark	Plasma
PT I	2368.276	32	25	
RE	2368.540	6	40	
FE II	2368.595	16	25	
IR	2369.220		32	
CU II	2369.887	20	32	
NB	2369.952	5	25	
W	2370.040	5	25	
TA	2370.760	5	25	
RE	2370.780	13	32	
RH	2371.070	3	40	
V	2371.070		500	
RH	2371.480	3	25	
PD II	2372.154	5	50	
V	2372.175		100	
CE	2372.370		50	
NB	2372.725	2	50	
IR I	2372.774	100	40	12 VAF
TA	2372.800	3	32	
MO	2372.960		32	
V	2373.060		200	
*AL I	2373.132	100	32	10 VAF
AL I	2373.362	200	100	10 VAF
MN II	2373.370	4	50	
SB	2373.623	79	25	
TA	2373.940		32	
MN	2374.300		32	
AG	2375.060	320	320	
OS	2375.060	10	25	
NI II	2375.421	10	32	
RU	2375.631	50	79	
CU	2376.270		25	
CU II	2376.380	3	32	
NB	2376.400	8	63	
FE II	2376.435	3	50	
OS	2377.030	50	32	
MO	2377.120		32	
PT II	2377.275	16	25	
RH	2377.810		50	
PD II	2377.916		32	

Table IIa. AS I 2288.120 A

	Element	Wavelength (Angstrom)	Intensity Arc	Intensity Spark	Intensity Plasma
	CR 3	2287.160		2	
	AR	2287.190			20
	W	2287.190		3	
	FE I	2287.248	20	6	
	NI I	2287.323	5		
---	W	2287.470		3	
	P II	2287.492			20
	PT II	2287.499		25	
	RE I	2287.510			4 VAF
	RU	2287.525	3		
	PD	2287.550			10
---	FE I	2287.630	5		
	NI II	2287.650	2	20	
	W	2287.670	10		
	RU	2287.680	63		
	KR II	2287.790			32
	AR	2287.800			63
	U	2287.800		2	
	CO	2287.809	13	4	
	TA	2287.840	5		
	TE	2287.840			5
	CE	2287.850		10	
--	IR	2287.878	20	5	
	I	2287.890			13
	V	2287.927		16	
	CD I	2288.018	1600	320	11 VAF
	IR	2288.090		5	
	V	2288.095	3	16	
	*AS I	2288.120	250	5	36 VAF
	SB	2288.139	5		
	PR	2288.150		6	
	CL	2288.160			4
	PT II	2288.192	16	32	
	AU II	2288.250		2	
	RH	2288.250		40	
--	U	2288.300		2	
	NI I	2288.390	13	2	
	AR	2288.400			2
	MN	2288.420	6	3	
	N II	2288.470			16

Table IIa. AS I 2288.120 A
(con't)

	Element	Wavelength (Angstrom)	Intensity Arc	Spark	Plasma
	W	2288.524		6	
	CO	2288.560		2	
	RH I	2288.570	25	10	21 VAF
	U	2288.580	2	2	
	PD	2288.590		4	
---	ZR II	2288.614	2	2	
	V	2288.628		79	
	AU	2288.630		3	
	HF II	2288.630	5	6	
	FE	2288.678		3	
	W	2288.680	6	4	
	I II	2288.720			2
	CD	2288.740		20	
	CO I	2288.786	16		
----	NB	2288.866	2	5	
	TE	2288.900			5
	IR	2288.920	2		
	YB	2288.930		6	
	SB	2288.986	20	10	
	FE I	2289.031	25		
	CO	2289.060		2	

Table IIb.(3) AS I 2288.120 A

Element	Wavelength (Angstrom)	Intensity Arc	Spark	Plasma
OS	2283.170		32	
TA	2283.190	2	16	
W	2283.270	10	5	
MO	2283.280		10	
AU II	2283.320		16	
U	2283.340		10	
V I	2283.376	5	5	
W	2283.380		3	
YB	2283.380	8	10	
ER	2283.410	3	5	
CO II	2283.519	10	16	

Table IIb.(3) AS I 2288.120 A
(con't)

Element	Wavelength (Angstrom)	Intensity Arc	Spark	Plasma
OS	2283.670	50	16	
U	2283.720		16	
V	2283.766		6	
YB	2283.980		8	
W	2283.990		5	
RH	2284.080		200	
FE I	2284.085	25	20	
NB	2284.348		16	
MO	2284.430	6	32	
CR 3	2284.440	2	10	
W	2284.440	4	5	
CR	2284.530	2	10	
HF II	2284.600	20	32	
IR	2284.605	20	5	
W	2284.620		8	
PR	2284.640		10	
V	2284.748		3	
TM	2284.800	20	10	
U	2284.830		4	
OS	2284.900	20	5	
V I	2284.984	2	4	
RH	2285.011		32	
TA	2285.020	3	13	
ND	2285.080		3	
MO	2285.140		20	
W	2285.170	10	3	
GA	2285.190		3	
NB	2285.221	8	16	
TA	2285.250	6	20	
RE	2285.310	25	6	
HF II	2285.430	6	6	
V	2285.455		100	
PT	2285.650		8	
U	2285.660		10	
W	2285.730		3	
IR	2285.740		10	
YB	2285.800	3	8	
ND	2285.860		3	
CO II	2286.156	40	320	43 VAF
TI II	2286.180	2	6	

Table IIb.(3) AS I 2288.120 A
(con't)

Element	Wavelength (Angstrom)	Intensity Arc	Spark	Plasma
PT II	2286.190		3	
W	2286.290	2	5	
NB	2286.352		4	
MO	2286.420	2	10	
AG	2286.460		13	
W	2286.500		3	
OS	2286.517	32	10	
PT II	2286.572		10	
TA	2286.590		6	
PR	2286.610		6	
RE	2286.620	4	25	
SN	2286.680	63	40	
CU II	2286.733		25	
NB	2286.750		4	
NB	2286.888	2	3	
NI II	2287.084	100	500	
W	2287.190		3	
FE I	2287.248	20	6	
W	2287.470		3	
PT II	2287.499		25	
RE I	2287.510			4 VAF
NI II	2287.650	2	20	
CO	2287.809	13	4	
CE	2287.850		10	
IR	2287.878	20	5	
V	2287.927		16	
CD I	2288.018	1600	320	11 VAF
IR	2288.090		5	
V	2288.095	3	16	
*AS I	2288.120	250	5	36 VAF
PR	2288.150		6	
PT II	2288.192	16	32	
RH	2288.250		40	
MN	2288.420	6	3	
W	2288.524		6	
RH I	2288.570	25	10	21 VAF
PD	2288.590		4	
V	2288.628		79	
AU	2288.630		3	
HF II	2288.630	5	6	

Table IIb.(3) AS I 2288.120 A
(con't)

Element	Wavelength (Angstrom)	Intensity Arc	Spark	Plasma
FE	2288.678		3	
W	2288.680	6	4	
CD	2288.740		20	
NB	2288.866	2	5	
YB	2288.930		6	
SB	2288.986	20	10	
TA II	2289.160	8	25	95 VAF
MO II	2289.190	4	20	
V	2289.219		32	
CR 3	2289.260		8	
PT I	2289.270	25	13	
W	2289.390	2	4	
IR	2289.393	5	40	
CA	2289.410		3	
RH	2289.580		16	
MO	2289.890	5	16	
MN	2289.970		3	
NI I	2289.976	20	20	
W	2289.980		3	
RH	2290.030	25	500	
MO	2290.060		20	
PR	2290.260		8	
MO	2290.320		20	
CO	2290.330		5	
NB	2290.380		25	
V	2290.544		50	
TM	2290.550		5	
W II	2290.560	2	10	
CR 3	2290.680		13	
IR	2290.800	3	50	
RE	2290.810	10	6	
MO	2290.870	3	20	
AG	2290.990		3	
CU II	2290.998		25	
ZR II	2291.113	79	16	
RE	2291.330	20	5	
W	2291.368		4	
PD	2291.440		20	
CO I	2291.455	13	5	
AU II	2291.520		16	

Table IIb.(3) AS I 2288.120 A
(con't)

Element	Wavelength (Angstrom)	Intensity Arc	Spark	Plasma
YB	2291.630		3	
HF II	2291.640	32	40	
U	2291.642		8	
NB	2291.645		5	
TH	2291.650		10	
PT	2291.720		20	
IR	2291.780		5	
TI II	2291.860		3	
HG	2291.920		5	
TA	2292.130	4	8	
OS	2292.210	16	5	
YB	2292.270		3	
LA II	2292.320		3	
W	2292.393		3	
PT I	2292.395	400	100	
TA	2292.544		20	
V	2292.589		25	
CU II	2292.681		5	
ND	2292.690		8	
HF II	2292.770	3	3	
FE II	2292.770		4	
YB	2292.810		3	
V	2292.855		250	
MO	2292.983		3	
CO II	2292.990	10	32	
NI I	2293.116	10	3	

Table IIa.   AU I   2675.950 A

	Element	Wavelength (Angstrom)	Intensity Arc	Spark	Plasma
	MN	2674.979		16	
	HG I	2674.990	5	5	
	NB	2675.101		4	
	U	2675.119	16	10	
	W	2675.128	10		
	RU	2675.185		32	
--	W	2675.204		8	
	NE I	2675.240			160
	S	2675.240			8
	SI	2675.250			5
	FE	2675.283	32	16	
	TA	2675.284		13	
	BH B	2675.300	63		
	KR II	2675.310			4
	ER	2675.347	5		
	CD	2675.360			2
	W	2675.400	10	3	
--	MN	2675.506		25	
	RU	2675.523		50	
	TA	2675.540	2		
	NE I	2675.640			160
	LA II	2675.655	2	5	
	TH	2675.670	8	4	
	CR	2675.682		16	
--	W	2675.730		13	
	CE	2675.733	2		
	TL II	2675.760			32
	V I	2675.761	13	2	
	W	2675.869	13	6	
	U	2675.880	16	10	
	TA II	2675.901	160	200	68 VAF
	NB	2675.944	10	100	
	*AU I	2675.950	250	100	96 VAF
	V I	2675.973	6		
	CO I	2675.982	10	5	
	CO II	2676.010		10	
	TL II	2676.030			32
	V	2676.043		16	
	EU	2676.050		5	
	TI I	2676.080	6		

Table IIa. AU I 2675.950 A
(con't)

	Element	Wavelength (Angstrom)	Intensity Arc	Spark	Plasma
	FE	2676.110	16	6	
	RH I	2676.110	10	5	
	CE	2676.121	2		
	NB	2676.125	2	10	
	YB	2676.130		2	
--	RU	2676.190	8	100	
	RH	2676.250		100	
	P	2676.280			10
	W	2676.310		6	
	MN	2676.331	16		
	V	2676.350		20	
	RU	2676.353	50	3	
	CE	2676.355	2		
	U	2676.410	4	4	
---	W	2676.412	10		
	CE	2676.467	3		
	RH I	2676.483	2	2	
	MO	2676.484	3	25	
	ZR I	2676.538	3		
	CR	2676.542		3	
	CE	2676.623	4		
	HF II	2676.630	10	13	
	V I	2676.634	6		
----	U	2676.687	6	4	
	MN	2676.747		10	
	CE	2676.755	4		
	RU	2676.762		20	
	IR I	2676.828	32	10	
	DY	2676.840	3		
	FE	2676.877	2	25	
	CL II	2676.950			160

Table IIb.(3) AU I 2675.950 A

Element	Wavelength (Angstrom)	Intensity Arc	Spark	Plasma
MN	2671.805		50	
MO	2671.834		100	
NB	2671.931	20	200	
V	2672.004	50	320	
MN	2672.586	16	130	
YB	2672.650	20	79	
MO	2672.843	16	100	
RU	2673.013	8	50	
V	2673.230	10	63	
MO	2673.273		100	
MN	2673.368		50	
EU	2673.409	130	50	
NB	2673.567	10	500	
RU	2674.195		50	
RH	2674.441		200	
RU	2675.523		50	
TA II	2675.901	160	200	68 VAF
NB	2675.944	10	100	
*AU I	2675.950	250	100	96 VAF
RU	2676.190	8	100	
RH	2676.250		100	
PT I	2677.149	790	200	
CR II	2677.159	32	320	42 VAF
NB	2677.660	2	50	
V	2677.802	63	320	
EU	2678.280	160	100	
V	2678.568	32	160	
ZR II	2678.632	79	100	20 VAF
RU II	2678.758	100	320	83 VAF
CR II	2678.792	10	79	
FE I	2679.062	200	200	
PT II	2679.129	6	50	
NI II	2679.241		500	
V	2679.322	63	320	
TM	2679.570	32	50	
PD II	2679.578		100	
NB	2680.057	4	79	
TA	2680.060	32	50	
RU	2680.543	5	50	

Table IIa. AU I 2427.950 A

	Element	Wavelength (Angstrom)	Intensity Arc	Spark	Plasma
	NB	2426.956	2		
	MO	2426.974		2	
	CO	2427.001	13		
	W	2427.068		2	
	CD II	2427.076		3	
	RU	2427.082		6	
	IR I	2427.086	2		
	RH II	2427.110			24 VAF
	FE II	2427.197		5	
----	IN II	2427.205			40
	LU	2427.210		3	
	W	2427.287	10		
	MO	2427.295	2	20	
	V	2427.327		32	
	RH II	2427.339	2	50	24 VAF
	MN	2427.410		50	
---	U	2427.454	13	2	
	W II	2427.490	10	13	
	NB	2427.539	5	2	
	IR I	2427.613	25	2	
	U	2427.622	13	2	
	TA	2427.642	160		
	CS	2427.650			20
--	AL II	2427.700			16
	RU	2427.741	2	13	
	V I	2427.745	6	6	
	MN	2427.753		50	
	CL II	2427.790			20
	W II	2427.813		6	
	OS	2427.900	8	40	
	*AU I	2427.950	400	100	170 VAF
	IR	2427.963	5		
	MN	2427.978		50	
	TH	2427.990	3	25	
	CL II	2428.020			10
	PT I	2428.035	100	10	
	SR I	2428.095	10		
	W	2428.173	5	13	
	MO	2428.180		16	
	AG II	2428.196		40	

Table IIa. AU I 2427.950 A
(con't)

	Element	Wavelength (Angstrom)	Intensity Arc	Intensity Spark	Intensity Plasma
--	FE	2428.200	10		
	PT I	2428.203	100	20	
	TI I	2428.228	16		
	V I	2428.279	32	20	
	FE II	2428.286		10	
	CO II	2428.293	10	20	
	KR	2428.350			20
	TI I	2428.359	8		
	IR I	2428.360	2		
	FE II	2428.361	3	16	
	IN	2428.380		3	
	MN	2428.422	13		
--	TM	2428.430	5	2	
	U	2428.476	5	2	
	TA	2428.480		20	
	CO	2428.596	10		
	W	2428.596	3		
	NB	2428.603		16	
--	PB	2428.644	10	8	
	RU	2428.718	13	3	
	W	2428.773		10	
	MO	2428.780		20	
	FE II	2428.795		16	
	ZN II	2428.860			10
	NB	2428.882		13	
	CU II	2428.911		5	

Table IIb.(2) AU I 2427.950 A

Element	Wavelength (Angstrom)	Intensity Arc	Intensity Spark	Intensity Plasma
OS	2423.071	25	79	
FE II	2423.210	2	40	
MO	2423.995	6	40	
FE II	2424.143	25	63	
CU II	2424.436	4	200	
PD II	2424.480		100	

Table IIb.(2) AU I 2427.950 A
(con't)

Element	Wavelength (Angstrom)	Intensity Arc	Spark	Plasma
PT II	2424.869	50	100	
CO I	2424.930	250		
FE II	2425.684		40	
TA	2425.906	10	25	
HF II	2425.975	100	200	
SB	2426.352	79	25	
NB	2426.791		40	
PD II	2426.868		50	
RH II	2427.110			24 VAF
V	2427.327		32	
RH II	2427.339	2	50	24 VAF
MN	2427.410		50	
MN	2427.753		50	
OS	2427.900	8	40	
*AU I	2427.950	400	100	170 VAF
MN	2427.978		50	
TH	2427.990	3	25	
AG II	2428.196		40	
HF II	2428.993	50	200	
MO	2429.410	4	25	
SN I	2429.495	2000	2500	31 VAF
AG II	2429.635		160	
TA	2429.710	20	40	
V	2430.040		63	
FE II	2430.072	16	63	
PD II	2430.538		25	
PD II	2430.929		32	
NB	2431.676		40	
TH	2431.741	2	32	
IR I	2431.938	50	50	
FE II	2432.267	25	63	
B II	2432.290		32	
NB	2432.322	2	40	
CO	2432.520		79	
TA	2432.701	320	400	
FE	2432.868	5	40	

Table IIa.   AU II 2082.090 A

	Element	Wavelength (Angstrom)	Intensity Arc	Spark	Plasma
	V	2081.110		5	
	PD I	2081.120	20		
	I	2081.290			40
---	W	2081.390	8		
	PD II	2081.420		25	
	OS	2081.460	20	10	
	AL II	2081.500			8
	RH	2081.520		32	
--	V	2081.670	4	3	
	OS	2081.720	3		
	IR	2081.830	2	10	
--	TE	2081.880			5
	SI I	2082.010	16		
	TA	2082.040	4		
	*AU II	2082.090			70 VAF
	K	2082.100			10
	IR	2082.220	6	5	
	RE	2082.260	6	20	
	PD I	2082.270	20	2	
--	RH	2082.340	2	10	
	PT I	2082.510	16	6	
	V I	2082.510	25		
--	OS	2082.540	20	5	
	KR	2082.600			2
	CO	2082.680	3	25	
	CA I	2082.730	2		
	IR	2082.790		13	
--	HF II	2082.800	16	8	
	NI I	2082.856	25	2	
	NB	2082.890	6	25	
	CU II	2082.920		6	
	RH	2082.930		8	
	TA	2083.030		5	

Table IIb.(2) AU II 2082.090 A

Element	Wavelength (Angstrom)	Intensity Arc	Spark	Plasma
W II	2079.106	13	32	100 VAF
IN II	2079.260		16	4 VAF
IR	2080.640	2	63	
TE I	2081.030	400		
*AU II	2082.090			70 VAF
IR	2084.490		79	
RH	2085.580		100	
RE I	2085.620	32	6	2 VAF

Table IIa.   B I   2497.733 A

	Element	Wavelength (Angstrom)	Intensity Arc	Spark	Plasma
	SN	2496.768	10		
	B I	2496.778	320	320	53 VAF
	HG	2496.780		3	
	RU	2496.846		10	
	N II	2496.880			25
	CS	2496.900			2
	CE	2496.914		2	
	TA	2496.957	2		
	NB	2496.973		3	
---	HF II	2496.986	32	40	
	FE	2496.991	20		
	V	2497.002		6	
	I	2497.010			20
	MO	2497.020		8	
	NA	2497.059			5
	V I	2497.099	2		
	O	2497.100			8
	IN	2497.200		3	
---	TH	2497.260		10	
	W	2497.292	5		
	FE II	2497.300		16	
	P II	2497.328			100
	MO	2497.376		16	
	BR	2497.420			2
	KR	2497.420			2
--	W II	2497.484	10	16	
	CO	2497.501		40	
	MO	2497.580	32		
	TH	2497.590	3	13	
	V I	2497.655	4		
	CA	2497.660		5	
	RU	2497.678	50		
	FE II	2497.717		3	
	SN	2497.720	8		
	*B I	2497.733	500	400	63 VAF
	HG	2497.770		3	
	MN	2497.780		3	
	NI II	2497.817		5	
	FE II	2497.820	16	50	
	AL II	2497.850			8

Table IIa. B I 2497.733 A
(con't)

	Element	Wavelength (Angstrom)	Intensity Arc	Spark	Plasma
	MO	2497.863	20	16	
	PT II	2497.922		3	
--	GE I	2497.963	8	10	
	V	2498.044	6		
	MO	2498.092		8	
	W	2498.097		3	
	FE	2498.208	3		
	RH	2498.210		3	
	PD II	2498.220		3	
	RE	2498.220	32		
---	V I	2498.232	10	4	
	AS	2498.250		5	
	NB	2498.253		25	
	W	2498.260		8	
	U	2498.267	3	8	
	MO	2498.281	10	25	
	TA	2498.330	32		
	YB	2498.350		3	
	BR	2498.400			3
	TH	2498.402	8	3	
	OS	2498.414	20	5	
----	RU II	2498.419	63	40	31 VAF
	PT I	2498.500	400	50	
	CL II	2498.530			32
	HF II	2498.570		2	
	RU II	2498.572	63	40	31 VAF
	IN II	2498.590			50
	TB	2498.630		10	

Table IIb.(2) B I 2497.733 A

Element	Wavelength (Angstrom)	Intensity Arc	Spark	Plasma
FE II	2493.180	10	100	
FE II	2493.261	10	100	
BE I	2494.733	32	20	8 VAF
SN	2495.719	100	100	
PD II	2496.686		100	
B I	2496.778	320	320	53 VAF
*B I	2497.733	500	400	63 VAF
RU II	2498.419	63	40	31 VAF
PT I	2498.500	400	50	
RU II	2498.572	63	40	31 VAF
PD II	2498.784	4	160	
NB	2499.750		200	
GA	2500.173	13	10	16 VAF
GA I	2500.705	5	5	16 VAF
NB	2501.405		160	
ZN II	2502.001	20	400	

Table IIa. B I 2496.778 A

	Element	Wavelength (Angstrom)	Intensity Arc	Spark	Plasma
	V I	2495.788	6	6	
	SI	2495.790			2
	PT I	2495.817	40	10	
	FE I	2495.863	25	32	
	RB	2495.900			20
	W	2495.938	4		
	RH	2495.951	2	50	
	IN	2495.970		2	
	P II	2496.003			50
----	CL II	2496.040			20
	SE	2496.050			100
	MO	2496.060		5	
	FE	2496.071	4		
	IN II	2496.215			10
	TA	2496.237	2	2	
	IR I	2496.271	10	2	
---	CR I	2496.309	130	2	
	IN II	2496.333			20
	GD	2496.400	2		
	ZR II	2496.480	2	5	
--	CD II	2496.533		4	
	FE	2496.533	40	16	
	TA	2496.635	20		
	W II	2496.638	10	20	
	PD II	2496.686		100	
	RE	2496.700	20		
	SN	2496.768	10		
	*B I	2496.778	320	320	53 VAF
	HG	2496.780		3	
	RU	2496.846		10	
	N II	2496.880			25
	CS	2496.900			2
	CE	2496.914		2	
	TA	2496.957	2		
	NB	2496.973		3	
	HF II	2496.986	32	40	
	FE	2496.991	20		
	V	2497.002		6	
	I	2497.010			20
--	MO	2497.020		8	

Table IIa. B I 2496.778 A
(con't)

Element	Wavelength (Angstrom)	Intensity Arc	Spark	Plasma
NA	2497.059			5
V I	2497.099	2		
O	2497.100			8
IN	2497.200		3	
TH	2497.260		10	
W	2497.292	5		
FE II	2497.300		16	
P II	2497.328			100
MO	2497.376		16	
BR	2497.420			2
KR	2497.420			2
W II	2497.484	10	16	
CO	2497.501		40	
MO	2497.580	32		
TH	2497.590	3	13	
V I	2497.655	4		
CA	2497.660		5	
RU	2497.678	50		
FE II	2497.717		3	
SN	2497.720	8		
B I	2497.733	500	400	63 VAF
HG	2497.770		3	

Table IIb.(2) B I 2496.778 A

Element	Wavelength (Angstrom)	Intensity Arc	Spark	Plasma
RH	2491.855	2	400	
CU I	2492.146	200	50	
FE II	2493.180	10	100	
FE II	2493.261	10	100	
BE I	2494.733	32	20	8 VAF
SN	2495.719	100	100	
PD II	2496.686		100	
*B I	2496.778	320	320	53 VAF
B I	2497.733	500	400	63 VAF
RU II	2498.419	63	40	31 VAF

Table IIb.(2) B I 2496.778 A
(con't)

Element	Wavelength (Angstrom)	Intensity Arc	Spark	Plasma
PT I	2498.500	400	50	
RU II	2498.572	63	40	31 VAF
PD II	2498.784	4	160	
NB	2499.750		200	
GA	2500.173	13	10	16 VAF
GA I	2500.705	5	5	16 VAF
NB	2501.405		160	

Table IIa. B I 2089.590 A

	Element	Wavelength (Angstrom)	Intensity Arc	Spark	Plasma
	TA	2088.670	8	32	
	PT II	2088.712	2	25	
	HF II	2088.770	50	50	
	IR I	2088.820	50	50	28 VAF
---	RU	2088.860	2		
	ZR	2088.890	2		
	B I	2088.930	100	16	25 VAF
	PT	2088.930	10		
	ND	2088.960		13	
	NI I	2088.980	32		
	OS	2089.030	20	4	
	PT II	2089.048		16	
	NI I	2089.070	16	16	
---	W	2089.143	10	20	
	CR	2089.160		16	
	OS	2089.210	20	4	
--	HF	2089.380	3		
	RH	2089.480		79	
	MO	2089.520	3	20	
	ZR	2089.560	4		
	*B I	2089.590	160	20	30 VAF
	PD II	2089.590	2	25	
	ND	2089.600		10	
	AS I	2089.790	5		
--	K	2089.800			40
	S	2089.900			320
	IR	2089.930		5	
	V I	2089.940	4		
	HF II	2089.950	32	32	
	ER	2089.960	8		
	PD I	2089.990	6		
---	BH C	2090.000	32		
	NI II	2090.100	3	25	
	PD II	2090.135		20	
	W	2090.183	3	13	
	IR	2090.320		25	
--	V	2090.330		16	
	NI I	2090.389	16		
	CO	2090.470		2	
	GA	2090.500		5	11 VAF

Table IIa. B I 2089.590 A
(con't)

Element	Wavelength (Angstrom)	Intensity Arc	Spark	Plasma
IR	2090.520	8	6	

Table IIb.(3) B I 2089.590 A

Element	Wavelength (Angstrom)	Intensity Arc	Spark	Plasma
PT I	2084.590	25	32	
NI II	2084.860		25	
W	2085.090	3	10	
ND	2085.150		20	
RH	2085.180		32	
CU II	2085.303	2	25	
PT II	2085.418		16	
RH	2085.580		100	
RE I	2085.620	32	6	2 VAF
IR	2085.740	32	20	
RE	2085.800	6	25	
TA	2085.920	3	25	
PD II	2086.490		25	
IR	2086.580		20	
MO	2086.780	2	10	
PT	2086.866	3	20	
IR	2086.920	3	20	
TA	2087.270	4	20	
W	2087.470	4	13	
RH	2087.480	32	25	
RE	2087.660	2	20	
CU II	2087.920		32	
YB	2088.050		16	
W II	2088.192	13	32	41 VAF
MO	2088.280	3	10	
PB	2088.430	32	40	
IR	2088.510		20	
CD	2088.520		10	
TA	2088.670	8	32	
PT II	2088.712	2	25	
HF II	2088.770	50	50	

Table IIb.(3) B I 2089.590 A
(con't)

Element	Wavelength (Angstrom)	Intensity Arc	Spark	Plasma
IR I	2088.820	50	50	28 VAF
B I	2088.930	100	16	25 VAF
ND	2088.960		13	
PT II	2089.048		16	
NI I	2089.070	16	16	
W	2089.143	10	20	
CR	2089.160		16	
RH	2089.480		79	
MO	2089.520	3	20	
*B I	2089.590	160	20	30 VAF
PD II	2089.590	2	25	
ND	2089.600		10	
HF II	2089.950	32	32	
NI II	2090.100	3	25	
PD II	2090.135		20	
W	2090.183	3	13	
IR	2090.320		25	
V	2090.330		16	
GA	2090.500		5	11 VAF
TA	2090.610	4	20	
PR	2090.750		20	
IR	2090.800		32	
HF II	2090.830	40	40	
RH	2091.070		100	
MO	2091.210	3	16	
YB	2091.280		20	
TA	2091.330	8	20	
GA II	2091.340			11 VAF
IR	2091.440		13	
RE II	2091.535		32	
IR	2091.860	10	20	
OS	2091.900		10	
ND	2091.980		13	
PD II	2092.010		32	
W	2092.077		25	
MN	2092.130	32	20	
YB	2092.310		32	
PD II	2092.380		25	
RE II	2092.410	5	100	3 VAF
MO	2092.500	6	20	

Table IIb.(3) B I 2089.590 A
(con't)

Element	Wavelength (Angstrom)	Intensity Arc	Spark	Plasma
IR I	2092.630	20	20	28 VAF
RH	2092.640	10	25	
MO	2093.110	6	25	
MN	2093.390	32	20	
NI II	2093.560	6	20	
CU II	2093.606	2	20	
W	2093.800	8	16	
IR	2093.970		10	
IR	2094.180	8	16	
GE I	2094.230	8		75 VAF

Table IIa. B I 2088.930 A

	Element	Wavelength (Angstrom)	Intensity Arc	Spark	Plasma
	PD I	2087.940	16	5	
	YB	2088.050		16	
	KR II	2088.160			20
---	W II	2088.192	13	32	41 VAF
	MO	2088.280	3	10	
	PT	2088.290	13	4	
	OS	2088.310	4		
	RU	2088.380	5		
	CD	2088.390			2
---	PB	2088.430	32	40	
	PD I	2088.490	6		
	IR	2088.510		20	
	CD	2088.520		10	
	CL	2088.520			8
	V	2088.560	16		
	TA	2088.670	8	32	
--	PT II	2088.712	2	25	
	HF II	2088.770	50	50	
	IR I	2088.820	50	50	28 VAF
	RU	2088.860	2		
	ZR	2088.890	2		
	*B I	2088.930	100	16	25 VAF
	PT	2088.930	10		
	ND	2088.960		13	
	NI I	2088.980	32		
	OS	2089.030	20	4	
	PT II	2089.048		16	
	NI I	2089.070	16	16	
	W	2089.143	10	20	
--	CR	2089.160		16	
	OS	2089.210	20	4	
---	HF	2089.380	3		
	RH	2089.480		79	
	MO	2089.520	3	20	
	ZR	2089.560	4		
	B I	2089.590	160	20	30 VAF
	PD II	2089.590	2	25	
---	ND	2089.600		10	
	AS I	2089.790	5		
	K	2089.800			40

Table IIa. B I 2088.930 A
(con't)

Element	Wavelength (Angstrom)	Intensity Arc	Spark	Plasma
S	2089.900			320
IR	2089.930		5	

Table IIb.(3) B I 2088.930 A

Element	Wavelength (Angstrom)	Intensity Arc	Spark	Plasma
IR	2084.220		13	
W	2084.270	2	10	
W	2084.480	2	8	
IR	2084.490		79	
PT I	2084.590	25	32	
NI II	2084.860		25	
W	2084.890	2	8	
W	2085.090	3	10	
ND	2085.150		20	
RH	2085.180		32	
CU II	2085.303	2	25	
PT II	2085.418		16	
RH	2085.580		100	
RE I	2085.620	32	6	2 VAF
IR	2085.740	32	20	
RE	2085.800	6	25	
TA	2085.920	3	25	
PD II	2086.490		25	
IR	2086.580		20	
MO	2086.780	2	10	
PT	2086.866	3	20	
IR	2086.920	3	20	
TA	2087.270	4	20	
W	2087.470	4	13	
RH	2087.480	32	25	
RE	2087.660	2	20	
CU II	2087.920		32	
YB	2088.050		16	
W II	2088.192	13	32	41 VAF
MO	2088.280	3	10	

Table IIb.(3) B I 2088.930 A
(con't)

Element	Wavelength (Angstrom)	Intensity Arc	Spark	Plasma
PB	2088.430	32	40	
IR	2088.510		20	
CD	2088.520		10	
TA	2088.670	8	32	
PT II	2088.712	2	25	
HF II	2088.770	50	50	
IR I	2088.820	50	50	28 VAF
*B I	2088.930	100	16	25 VAF
ND	2088.960		13	
PT II	2089.048		16	
NI I	2089.070	16	16	
W	2089.143	10	20	
CR	2089.160		16	
RH	2089.480		79	
MO	2089.520	3	20	
B I	2089.590	160	20	30 VAF
PD II	2089.590	2	25	
ND	2089.600		10	
HF II	2089.950	32	32	
NI II	2090.100	3	25	
PD II	2090.135		20	
W	2090.183	3	13	
IR	2090.320		25	
V	2090.330		16	
GA	2090.500		5	11 VAF
TA	2090.610	4	20	
PR	2090.750		20	
IR	2090.800		32	
HF II	2090.830	40	40	
RH	2091.070		100	
MO	2091.210	3	16	
YB	2091.280		20	
TA	2091.330	8	20	
GA II	2091.340			11 VAF
IR	2091.440		13	
RE II	2091.535		32	
IR	2091.860	10	20	
OS	2091.900		10	
ND	2091.980		13	
PD II	2092.010		32	

Table IIb.(3) B I 2088.930 A
(con't)

Element	Wavelength (Angstrom)	Intensity Arc	Spark	Plasma
W	2092.077		25	
MN	2092.130	32	20	
YB	2092.310		32	
PD II	2092.380		25	
RE II	2092.410	5	100	3 VAF
MO	2092.500	6	20	
IR I	2092.630	20	20	28 VAF
RH	2092.640	10	25	
MO	2093.110	6	25	
NB	2093.280		8	
MN	2093.390	32	20	
NI II	2093.560	6	20	
CU II	2093.606	2	20	
W	2093.800	8	16	

Table IIa. BA II 4554.042 A

	Element	Wavelength (Angstrom)	Intensity Arc	Spark	Plasma
	V I	4553.051	20	16	
	CE	4553.062	8		
	BH C	4553.100			
	ER	4553.129	5		
	NE II	4553.160			50
	NI I	4553.175	16		
	MO	4553.220	13	6	
	HF	4553.239	5		
	PR	4553.256	5		
	CA	4553.274		4	
---	EU	4553.300	3		
	MN	4553.310	13		
	MO	4553.320	13	4	
	CO I	4553.327	25		
	TI I	4553.415	6		
	CE	4553.419	2		
	PR	4553.498	4		
	MO	4553.503		25	
---	YB	4553.561	20	63	
	W	4553.662	6		
	TA	4553.694	200	2	
	BH PB	4553.700	6		
	CE	4553.755	2		
	HF	4553.776	10		
--	MO	4553.798	20	20	
	NB	4553.836	5	8	
	TH	4553.850	3	2	
	U	4553.858	4		
	CR	4553.949	20	3	
	ZR II	4553.967	4	13	
	MO	4554.028		4	
	CE	4554.035	32		
	*BA II	4554.042	1000	200	230 VAF
--	DY	4554.234	2		
	AR I	4554.319			16
	CE	4554.333	2		
	NE I	4554.415			10
	SM II	4554.443	63		
	Y I	4554.459	3	6	
	PR	4554.498	2		

Table IIa. BA II 4554.042 A
(con't)

	Element	Wavelength (Angstrom)	Intensity Arc	Spark	Plasma
---	RU I	4554.509	1000	200	
	CE	4554.557	6		
	NE I	4554.561			5
	CA	4554.590		2	
	PT	4554.593	10	5	
	TM	4554.650	5		
	W	4554.683	4		
	IR	4554.778	4		
----	PR	4554.790	4		
	P II	4554.800			100
	NE I	4554.824			40
	CR	4554.830	25	2	
	ND	4554.967	5		
	GD	4554.990	6	2	
	CR II	4555.029		40	

Table IIb.(1) BA II 4554.042 A

Element	Wavelength (Angstrom)	Intensity Arc	Spark	Plasma
AS II	4549.230		130	
FE II	4549.470	100	100	
TI II	4549.628	100	200	
V	4549.647	32	20	
CO I	4549.656	630		
U	4549.852	13	20	
AS II	4549.870		10	
NB	4550.047	3	10	
OS	4550.410	160	10	
GD	4550.964	10	10	
OS	4551.298	160	8	
RH I	4551.638	25	10	
W	4551.850	32	10	
TA	4551.950	400	8	
AS II	4552.370		50	
PT I	4552.423	63	10	
TI	4552.459	160	50	

Table IIb.(1) BA II 4554.042 A
(con't)

Element	Wavelength (Angstrom)	Intensity Arc	Spark	Plasma
SI	4552.500		40	
SM II	4552.661	79	40	
V I	4553.051	20	16	
MO	4553.503		25	
YB	4553.561	20	63	
TA	4553.694	200	2	
MO	4553.798	20	20	
ZR II	4553.967	4	13	
*BA II	4554.042	1000	200	230 VAF
RU I	4554.509	1000	200	
CR II	4555.029		40	
CR	4555.092	16	50	
U	4555.095	20	40	
TM	4555.260	25	50	
CS I	4555.355	2000	100	0.3 VAF
TI I	4555.489	130	63	
TH	4555.610		10	
FE II	4555.895	13	13	
CU II	4555.922	2	63	
FE I	4556.125	160	32	
CR	4556.169	40	13	
TA	4556.354	200	5	
TM	4556.670	32	63	
U	4557.805	3	16	
MO	4558.107	32	32	
LA II	4558.464	100	200	
CR II	4558.659	20	630	
MO	4558.744	13	16	
TM	4559.010	16	25	

Table IIa. BA I 5535.551 A

	Element	Wavelength (Angstrom)	Intensity Arc	Spark	Plasma
	AR I	5528.930			40
	TB	5534.580	10		
	FE I	5534.660	20		
----	SR	5534.807	20	16	
	FE II	5534.860		10	
	CU II	5534.980		3	
	RH	5535.040	79		
---	GD	5535.160	8		
	PR	5535.176	10	2	
	CE	5535.240	16		
	ND	5535.271	2		
--	TB	5535.380	16		
	V I	5535.382	2	2	
	N II	5535.390			63
	FE	5535.411	50		
	ND	5535.476	3		
	SM	5535.504	3		
	*BA I	5535.551	1000	200	
	LA II	5535.671	50	100	
--	U	5535.796	6	2	
	IN II	5535.940			63
	C II	5536.000		5	
---	K II	5536.010			10
	EU	5536.130	32		
	SM	5536.198	2		
	ER	5536.260	8		
	TB	5536.273	25		
----	BR	5536.300			50
	BH LA	5536.400	10		
	BR I	5536.400			20
	ND	5536.444	2		

Table IIb.(1) BA I 5535.551 A

Element	Wavelength (Angstrom)	Intensity Arc	Spark	Plasma
CO I	5530.769	500		
LA I	5532.170	2	10	
RE	5532.663	100		
MO	5533.046	200	100	
V I	5533.838	20	20	
SR	5534.807	20	16	
FE II	5534.860		10	
*BA I	5535.551	1000	200	
LA II	5535.671	50	100	
YB	5539.064	200	5	
MO	5539.410	25	13	
SR	5540.052	20	32	
BI II	5540.200		20	

Table IIa. BA II 2335.269 A

	Element	Wavelength (Angstrom)	Intensity Arc	Spark	Plasma
	W	2334.300	4		
	TI	2334.330		5	
	IR I	2334.330	10	2	
	RE	2334.360	20	8	
	MO	2334.380		3	
	CR	2334.390	5	10	
	V I	2334.446	25	40	
	SI	2334.480			2
	RE	2334.490	10		
	IR I	2334.505	25	5	
----	FE	2334.528	3		
	TI	2334.530	4	10	
	IN II	2334.570			50
	NI II	2334.580	8	20	
	ZR	2334.670	6		
---	RH II	2334.770	25	500	67 VAF
	MO	2334.800		10	
	SN I	2334.808	100	100	
	NB	2334.810	3	25	
	CO	2334.880		4	
	TA	2334.880	3	20	
	TE	2334.910			5
	MO	2334.920		20	
	MN	2334.940		25	
	IR	2334.958		25	
	TM	2335.000		8	
--	RE	2335.030	10		
	CO I	2335.107	16		
	PT II	2335.190	8	25	
	W II	2335.200	6	8	
	HF II	2335.250		2	
	*BA II	2335.269	63	100	75 VAF
	RH	2335.280		16	
	IR I	2335.285	5		
	NB	2335.314	3	10	
	V	2335.337	2	3	
	MO	2335.380		5	
	YB	2335.419	3	16	
	RB	2335.420			5
	XE II	2335.420			2

Table IIa. BA II 2335.269 A
(con't)

	Element	Wavelength (Angstrom)	Intensity Arc	Spark	Plasma
	PD	2335.430			3
	I II	2335.450			63
	TH	2335.490		3	
--	V	2335.492	3	50	
	I II	2335.520			16
	W	2335.560		3	
	PD	2335.570			5
	NB	2335.617	2	16	
	ZR	2335.630		2	
	RU	2335.740	13		
	RE	2335.750	40	10	
---	TA	2335.750	2	20	
	MO	2335.800		10	
	RU	2335.840		8	
	U	2335.842	2	2	
	W	2335.930		6	
	RU	2335.984	2	16	
--	CO I	2335.992	10	3	
	HG	2336.050			2
	V	2336.105	2	50	
	RE	2336.130	25	10	
	TE	2336.150			16
	MO	2336.190		10	
	OS	2336.190	5	8	
	CU II	2336.200	3	20	
	NI	2336.220		2	
	PR	2336.230		10	
	CO II	2336.241	6	16	
	S	2336.250			25

Table IIb.(3) BA II 2335.269 A

Element	Wavelength (Angstrom)	Intensity Arc	Spark	Plasma
V	2330.460	8	320	
IR	2330.500		50	
TA	2330.790		63	
V	2331.290		50	
AG II	2331.370	20	160	1 VAF
ZR	2331.570		100	
V	2331.771		320	
TA II	2331.980	10	20	96 VAF
MO	2332.120	8	79	
PB	2332.426	63	32	
FE II	2332.796	16	40	15 VAF
HF II	2332.968	40	50	
RH I	2333.310	8	130	
V	2333.603		50	
V	2334.213		250	
RH II	2334.770	25	500	67 VAF
SN I	2334.808	100	100	
*BA II	2335.269	63	100	75 VAF
V	2335.492	3	50	
V	2336.105	2	50	
NI II	2336.586		160	
OS II	2336.800	50	79	24 VAF
RH	2336.840		130	
V	2337.130		100	
V	2337.324		79	
GA I	2338.283	3	4	4 VAF
TM	2338.360	2	50	

Table IIa. BE I 2348.610 A

	Element	Wavelength (Angstrom)	Intensity Arc	Spark	Plasma
	RE	2347.640	3	10	
	W II	2347.640		3	
	CD	2347.650	2		
	CO I	2347.660	4		
	RH	2347.690		10	
	PD II	2347.760			25
	MO	2347.790		20	
	GA	2347.820		2	
	CO	2347.826		8	
	RH	2347.850	8		
---	CR	2347.900	2		
	CU	2347.900	4		
	RH	2347.900		16	
	BI	2347.909	5		
	W	2347.967	10	10	
	TI II	2347.990		20	
	FE II	2348.099	5	20	23 VAF
---	W	2348.151	8		
	K	2348.300			20
	V	2348.300		25	
	IR	2348.301	20		
	FE II	2348.303	5	20	23 VAF
	RU	2348.327	50		
--	CO	2348.460		2	
	ND	2348.460		20	
	PT II	2348.548	3	20	
	I	2348.560			13
	W	2348.560	2		
	MO	2348.580		10	
	ZR	2348.586	16		
	TA	2348.590		13	
	LU	2348.600		2	
	OS	2348.610	3		
	*BE I	2348.610	2000	50	96 VAF
	NB	2348.660		5	
	NI I	2348.740	10		
	TE	2348.740			5
	NB	2348.748	3		
	RE	2348.800	10		
	CU II	2348.820	16	20	

Table IIa. BE I 2348.610 A
(con't)

	Element	Wavelength (Angstrom)	Intensity Arc	Spark	Plasma
	MN	2348.830	3	16	
	MO	2348.840	3	10	
--	LA II	2348.860	2	2	
	TM	2348.890		8	
	U	2348.910	2	2	
	MO	2348.980		6	
	PD	2349.020		16	
	TM	2349.040		6	
---	BI I	2349.100	10		
	CO	2349.160		2	
	NB	2349.213		16	
	MN	2349.220	3	16	
	ER	2349.260	3		
	SE	2349.290			5
	ND	2349.300		16	
	IR	2349.309	5	10	
	HF II	2349.320		2	
	W	2349.320	8	13	
----	RU	2349.336	63	4	
	PD	2349.380			2
	YB	2349.390		8	
	RE	2349.400	13	6	
	NB	2349.421		2	
	OS	2349.480		4	
	SI	2349.540			2
	ZR	2349.590	10		
	U	2349.600		16	
	RE	2349.610	13	10	

Table IIb.(3) BE I 2348.610 A

Element	Wavelength (Angstrom)	Intensity Arc	Spark	Plasma
TA	2343.640		25	
CO II	2344.260	16	16	14 VAF
CR	2345.330	8	63	
NI I	2345.540	20		10 VAF
HG II	2345.550	10	32	
CU	2346.130	2	25	
V	2346.343		130	
RH	2346.440	2	100	
PD II	2346.463	2	25	
RH	2346.770	2	32	
V	2346.879		25	
V	2347.151		160	
PD II	2347.306	3	25	
CO II	2347.390	10	25	13 VAF
HF II	2347.444	79	130	
BA II	2347.577	32	40	8 VAF
FE II	2348.099	5	20	23 VAF
V	2348.300		25	
FE II	2348.303	5	20	23 VAF
*BE I	2348.610	2000	50	96 VAF
RH	2349.680	3	130	
V	2349.806	3	160	
AS I	2349.840	250	20	21 VAF
MO	2349.890		25	
CE	2350.150		50	
OS	2350.230	32	50	
RH	2350.350		50	
IR	2350.610		25	
PD	2350.700	6	25	
HF II	2351.215	100	160	13 VAF
V	2351.260	2	50	
PD II	2351.338	10	63	17 VAF
V	2351.537		50	
PD II	2351.857		32	
V	2352.177	5	200	
TA	2352.970	2	32	
TM	2353.100		40	
IR	2353.117	4	50	
CO II	2353.420	10	32	17 VAF

Table IIa. BE II 3130.416 A

	Element	Wavelength (Angstrom)	Intensity Arc	Spark	Plasma
	CO I	3129.482	40	2	
	TA	3129.549	50	6	
	HF	3129.584	16		
	RU	3129.604	50		
	TI I	3129.624	2		
	PB I	3129.630		3	
	NB	3129.644		5	
----	U	3129.728	8	16	
	ZR II	3129.761	10	10	
	RU I	3129.837	63	4	
	TH	3129.911	10	8	
---	Y II	3129.933	8	50	
	TA	3129.946	50	8	
	SM	3129.950	5		
	GD	3129.965	2		
	OS	3130.002	32	8	
	AG	3130.009	25	16	
	MO	3130.060		20	
	ZR I	3130.063	3		
	SE	3130.120			8
	W	3130.155	5	4	
	DY	3130.159	4	2	
--	TI I	3130.175	2		
	CE	3130.197	16		
	SM	3130.235	5		
	LA II	3130.250	3	4	
	V II	3130.267	50	200	
	FE	3130.278	5	4	
	IR	3130.285	4		
	TA	3130.290	16		
	CE	3130.334	32		
	U	3130.373	3		
	TI I	3130.376	2		
	HO	3130.380		4	
	P II	3130.380			32
	XE II	3130.400			2
	*BE II	3130.416	200	200	110 VAF
	W	3130.456	10	8	
	SI	3130.480		5	
	CE	3130.516	5		

Table IIa. BE II 3130.416 A
(con't)

	Element	Wavelength (Angstrom)	Intensity Arc	Spark	Plasma
	U	3130.564		16	
	CR	3130.565		13	
	FE II	3130.567	4	4	
	BA I	3130.570	2	3	
	IR	3130.578	3		
--	TA	3130.578	100	32	
	CS	3130.700			4
	S	3130.710			16
	U	3130.732	10	6	
	EU	3130.740	100	100	
	NB II	3130.786	1000	100	60 VAF
	RH I	3130.790	63	2	
	AR I	3130.800			20
	TI II	3130.800	25	100	
	GD	3130.814	3		
--	CE	3130.872	32	2	
	CE	3130.919	5		
	HO	3130.990	6	6	
	ER	3131.070	3		
	TH	3131.070	13	10	
	BE II	3131.072	200	160	41 VAF
	ZR I	3131.110	6		
--	OS	3131.115	130	32	
	MO	3131.195		5	
	CR	3131.211	20	6	
	TA	3131.216	5	25	
	TM II	3131.260	400	500	58 VAF
	U	3131.325	8	6	
	TB	3131.350	8	3	

Table IIb.(2) BE II 3130.416 A

Element	Wavelength (Angstrom)	Intensity Arc	Spark	Plasma
FE I	3125.654	400	320	
HG I	3125.663	200	160	
FE	3126.173	160	63	
V II	3126.215	63	100	
RU	3126.608	13	50	
CO I	3127.249	100		
NB	3127.526	10	50	
TA	3127.765	20	100	
RU	3127.913	10	100	
V	3128.284	2	63	
TI I	3128.640	13	63	
TI II	3128.640	13	63	
V	3128.686	3	63	
CR II	3128.699	32	160	
RE I	3128.952	100		
NI I	3129.314	130		
FE I	3129.330	100	63	
Y II	3129.933	8	50	
V II	3130.267	50	200	
*BE II	3130.416	200	200	110 VAF
TA	3130.578	100	32	
EU	3130.740	100	100	
NB II	3130.786	1000	100	60 VAF
TI II	3130.800	25	100	
BE II	3131.072	200	160	41 VAF
OS	3131.115	130	32	
TM II	3131.260	400	500	58 VAF
HG I	3131.546	400	320	
HG I	3131.833	200	100	
CR II	3132.058	25	130	
MO	3132.594	1000	320	
YB II	3132.628	16	160	
TA	3132.643	250	25	
CD I	3133.167	200	320	
V II	3133.328	50	200	
TM II	3133.890	200	200	22 VAF
NI I	3134.108	1000	160	
FE I	3134.111	200	130	
CR	3134.307	3	50	
HF	3134.718	79	130	

Table IIb.(2) BE II 3130.416 A
(con't)

Element	Wavelength (Angstrom)	Intensity Arc	Spark	Plasma
RU I	3134.802	10	100	
V	3134.931	32	160	
FE II	3135.364		100	
DY	3135.370	100	50	

Table IIa. BE II 3131.072 A

	Element	Wavelength (Angstrom)	Intensity Arc	Spark	Plasma
	SE	3130.120			8
	W	3130.155	5	4	
	DY	3130.159	4	2	
	TI I	3130.175	2		
	CE	3130.197	16		
	SM	3130.235	5		
	LA II	3130.250	3	4	
	V II	3130.267	50	200	
	FE	3130.278	5	4	
	IR	3130.285	4		
	TA	3130.290	16		
----	CE	3130.334	32		
	U	3130.373	3		
	TI I	3130.376	2		
	HO	3130.380		4	
	P II	3130.380			32
	XE II	3130.400			2
	BE II	3130.416	200	200	110 VAF
	W	3130.456	10	8	
	SI	3130.480		5	
	CE	3130.516	5		
	U	3130.564		16	
	CR	3130.565		13	
	FE II	3130.567	4	4	
	BA I	3130.570	2	3	
---	IR	3130.578	3		
	TA	3130.578	100	32	
	CS	3130.700			4
	S	3130.710			16
	U	3130.732	10	6	
	EU	3130.740	100	100	
	NB II	3130.786	1000	100	60 VAF
	RH I	3130.790	63	2	
	AR I	3130.800			20
	TI II	3130.800	25	100	
	GD	3130.814	3		
--	CE	3130.872	32	2	
	CE	3130.919	5		
	HO	3130.990	6	6	
	ER	3131.070	3		

Table IIa. BE II 3131.072 A
(con't)

	Element	Wavelength (Angstrom)	Intensity Arc	Intensity Spark	Intensity Plasma
	TH	3131.070	13	10	
	*BE II	3131.072	200	160	41 VAF
	ZR I	3131.110	6		
	OS	3131.115	130	32	
	MO	3131.195		5	
	CR	3131.211	20	6	
	TA	3131.216	5	25	
--	TM II	3131.260	400	500	58 VAF
	U	3131.325	8	6	
	TB	3131.350	8	3	
	FE	3131.455	2		
	OS	3131.481	20	10	
	CE	3131.505	2		
	U	3131.534	4	2	
--	HG I	3131.546	400	320	
	U	3131.618		2	
	CE	3131.679	4		
	FE II	3131.718		32	
	EU	3131.750	20	5	
	SR II	3131.750		10	
--	HF	3131.814	40	10	
	CO I	3131.830	8		
	HG I	3131.833	200	100	
	U	3131.987	8	6	
	NB	3132.011		5	
	RE I	3132.025	2		
	ER	3132.030	6	2	
	CE	3132.043	16		
	CR II	3132.058	25	130	
	ZR I	3132.066	10	2	

Table IIb.(2) BE II 3131.072 A

Element	Wavelength (Angstrom)	Intensity Arc	Spark	Plasma
FE	3126.173	160	63	
V II	3126.215	63	100	
RU	3126.608	13	50	
CO I	3127.249	100		
NB	3127.526	10	50	
TA	3127.765	20	100	
RU	3127.913	10	100	
V	3128.284	2	63	
TI I	3128.640	13	63	
TI II	3128.640	13	63	
V	3128.686	3	63	
CR II	3128.699	32	160	
Y II	3128.789	10	40	
RE I	3128.952	100		
NI I	3129.314	130		
FE I	3129.330	100	63	
Y II	3129.933	8	50	
V II	3130.267	50	200	
BE II	3130.416	200	200	110 VAF
TA	3130.578	100	32	
EU	3130.740	100	100	
NB II	3130.786	1000	100	60 VAF
TI II	3130.800	25	100	
*BE II	3131.072	200	160	41 VAF
OS	3131.115	130	32	
TM II	3131.260	400	500	58 VAF
HG I	3131.546	400	320	
HG I	3131.833	200	100	
CR II	3132.058	25	130	
FE	3132.514	63	40	
MO	3132.594	1000	320	
YB II	3132.628	16	160	
TA	3132.643	250	25	
CD I	3133.167	200	320	
V II	3133.328	50	200	
TM II	3133.890	200	200	22 VAF
NI I	3134.108	1000	160	
FE I	3134.111	200	130	
CR	3134.307	3	50	
HF	3134.718	79	130	

Table IIb.(2) BE II 3131.072 A
(con't)

Element	Wavelength (Angstrom)	Intensity Arc	Spark	Plasma
RU I	3134.802	10	100	
V	3134.931	32	160	
FE II	3135.364		100	
DY	3135.370	100	50	
RU	3135.802	10	79	
TA	3135.893	32	100	

Table IIa. BI I 2897.975 A

	Element	Wavelength (Angstrom)	Intensity Arc	Spark	Plasma
	ND	2896.980		5	
	CL	2897.000			6
	NB	2897.002	2	3	
	NE II	2897.030			5
	MN	2897.065	5	2	
	TH	2897.071	13	5	
	IR I	2897.152	20	10	
	CE	2897.172	4		
	W	2897.200	4	3	
----	FE II	2897.262		200	
	AR	2897.300			10
	NB	2897.351	3	2	
	HO	2897.360		20	
	MO	2897.421		20	
	MN	2897.427	8		
	U	2897.458	6	4	
	TB	2897.460	3	20	
---	N	2897.490			16
	ER	2897.518	13	2	
	RE	2897.592	10		
	MO	2897.628	20	25	
	RH	2897.633	2	10	
	U	2897.636	3	6	
	FE	2897.640	4		
	IR	2897.653	3		
	Y II	2897.676	3	13	
	CR	2897.704	3	25	
	RU	2897.715	6	63	
--	LA II	2897.760	2	5	
	MN	2897.797	16		
	NB	2897.812	16	160	
	I	2897.830			20
	FE	2897.850	2		
	PT I	2897.873	400	16	
	V	2897.896		25	
	*BI I	2897.975	500	500	9 VAF
	MN	2897.990	16		
	U	2898.013	6	6	
	FE	2898.060	2		
	W II	2898.090		3	

Table IIa. BI I 2897.975 A
(con't)

	Element	Wavelength (Angstrom)	Intensity Arc	Spark	Plasma
	BE I	2898.190	16		
	TB	2898.200		10	
--	RU	2898.219		63	
	W	2898.253	8	3	
	HF I	2898.256	500	50	
	ZR	2898.256	4		
	TH	2898.267	4	3	
	BE I	2898.270	20		
	YB	2898.330		10	
	CE	2898.336	6		
	IR I	2898.350	10		
	FE	2898.355	100	32	
	U	2898.366	4	4	
	MO	2898.386	2		
	TA	2898.425	32	5	
	CL	2898.450			8
--	FE	2898.470	2		
	MO	2898.481		16	
	CR	2898.536	13	40	
	RU	2898.538	20	4	
	U	2898.563	8	4	
	MO	2898.652	20	6	
	MN	2898.693	13	3	
	HF II	2898.709	50	200	
	AS I	2898.710	25	40	
	U	2898.711	8	4	
--	ZR II	2898.712	6	6	
	FE II	2898.735		3	
	RE	2898.787	16		
	V I	2898.817	10		
	TB	2898.840	16	10	
	FE	2898.860	6	4	
	FE	2898.930	2		
	TH	2898.930		25	
	Y II	2898.940	8	10	
	IR I	2898.942	8		
	CE	2898.944	6		

Table IIb.(2) BI I 2897.975 A

Element	Wavelength (Angstrom)	Intensity Arc	Spark	Plasma
CR II	2893.254	79	10	35 VAF
V II	2893.320	50	320	29 VAF
PT I	2893.865	500	25	
FE	2894.505	160	160	
LU II	2894.840	63	200	15 VAF
V II	2896.211	32	160	
AG II	2896.490	2	160	
FE II	2897.262		200	
NB	2897.812	16	160	
PT I	2897.873	400	16	
*BI I	2897.975	500	500	9 VAF
HF I	2898.256	500	50	
HF II	2898.709	50	200	
NB	2899.239	20	500	
LU II	2900.300	50	160	12 VAF
TA	2902.046	1000	200	
AG II	2902.070	5	200	

Table IIa. BI I 2230.608 A

	Element	Wavelength (Angstrom)	Intensity Arc	Spark	Plasma
	W II	2229.626	10	20	
	NB	2229.660	13		
	NB	2229.720	32	20	
	TI	2229.730	10		
	CO I	2229.732	10	2	
	V I	2229.739	13	79	
	PD II	2229.760			4
	IR I	2229.780	20	5	
	IN	2229.800	2		
	CU II	2229.850		4	
--	NI II	2229.858		6	
	HG II	2229.860		3	
	I	2229.970			400
	MN	2229.970		63	
	V	2229.988	3	10	
	HG	2230.040		5	
	MO	2230.080	4	25	
	CU I	2230.084	32	20	23 VAF
	CU II	2230.084	32	20	
--	TI I	2230.221	20		
	OS	2230.250	20		
	RU	2230.310	32		
--	CD	2230.370	2		
	PR	2230.370		40	
	V I	2230.372	8	20	
	CU II	2230.400			10
	W	2230.420		6	
	PD II	2230.430			4
	IR	2230.441	20	20	
	TI I	2230.473	5		
	CO	2230.520		10	
	TA	2230.530	10		
	MO	2230.560		10	
	*BI I	2230.608	100	32	87 VAF
	RE I	2230.610	5		
	U	2230.620		4	
	RH	2230.650		25	
	IN I	2230.699	5		
	LA II	2230.740	2	6	
	PD	2230.750			2

Table IIa. BI I 2230.608 A
(con't)

	Element	Wavelength (Angstrom)	Intensity Arc	Intensity Spark	Intensity Plasma
	XE II	2230.790			3
	RU	2230.804	8	2	
	RE I	2230.830	10		
--	NB	2230.853		4	
	ZR	2230.927	16		
	TI II	2230.945	6	40	
	CU II	2230.948		16	
	NI I	2230.954	10		
	IR	2230.990		25	
	MO	2231.080		16	
---	W	2231.080	3	6	
	OS	2231.160	13	50	
	FE I	2231.211	8		
	CL	2231.220			4
	W	2231.240		10	
	AU II	2231.310		20	
----	PD II	2231.317		10	
	TE	2231.360			5
	HF II	2231.410	3	3	
	V	2231.418	16		
	NB	2231.434	3		
	MO	2231.483		6	
	YB	2231.550		2	
	TM	2231.570	8	5	
	CU II	2231.571		16	
	PD II	2231.590	10	63	25 VAF

Table IIb.(3) BI I 2230.608 A

Element	Wavelength (Angstrom)	Intensity Arc	Intensity Spark	Intensity Plasma
CU I	2225.697	160	20	
W II	2225.893	10	20	50 VAF
AG II	2226.110		20	
NI II	2226.332	13	16	
TA	2226.525		25	
CR	2226.700		32	

Table IIb.(3) BI I 2230.608 A
(con't)

Element	Wavelength (Angstrom)	Intensity Arc	Spark	Plasma
CU II	2226.864		32	
CU II	2227.157	4	32	
CU I	2227.775	40	25	19 VAF
CU I	2227.840	13	32	
TA	2227.840	13	32	
CE	2227.870		20	
OS I	2227.980	32	32	11 VAF
CE	2228.090		25	
BI I	2228.251	100	50	36 VAF
V	2228.299		100	
PT II	2228.498		16	
CU II	2228.863	10	40	
AU II	2229.030		50	
PT II	2229.218	3	25	
PD II	2229.220		32	
PT	2229.310		25	
AG II	2229.523		25	
W II	2229.626	10	20	
NB	2229.720	32	20	
V I	2229.739	13	79	
MN	2229.970		63	
MO	2230.080	4	25	
CU I	2230.084	32	20	23 VAF
CU II	2230.084	32	20	
PR	2230.370		40	
V I	2230.372	8	20	
IR	2230.441	20	20	
*BI I	2230.608	100	32	87 VAF
RH	2230.650		25	
TI II	2230.945	6	40	
CU II	2230.948		16	
IR	2230.990		25	
MO	2231.080		16	
OS	2231.160	13	50	
AU II	2231.310		20	
CU II	2231.571		16	
PD II	2231.590	10	63	25 VAF
MN	2231.662		63	
SN I	2231.720	32	63	
CO II	2232.066	6	20	

Table IIb.(3) BI I 2230.608 A
(con't)

Element	Wavelength (Angstrom)	Intensity Arc	Spark	Plasma
IR	2232.248	32	50	
RE	2232.310	6	25	
CO I	2232.468	8	16	
V	2232.912		500	
RH	2232.980		25	
IR I	2233.370	10	100	
FE II	2233.914		32	
IR	2234.380	5	79	
PT I	2234.915	25	16	
RH	2235.300	2	25	
PT II	2235.301	5	25	
HF II	2235.410	3	20	

Table IIa. BR II 4704.860 A

	Element	Wavelength (Angstrom)	Intensity Arc	Spark	Plasma
	ND	4703.864	16		
	EU	4703.887	8		
	NB	4703.930	3	5	
	SR I	4703.984	2		
	TH	4703.992	3		
	PR	4703.998	4		
	CE	4704.012	8		
	IR	4704.054	13		
	RH I	4704.081	10	4	
---	TB	4704.150	2		
	GD	4704.289	4		
---	CO I	4704.386	3		
	NE I	4704.395			1600
	SM II	4704.408	200		
	U	4704.476	3	3	
	SB II	4704.480		10	
	ER	4704.493	2		
	TB	4704.580	2		
	CU I	4704.596	200	50	
	EU	4704.601	32		
--	HG II	4704.630			200
	XE II	4704.670			8
	*BR II	4704.860			250
	YB	4704.870	4		
	TB	4704.900	2		
	FE	4704.963	10		
	SM II	4705.000	3		
	RE	4705.034	40		
--	V I	4705.087	16	13	
	RU I	4705.159	5		
	ER	4705.305	4		
	O II	4705.320			320
---	BI II	4705.350		50	
	KR II	4705.440			2
	TM	4705.443	3		
	FE	4705.464	2		
	TB	4705.480	2		
	NI I	4705.500	2		
	PR	4705.573	5		
--	CE	4705.579	3		

Table IIa. BR II 4704.860 A
(con't)

Element	Wavelength (Angstrom)	Intensity Arc	Spark	Plasma
DY	4705.730	3		
TH	4705.764	10	6	
GD	4705.767	25		
ER	4705.808	6		

Table IIb.(2) BR II 4704.860 A

Element	Wavelength (Angstrom)	Intensity Arc	Spark	Plasma
TA	4701.324	160	2	
NI I	4701.536	160		
AR I	4702.316			7 VAF
GD	4702.323	50	100	
LA II	4703.278	200	320	
NI I	4703.808	200		
SM II	4704.408	200		
CU I	4704.596	200	50	
*BR II	4704.860			250
TA	4706.091	200	2	
MO	4707.255	130	130	
AS II	4707.820		200	
CR I	4708.040	200	160	
RU I	4709.484	160	79	
MN	4709.715	160	16	

Table IIa. CA II 3933.666 A

	Element	Wavelength (Angstrom)	Intensity Arc	Spark	Plasma
	CE	3932.794	3		
	TH	3932.915	10	3	
---	CU I	3932.917	10		
	FE	3932.921	8	4	
	SM II	3932.966	6	3	
	GD	3932.970	10	10	
	IR I	3932.971	6		
	PR	3932.978	25	8	
	CE	3932.979	6		
	DY	3932.980	10		
	NB	3933.013	3	3	
	U	3933.030	5	10	
---	ZR I	3933.178	10		
	S II	3933.294			79
	PR	3933.300	10	2	
	P	3933.370			50
	SC I	3933.381	63	63	
	NB	3933.394	3	3	
--	TB	3933.469	6		
	SM II	3933.592	200	200	
	FE I	3933.605	200	200	
	CO	3933.654	79		
	U	3933.662	2	10	
	CA II	3933.664	630	630	89 VAF
	HF II	3933.664	20	16	
	IR	3933.664	20		
	*CA II	3933.666	630	630	89 VAF
	EU	3933.677	10		
	RU	3933.680	5	200	
	CE	3933.731	63	63	
	IR	3933.901	20	8	
	TB	3933.905	4		
--	CO I	3933.914	63		
	U	3933.985	6		
	V I	3934.013	100	32	
	CE	3934.076	6		
	ND	3934.093	20		
	ND	3934.107	5		
	BR I	3934.110			2
	ZR II	3934.121	20	13	

Table IIa. CA II 3933.666 A
(con't)

	Element	Wavelength (Angstrom)	Intensity Arc	Spark	Plasma
	IN II	3934.123			10
---	NB	3934.139	5	16	
	DY	3934.170	13		
	RH I	3934.229	100	2	
	FE	3934.230	2		
	TI I	3934.236	32	2	
	MO	3934.260	5	5	
	YB	3934.300	2	4	
	EU	3934.387	5		
	TB	3934.400	10		
----	NB	3934.408	5	5	
	IN II	3934.431			5
	K	3934.460			20
	U	3934.470	2		
	CE	3934.509	2		
	W	3934.601	5	5	

Table IIb.(1) CA II 3933.666 A

Element	Wavelength (Angstrom)	Intensity Arc	Spark	Plasma
LA II	3929.216	400	320	12 VAF
MN	3929.241	32	32	
PR	3929.256	40	32	
ZR I	3929.530	100	6	
ZR II	3929.530	100	6	
TM	3929.584	63	25	
V II	3929.730	3	32	
RE I	3929.843	100		
TI I	3929.875	63	32	
OS	3929.997	79	13	
FE I	3930.299	630	400	
EU II	3930.440	100		53 VAF
EU II	3930.503	1000	400	
ND	3930.506	20	32	
U	3930.982	13	32	
CE II	3931.088	32	8	5 VAF

Table IIb.(1) CA II 3933.666 A
(con't)

Element	Wavelength (Angstrom)	Intensity Arc	Spark	Plasma
DY	3931.537	200		
RU	3931.759	50	63	
TI II	3932.017	20	32	
U II	3932.026	32	50	8 VAF
FE I	3932.630	79	40	
SC I	3933.381	63	63	
SM II	3933.592	200	200	
FE I	3933.605	200	200	
CO	3933.654	79		
CA II	3933.664	630	630	89 VAF
*CA II	3933.666	630	630	89 VAF
RU	3933.680	5	200	
CE	3933.731	63	63	
CO I	3933.914	63		
V I	3934.013	100	32	
RH I	3934.229	100	2	
GD I	3934.801	100	50	
ND	3934.823	63	32	
IR I	3934.844	200	50	
BA I	3935.725	79	32	
FE I	3935.815	100	8	
PR	3935.823	130	50	
CO I	3935.970	400	16	
NB	3936.023	5	200	
LA II	3936.221	100	50	
MN	3936.761	25	50	
FE I	3937.332	79	32	
NB	3937.438	32	32	
YB II	3938.250	40	500	
PR	3938.312	40	32	
NB	3938.547		100	
OS	3938.593	130	20	

Table IIa. CA II 3968.468 A

	Element	Wavelength (Angstrom)	Intensity Arc	Spark	Plasma
	U	3967.480	10		
	IR I	3967.507	3		
	DY	3967.517	8		
	CE	3967.530	5		
	P	3967.540			16
	XE I	3967.541			200
	U	3967.639		2	
	CE	3967.644	3		
	TB	3967.649	6		
	PR	3967.661	10	3	
	SM II	3967.675	8	6	
	Y II	3967.690	3	10	
	ND	3967.705	20	6	
----	SM	3967.779	6	2	
	CE	3967.915	3		
---	FE I	3967.969	63	16	
	U	3968.007	6	4	
	HF	3968.010	5		
	HG	3968.030			50
	YB	3968.030		3	
	V II	3968.094	25	40	
	TB	3968.146	2		
	PR	3968.158	25	10	
	TA	3968.160	4		
	RH I	3968.164	2		
	W	3968.171	8	8	
--	ZR I	3968.257	100	4	
	GD	3968.350	20		
	AR	3968.360			200
	FE	3968.370	2		
	U	3968.374		2	
	DY II	3968.395	320		10 VAF
	Y	3968.430	10	32	
	RU	3968.461	13	200	
	LU	3968.464	50		
	*CA II	3968.468	500	500	30 VAF
	CE	3968.469	32	32	
	NB	3968.471	3	10	
	IR	3968.475	25		
	W	3968.590	6	6	

Table IIa. CA II 3968.468 A
(con't)

	Element	Wavelength (Angstrom)	Intensity Arc	Spark	Plasma
--	BR	3968.660			8
	ZR	3968.723	3		
	TB	3968.730	2		
	MO	3968.748	8	50	
	EU	3968.870	3		
---	ND	3968.876	20	4	
	TH	3969.003	10	5	
	GD I	3969.005	40		
	MO	3969.015	3	3	
	U	3969.022	5	8	
	CR I	3969.061	79	50	
	CO I	3969.117	100	6	
	IN	3969.130	16		
	NB	3969.135		20	
	CE	3969.156	2		
	IR	3969.172	32	10	
	TE	3969.180			10
---	W	3969.199	13	10	
	EU	3969.230	20		
	DY	3969.233	6		
	FE I	3969.261	630	400	
	GD	3969.261	200		
	SR I	3969.261	32		
	TH	3969.343	3	2	
	U	3969.422	5		
	ER	3969.434	6		

Table IIb.(2) CA II 3968.468 A

Element	Wavelength (Angstrom)	Intensity Arc	Spark	Plasma
OS I	3963.628	500	50	
CR I	3963.690	320	320	
DY II	3968.395	320		10 VAF
RU	3968.461	13	200	
*CA II	3968.468	500	500	30 VAF
FE I	3969.261	630	400	

Table IIb.(2) CA II 3968.468 A
(con't)

Element	Wavelength (Angstrom)	Intensity Arc	Spark	Plasma
H I	3970.074			1 VAF
FE I	3971.329	200	130	
EU II	3971.990		2000	32 VAF
TA	3973.178		400	

Table IIa. CA I 3644.410 A

Element	Wavelength (Angstrom)	Intensity Arc	Spark	Plasma
CE	3643.450	5		
MO	3643.470		20	
IR	3643.481	2		
DY	3643.500	3		
W	3643.514		6	
TH	3643.515	5		
NB	3643.522	2	2	
FE I	3643.626	20	8	
ND	3643.630	5		
CU I	3643.631	5		
TM	3643.652	63	40	
--- FE I	3643.713	6	4	
SM	3643.715	5	3	
NB	3643.721	16	16	
TB	3643.760	32	3	
FE	3643.815	6		
V I	3643.864	40	32	
DY	3643.890	16	6	
NE II	3643.890			20
--- NI I	3643.941	2		
SM	3643.991	8		
EU	3644.080	2		
TB	3644.130	16	3	
-- FE	3644.185	4		
U	3644.245	20	2	
EU	3644.270	5		
TE	3644.270			25
CE	3644.297	8		
HG	3644.320			40
TH	3644.347	5	4	
HF II	3644.355	25	50	
HO	3644.390		4	
*CA I	3644.410	200	16	
AG	3644.420		2	
ND	3644.426		16	
XE II	3644.430			3
TE	3644.460			10
TI I	3644.460	3		
EU	3644.500	6		
PR	3644.543	8	3	

Table IIa. CA I 3644.410 A
(con't)

	Element	Wavelength (Angstrom)	Intensity Arc	Spark	Plasma
	CE II	3644.548	5		
	FE	3644.587	2		
--	ND	3644.660	10	4	
	SC	3644.686	2		
	CR II	3644.693	5	20	
	TI	3644.699	32	5	
	V I	3644.706	79	50	
	TH	3644.717	5	4	
	ER	3644.730	3	2	
	CA I	3644.765	32		
	FE	3644.798	20	6	
	U	3644.851	4	6	
	NE II	3644.860			13
	PR	3644.897	5	2	
---	XE II	3644.910			3
	TB	3644.930	16		
	NB	3644.935	5	10	
	CE	3644.967	2		
	EU	3644.990	5		
	U	3645.030	8	16	
	CA I	3645.080	3	2	
	FE I	3645.080	20	8	
----	PR	3645.114	5	2	
	ND	3645.167	6	2	
	EU	3645.170	10	5	
	CO I	3645.195	63	3	
	FE	3645.222	10		
	CE II	3645.228	5		
	CU I	3645.232	20	5	
	SM II	3645.290	10	6	
	IR	3645.303	25	2	
	SC II	3645.311	50	50	35 VAF
	NB	3645.356	5	5	
	TB	3645.380	50	16	
	SM II	3645.387	8	6	
	ER	3645.399	25	13	
	Y	3645.403	4	4	

Table IIb.(3) CA I 3644.410 A

Element	Wavelength (Angstrom)	Intensity Arc	Spark	Plasma
CO I	3639.444	200	20	
TH	3639.449	10	10	
RH I	3639.512	130	63	
PB I	3639.580	320	50	5 VAF
AIR	3639.600		16	
MO	3639.640		10	
CR I	3639.802	63	25	
TB	3639.820	32	8	
W	3640.132	10	8	
GD	3640.189	10	8	
ND	3640.235	16	10	
OS I	3640.329	200	40	
FE I	3640.390	320	200	
NB	3640.636	8	10	
RU I	3640.640	3	13	
U	3640.757	8	20	
U	3640.948	8	20	
V I	3641.096	100	32	
OS	3641.227	32	10	
TI II	3641.331	63	160	
GD	3641.380	13	13	
MN	3641.391	50	50	
W II	3641.408	13	40	
CR I	3641.470	32	25	
ND	3641.502	20	10	
TB	3641.660	63	32	
CO I	3641.788	63	8	
CR I	3641.830	40	16	
TA	3642.057	130	20	
MO	3642.204	5	10	
U	3642.443	10	8	
OS	3642.505	20	10	
TI I	3642.675	320	130	
TB	3642.680	16	8	
SC II	3642.785	63	50	11 VAF
PT I	3643.172	63	8	
CO I	3643.184	79	16	
CR II	3643.199	5	32	
PR	3643.317	25	8	
NB	3643.337	5	10	

Table IIb.(3) CA I 3644.410 A
(con't)

Element	Wavelength (Angstrom)	Intensity Arc	Spark	Plasma
MO	3643.470		20	
FE I	3643.626	20	8	
TM	3643.652	63	40	
NB	3643.721	16	16	
V I	3643.864	40	32	
HF II	3644.355	25	50	
*CA I	3644.410	200	16	
ND	3644.426		16	
CR II	3644.693	5	20	
V I	3644.706	79	50	
NB	3644.935	5	10	
U	3645.030	8	16	
FE I	3645.080	20	8	
SC II	3645.311	50	50	35 VAF
TB	3645.380	50	16	
ER	3645.399	25	13	
LA II	3645.414	100	63	
DY II	3645.416	320	100	13 VAF
PR	3645.539	20	8	
W II	3645.599	4	20	
GD	3645.634	20	20	
PR	3645.660	32	20	
FE	3645.825	79	63	
V	3645.905		20	
RU	3646.114	2	8	
CR	3646.161	20	8	
GD II	3646.196	200	160	10 VAF
TI I	3646.200	63	25	
PR	3646.299	50	16	
W II	3646.525	10	32	
U	3646.847	4	10	
V	3646.847		20	
TB	3647.060	32	8	
TM	3647.240	10	20	
NB	3647.309	10	20	
CR II	3647.394	2	20	
FE I	3647.426	20	16	
W	3647.525	8	10	
TH	3647.648	8	8	
CO I	3647.659	100	8	

Table IIb.(3) CA I 3644.410 A
(con't)

Element	Wavelength (Angstrom)	Intensity Arc	Spark	Plasma
TM	3647.730	32	16	
FE I	3647.844	500	400	
ND	3648.188	13	8	
PR	3648.299	32	10	
HF II	3648.350	5	8	
CR I	3648.534	10	8	
MO	3648.607	10	10	
OS	3648.806	100	10	
DY	3648.807	50	32	
TI II	3648.860	3	10	
V I	3648.966	79	50	
CR I	3648.997	40	20	
TH	3649.251	10	10	
FE I	3649.304	63	25	
CO	3649.351	200	4	

Table IIa. CD II 2265.017 A

	Element	Wavelength (Angstrom)	Intensity Arc	Spark	Plasma
	TI	2264.022	16		
	NB	2264.074	2	3	
	ND	2264.080		16	
	RH	2264.140	50	25	
	CO	2264.150		2	
	IR I	2264.172	13	3	
	W II	2264.182	3	13	
	MO	2264.200		5	
	XE II	2264.200			3
----	PD II	2264.282		25	
	U	2264.300	5	2	
	MO	2264.350	4	10	
	FE I	2264.389	32	5	
	CO	2264.418	10	2	
	NI II	2264.457	160	400	
---	NB	2264.553	2	8	
	FE II	2264.591		20	
	IR I	2264.607	32	16	
	LU	2264.630		2	
	PD II	2264.680			3
	MO	2264.742		20	
--	CO	2264.874	16		
	CE	2264.890		20	
	CR	2264.920		10	
	MO	2264.950		4	
	*CD II	2265.017	25	320	89 VAF
	TE	2265.020			10
	K	2265.040			32
	FE I	2265.050	10		
	IN	2265.060		6	
	IR	2265.163	5	50	
--	MO	2265.170		2	
	W II	2265.350	5	10	
	CU II	2265.353		5	
	NI II	2265.354		5	
	OS	2265.390	5		
	IR I	2265.454	13	3	
	CU	2265.458		16	
---	ZN II	2265.510			10
	MO	2265.520		8	

Table IIa. CD II 2265.017 A
(con't)

Element	Wavelength (Angstrom)	Intensity Arc	Spark	Plasma
TE I	2265.550			3 VAF
LA II	2265.580		3	
FE I	2265.590	4		
NB	2265.592		50	
XE II	2265.620			3
YB	2265.650	3	50	
W	2265.666	8	6	
NB	2265.671	5	20	
I	2265.690			32
SE	2265.720			5
CO II	2265.742	3	5	
-- PD II	2265.750		3	
CD II	2265.810		32	
AG II	2265.880		2	
XE II	2265.940			3
SN	2265.980			40
W	2265.980	3		
AU II	2265.990		3	
FE II	2265.994	4	10	
ZN II	2266.000			250

Table IIb.(3) CD II 2265.017 A

Element	Wavelength (Angstrom)	Intensity Arc	Spark	Plasma
OS	2260.110	25	50	
CU I	2260.528	25	6	
GD	2261.195	25	3	
OS	2261.730	25	50	
TA II	2262.300	13	20	12 VAF
OS	2262.320	25	6	
SB	2262.538	40	25	
CU I	2263.079	32	10	
RH	2263.430	5	200	
AL I	2263.453	63	25	5 VAF
RH	2264.140	50	25	
FE I	2264.389	32	5	

Table IIb.(3) CD II 2265.017 A
(con't)

Element	Wavelength (Angstrom)	Intensity Arc	Spark	Plasma
NI II	2264.457	160	400	
IR I	2264.607	32	16	
*CD II	2265.017	25	320	89 VAF
TE I	2265.550			3 VAF
RE	2266.090	25	2	
IR I	2266.331	25	10	
HF II	2266.520	32	40	
HF II	2266.830	63	79	
OS	2268.280	40	10	
PT	2268.840	25	20	
IR I	2268.898	25	32	
SN I	2268.910	100	100	25 VAF
AL I	2269.093	63	25	9 VAF
AL I	2269.212	16		9 VAF
ZR	2269.392	32		

Table IIa. CD I 2288.018 A

	Element	Wavelength (Angstrom)	Intensity Arc	Spark	Plasma
	SI	2287.060			10
	NI II	2287.084	100	500	
	CR 3	2287.160		2	
	AR	2287.190			20
	W	2287.190		3	
	FE I	2287.248	20	6	
--	NI I	2287.323	5		
	W	2287.470		3	
	P II	2287.492			20
	PT II	2287.499		25	
	RE I	2287.510			4 VAF
--	RU	2287.525	3		
	PD	2287.550			10
	FE I	2287.630	5		
	NI II	2287.650	2	20	
	W	2287.670	10		
	RU	2287.680	63		
--	KR II	2287.790			32
	AR	2287.800			63
	U	2287.800		2	
	CO	2287.809	13	4	
	TA	2287.840	5		
	TE	2287.840			5
	CE	2287.850		10	
	IR	2287.878	20	5	
	I	2287.890			13
	V	2287.927		16	
	*CD I	2288.018	1600	320	110 VAF
	IR	2288.090		5	
	V	2288.095	3	16	
	AS I	2288.120	250	5	36 VAF
	SB	2288.139	5		
	PR	2288.150		6	
	CL	2288.160			4
	PT II	2288.192	16	32	
	AU II	2288.250		2	
--	RH	2288.250		40	
	U	2288.300		2	
	NI I	2288.390	13	2	
	AR	2288.400			2

Table IIa. CD I 2288.018 A
(con't)

	Element	Wavelength (Angstrom)	Intensity Arc	Spark	Plasma
	MN	2288.420	6	3	
---	N II	2288.470			16
	W	2288.524		6	
	CO	2288.560		2	
	RH I	2288.570	25	10	21 VAF
	U	2288.580	2	2	
	PD	2288.590		4	
	ZR II	2288.614	2	2	
	V	2288.628		79	
	AU	2288.630		3	
	HF II	2288.630	5	6	
	FE	2288.678		3	
	W	2288.680	6	4	
	I II	2288.720			2
----	CD	2288.740		20	
	CO I	2288.786	16		
	NB	2288.866	2	5	
	TE	2288.900			5
	IR	2288.920	2		
	YB	2288.930		6	
	SB	2288.986	20	10	

Table IIb.(3) CD I 2288.018 A

Element	Wavelength (Angstrom)	Intensity Arc	Spark	Plasma
OS	2283.170		32	
TA	2283.190	2	16	
W	2283.270	10	5	
MO	2283.280		10	
AU II	2283.320		16	
U	2283.340		10	
V I	2283.376	5	5	
W	2283.380		3	
YB	2283.380	8	10	
ER	2283.410	3	5	
CO II	2283.519	10	16	

Table IIb.(3) CD I 2288.018 A
(con't)

Element	Wavelength (Angstrom)	Intensity Arc	Spark	Plasma
OS	2283.670	50	16	
U	2283.720		16	
V	2283.766		6	
YB	2283.980		8	
W	2283.990		5	
RH	2284.080		200	
FE I	2284.085	25	20	
NB	2284.348		16	
MO	2284.430	6	32	
CR 3	2284.440	2	10	
W	2284.440	4	5	
CR	2284.530	2	10	
HF II	2284.600	20	32	
IR	2284.605	20	5	
W	2284.620		8	
PR	2284.640		10	
V	2284.748		3	
TM	2284.800	20	10	
U	2284.830		4	
OS	2284.900	20	5	
V I	2284.984	2	4	
RH	2285.011		32	
TA	2285.020	3	13	
ND	2285.080		3	
MO	2285.140		20	
W	2285.170	10	3	
GA	2285.190		3	
NB	2285.221	8	16	
TA	2285.250	6	20	
RE	2285.310	25	6	
HF II	2285.430	6	6	
V	2285.455		100	
PT	2285.650		8	
U	2285.660		10	
W	2285.730		3	
IR	2285.740		10	
YB	2285.800	3	8	
ND	2285.860		3	
CO II	2286.156	40	320	43 VAF
TI II	2286.180	2	6	

Table IIb.(3) CD I 2288.018 A
(con't)

Element	Wavelength (Angstrom)	Intensity Arc	Spark	Plasma
PT II	2286.190		3	
W	2286.290	2	5	
NB	2286.352		4	
MO	2286.420	2	10	
AG	2286.460		13	
W	2286.500		3	
OS	2286.517	32	10	
PT II	2286.572		10	
TA	2286.590		6	
PR	2286.610		6	
RE	2286.620	4	25	
SN	2286.680	63	40	
CU II	2286.733		25	
NB	2286.750		4	
NB	2286.888	2	3	
NI II	2287.084	100	500	
W	2287.190		3	
FE I	2287.248	20	6	
W	2287.470		3	
PT II	2287.499		25	
RE I	2287.510			4 VAF
NI II	2287.650	2	20	
CO	2287.809	13	4	
CE	2287.850		10	
IR	2287.878	20	5	
V	2287.927		16	
*CD I	2288.018	1600	320	110 VAF
IR	2288.090		5	
V	2288.095	3	16	
AS I	2288.120	250	5	36 VAF
PR	2288.150		6	
PT II	2288.192	16	32	
RH	2288.250		40	
MN	2288.420	6	3	
W	2288.524		6	
RH I	2288.570	25	10	21 VAF
PD	2288.590		4	
V	2288.628		79	
AU	2288.630		3	
HF II	2288.630	5	6	

Table IIb.(3) CD I 2288.018 A
(con't)

Element	Wavelength (Angstrom)	Intensity Arc	Spark	Plasma
FE	2288.678		3	
W	2288.680	6	4	
CD	2288.740		20	
NB	2288.866	2	5	
YB	2288.930		6	
SB	2288.986	20	10	
TA II	2289.160	8	25	95 VAF
MO II	2289.190	4	20	
V	2289.219		32	
CR 3	2289.260		8	
PT I	2289.270	25	13	
W	2289.390	2	4	
IR	2289.393	5	40	
CA	2289.410		3	
RH	2289.580		16	
MO	2289.890	5	16	
MN	2289.970		3	
NI I	2289.976	20	20	
W	2289.980		3	
RH	2290.030	25	500	
MO	2290.060		20	
PR	2290.260		8	
MO	2290.320		20	
CO	2290.330		5	
NB	2290.380		25	
V	2290.544		50	
TM	2290.550		5	
W II	2290.560	2	10	
CR 3	2290.680		13	
IR	2290.800	3	50	
RE	2290.810	10	6	
MO	2290.870	3	20	
AG	2290.990		3	
CU II	2290.998		25	
ZR II	2291.113	79	16	
RE	2291.330	20	5	
W	2291.368		4	
PD	2291.440		20	
CO I	2291.455	13	5	
AU II	2291.520		16	

Table IIb.(3) CD I 2288.018 A
(con't)

Element	Wavelength (Angstrom)	Intensity Arc	Spark	Plasma
YB	2291.630		3	
HF II	2291.640	32	40	
U	2291.642		8	
NB	2291.645		5	
TH	2291.650		10	
PT	2291.720		20	
IR	2291.780		5	
TI II	2291.860		3	
HG	2291.920		5	
TA	2292.130	4	8	
OS	2292.210	16	5	
YB	2292.270		3	
LA II	2292.320		3	
W	2292.393		3	
PT I	2292.395	400	100	
TA	2292.544		20	
V	2292.589		25	
CU II	2292.681		5	
ND	2292.690		8	
HF II	2292.770	3	3	
FE II	2292.770		4	
YB	2292.810		3	
V	2292.855		250	
MO	2292.983		3	
CO II	2292.990	10	32	

Table IIa.   CD II 2144.382 A

	Element	Wavelength (Angstrom)	Intensity Arc	Intensity Spark	Intensity Plasma
	BI II	2143.400		13	
	YB	2143.420		5	
	BI II	2143.460		13	
	RE	2143.480	6	2	
	TE	2143.540			25
	TI	2143.610	5		
--	TA	2143.650	2	5	
	CO I	2143.679	3		
	V II	2143.713	3	4	
	LU	2143.840		3	
	YB	2143.860		4	
--	FE	2143.900	8		
	RH	2143.970		25	
	ZR II	2144.010		3	
	IR	2144.070		5	
	MO	2144.070		20	
	RE II	2144.085	2	32	
	W	2144.090	10	10	
	AS I	2144.100	50		
--	HF II	2144.170	6	6	
	NB	2144.200		10	
	PT I	2144.231	32	100	
	PT II	2144.231	32	100	100 VAF
	PD	2144.250	2	25	
	IR	2144.280	10	3	
	*CD II	2144.382	50	1600	120 VAF
	BI	2144.410		5	
	FE	2144.450	25		
	NB	2144.492	4	2	
	K	2144.500			20
	W	2144.500	8	10	
--	IR	2144.580		16	
	HG	2144.670			5
	CU II	2144.730		3	
	HF II	2144.730	3	3	
	YB	2144.740		32	
	NB	2144.800		4	
	MN	2144.800		10	
-	SB I	2144.860			12 VAF
	PT	2145.026	13	4	

Table IIa. CD II 2144.382 A
(con't)

	Element	Wavelength (Angstrom)	Intensity Arc	Spark	Plasma
	CD	2145.040			20
	SB	2145.040	20	20	
----	KR II	2145.080			10
	NI I	2145.180	8	10	
	FE I	2145.188	13	2	
	OS	2145.260	10		

Table IIb.(2) CD II 2144.382 A

Element	Wavelength (Angstrom)	Intensity Arc	Spark	Plasma
SB I	2139.760	32	32	16 VAF
V II	2140.087	10	79	20 VAF
SB	2141.760	32	32	
TE I	2142.810	630		73 VAF
AS I	2144.100	50		
PT I	2144.231	32	100	
PT II	2144.231	32	100	100 VAF
*CD II	2144.382	50	1600	120 VAF
FE	2144.450	25		
SB I	2144.860			12 VAF
TE I	2147.190			14 VAF
IR	2148.220	25	50	
SN I	2148.740	50	20	
P I	2149.108	16		39 VAF

Table IIa.   CE II 4186.599 A

	Element	Wavelength (Angstrom)	Intensity Arc	Spark	Plasma
	CL II	4185.610			20
	IR	4185.658	25		
	NB	4185.670	2	2	
	RE	4185.722	2		
	ER	4185.724	16		
	ND	4185.766	16	8	
	U	4185.780	6	6	
	PR	4185.820	16	5	
	MO	4185.825	40	40	
--	TB	4185.890	8		
	S	4185.950			16
	W	4186.020	13	2	
	ND	4186.033	8	4	
	U	4186.037	6		
--	NB	4186.105	5	8	
	TI I	4186.123	100	40	
	K II	4186.226			63
	TB	4186.240	10		
	IR	4186.280	2		
	MO	4186.281	16	13	
	TM	4186.310	5	8	
	ND	4186.311	25	3	
	SB II	4186.340		4	
--	CR	4186.359	50	10	
	PR	4186.395	13	3	
	EU	4186.420	6		
	U	4186.477	6	5	
	*CE II	4186.599	79	25	6 VAF
	ZR II	4186.688	3	3	
	ER	4186.710	8	2	
	ZR I	4186.777	3		
	U	4186.790	6	5	
	DY	4186.810	100	13	
--	HO	4186.840	3	3	
	CE	4186.860	2		
	YB	4186.900		10	
	U	4186.977	10	4	
--	FE I	4187.044	250	200	
	TB	4187.160	16		
	CO I	4187.250	50	3	

Table IIa. CE II 4186.599 A
(con't)

	Element	Wavelength (Angstrom)	Intensity Arc	Spark	Plasma
	LA I	4187.316	50	40	
----	CE II	4187.323	32	16	
	ZR	4187.467	4		
	ZR I	4187.564	10	3	
	FE	4187.589	3		

Table IIb.(3) CE II 4186.599 A

Element	Wavelength (Angstrom)	Intensity Arc	Spark	Plasma
FE I	4181.757	200	160	
AR I	4181.883			28 VAF
FE I	4182.386	79	32	
RU	4182.457	20	32	
RU I	4182.644	16	13	
RE I	4182.979	160		
V II	4183.434	3	20	
SM II	4183.757	10	16	
LU	4184.250	100	200	
GD	4184.264	160	160	
TI II	4184.329	8	20	
NB	4184.444	20	50	
FE I	4184.895	100	79	
NB	4185.541		32	
MO	4185.825	40	40	
TI I	4186.123	100	40	
MO	4186.281	16	13	
*CE II	4186.599	79	25	6 VAF
DY	4186.810	100	13	
FE I	4187.044	250	200	
LA I	4187.316	50	40	
CE II	4187.323	32	16	
TM	4187.620	320	32	
FE I	4187.801	200	160	
SM II	4188.121	10	25	
MO	4188.321	100	79	
PR II	4189.518	100	50	5 VAF

Table IIb.(3) CE II 4186.599 A
(con't)

Element	Wavelength (Angstrom)	Intensity Arc	Spark	Plasma
MN	4189.988	79	40	
MO	4190.005	20	16	
CR	4190.131	40	16	
GD	4190.147	100		
YB	4190.300	6	32	
AR I	4190.712			32 VAF
GD I	4190.785	100	40	
NB	4190.884	20	32	
AR I	4191.028			32 VAF
MO	4191.030		32	
CR I	4191.271	63	16	
FE I	4191.436	200	100	

Table IIa. CE II 4012.388 A

	Element	Wavelength (Angstrom)	Intensity Arc	Spark	Plasma
	FE I	4011.412	5		
	U	4011.450	8	10	
	RE	4011.514	32		
	MN	4011.531	16	16	
	PR	4011.532	5		
	TI I	4011.534	4		
	CE II	4011.560	16	3	
----	TE	4011.680			32
	EU II	4011.683	25		
	SM II	4011.719	8	4	
	FE I	4011.728	2		
	RU	4011.729	6		
	TH	4011.752	16	16	
	CE	4011.773	2		
	U	4011.780	10	5	
	W	4011.809	4	3	
	SE I	4011.880			200
---	RB	4011.900			16
	MN	4011.905	13	10	
	HF	4011.944	3		
	MO	4011.966	25	25	
	W	4012.096		5	
	K II	4012.100			20
	TA	4012.111	5	2	
--	CE	4012.139	4		
	CO I	4012.160	2		
	U	4012.161	6	4	
	NB	4012.168		100	
	ND II	4012.250	79	40	58 VAF
	ZR I	4012.252	20		
	RE	4012.260	25		
	MO	4012.270	3	3	
	*CE II	4012.388	63	20	4 VAF
	CE I	4012.388	63	20	
	TI II	4012.391	32	50	
	TB	4012.450	5		
	CR	4012.469	63	63	
	TH	4012.497	16	16	
	MO	4012.510	3	3	
	DY	4012.523	4		

Table IIa. CE II 4012.388 A
(con't)

	Element	Wavelength (Angstrom)	Intensity Arc	Spark	Plasma
--	ER	4012.580	4	2	
	ND	4012.704	16	10	
	U	4012.708	2	2	
	MO	4012.800	3	2	
	TI I	4012.806	13		
	EU	4012.816	20	25	
	DY	4012.820	5	2	
---	TB	4012.870	6		
	PR	4012.910	8		
	IN II	4012.960			10
	SE I	4012.960			160
---	U	4013.029	8		
	W	4013.183	6	5	
	TA	4013.190	5		
	MO	4013.210		40	
	ND	4013.224	10	5	
	PR	4013.230	10	4	
	TH	4013.264	5	5	
	CE	4013.265	2		
	NB	4013.270	3	5	
	TB	4013.276	20		
	U	4013.288	2	2	
	ND	4013.359	3	3	

Table IIb.(3) CE II 4012.388 A

Element	Wavelength (Angstrom)	Intensity Arc	Spark	Plasma
ND	4007.435	20	20	
SM II	4007.486	50	25	
RU	4007.535	20	10	
V II	4008.169	2	10	
RU I	4008.269	20	20	
NB	4008.280	5	10	
SM	4008.330	8	10	
GD I	4008.331	16	10	
MO	4008.667		20	

Table IIb.(3) CE II 4012.388 A
(con't)

Element	Wavelength (Angstrom)	Intensity Arc	Spark	Plasma
PR II	4008.714	160	50	5 VAF
W I	4008.753	50	50	
ND	4008.754	13	10	
TI I	4008.928	79	32	
U	4009.170	8	16	
MO	4009.366	20	25	
TI I	4009.663	63	25	
NB	4009.714	5	10	
FE I	4009.717	130	100	
W	4009.806		10	
C II	4009.900		10	
MO	4010.300		10	
AS II	4010.660		10	
ND	4011.069	16	10	
U	4011.450	8	10	
MN	4011.531	16	16	
TH	4011.752	16	16	
MN	4011.905	13	10	
MO	4011.966	25	25	
NB	4012.168		100	
ND II	4012.250	79	40	58 VAF
*CE II	4012.388	63	20	4 VAF
CE I	4012.388	63	20	
TI II	4012.391	32	50	
CR	4012.469	63	63	
TH	4012.497	16	16	
ND	4012.704	16	10	
EU	4012.816	20	25	
MO	4013.210		40	
RU I	4013.505	16	13	
TI I	4013.584	63	6	
FE I	4013.795	79	40	
FE	4013.824	200		
CO I	4013.945	320		
FE	4014.534	200	100	
CE II	4014.899	63	13	
W	4015.216	25	32	
TI I	4015.384	63	10	
PR	4015.389	50	32	
LA I	4015.393	100	2	

Table IIb.(3) CE II 4012.388 A
(con't)

Element	Wavelength (Angstrom)	Intensity Arc	Spark	Plasma
AU II	4016.050	10	16	
MO	4016.062		10	
NB	4016.079	5	32	
W	4016.527	10	13	
PR	4016.748	25	20	
V	4016.821		16	
FE I	4017.153	79	50	
V	4017.296		16	

Table IIa. CE II 4137.646 A

	Element	Wavelength (Angstrom)	Intensity Arc	Spark	Plasma
	ND	4136.752	3	5	
	CE	4136.769	5		
	U	4136.807	16	8	
----	CE	4136.897	5	2	
	MO	4136.950	16	8	
	FE	4137.004	100	79	
	LA I	4137.025	16		
	TB	4137.040	5		
	EU	4137.076	10		
	GD	4137.083	20	40	
	NB I	4137.095	100	63	
---	PR	4137.160	3		
	RU I	4137.234	25	16	
	MN	4137.257	40	5	
	ER	4137.272	8		
	TI I	4137.288	50	16	
	P	4137.300			16
	SE II	4137.310			50
--	CE II	4137.473	6	2	
	PR	4137.490	2		
	TM	4137.490	3		
	W	4137.543	13	13	
	NB	4137.593	3	5	
	RE I	4137.606	16		
	N I	4137.630			50
	*CE II	4137.646	25	13	6 VAF
	U	4137.711	6		
--	OS	4137.840	100	3	
	U	4137.903	6	3	
	IR	4137.923	5		
	LA II	4137.925	4	2	
	KR	4137.960			50
	ND	4137.989	10	4	
	GD	4138.020	5		
	I	4138.022			8
	CE	4138.102	13	3	
---	DY	4138.140	2		
	MO	4138.185	20	16	
	PR	4138.190	20	8	
	ZR	4138.210	10	10	

Table IIa. CE II 4137.646 A
(con't)

Element	Wavelength (Angstrom)	Intensity Arc	Spark	Plasma
EU	4138.300	2		
NB	4138.300	4	2	
ER	4138.335	10		
CE	4138.352	13	4	
TM	4138.360	79	8	
--- BR	4138.380			8
RE	4138.530	2		
DY	4138.540	10		
MO	4138.543	20	13	
BR	4138.580			5
NA	4138.600	3		

Table IIb.(3) CE II 4137.646 A

Element	Wavelength (Angstrom)	Intensity Arc	Spark	Plasma
TM	4132.690	13	16	
MO	4132.750	13	10	
TH	4132.751	10	13	
IN	4132.904		16	
U	4133.201	8	8	
LA	4133.334	16	25	
ND	4133.361	16	10	
RE I	4133.417	200		
CE II	4133.800	32	8	6 VAF
FE I	4133.862	50	6	
TH	4134.117	10	10	
GD I	4134.167	25		
SB II	4134.172		8	
CR	4134.389	25	3	
V I	4134.488	40	16	
NB	4134.591	10	10	
FE I	4134.681	160	100	
CD II	4134.780		16	
MN	4135.036	50	32	
YB	4135.100	16	50	
SM II	4135.135	5	10	

Table IIb.(3) CE II 4137.646 A
(con't)

Element	Wavelength (Angstrom)	Intensity Arc	Spark	Plasma
RH I	4135.272	320	160	
ND	4135.325	10	16	
MO	4135.383	10	10	
OS	4135.784	200	50	
RB II	4136.125		16	
TA	4136.198	79	32	
HO	4136.240	40	25	
W	4136.351	10	8	
TH	4136.383	5	8	
RE I	4136.446	160		
U	4136.807	16	8	
MO	4136.950	16	8	
FE	4137.004	100	79	
GD	4137.083	20	40	
NB I	4137.095	100	63	
RU I	4137.234	25	16	
MN	4137.257	40	5	
TI I	4137.288	50	16	
W	4137.543	13	13	
*CE II	4137.646	25	13	6 VAF
OS	4137.840	100	3	
MO	4138.185	20	16	
PR	4138.190	20	8	
ZR	4138.210	10	10	
TM	4138.360	79	8	
MO	4138.543	20	13	
U	4138.664	10	10	
U	4139.140	10	10	
NB	4139.436	10	10	
NB	4139.707	50	50	
FE I	4139.923	40	32	
TH	4140.239	16	16	
GD	4140.454	20	20	
SB II	4140.607		16	
PD I	4140.827	100		
MN	4141.063	50	32	
U	4141.228	20	32	
PR	4141.257	160	50	
PB	4141.420		10	
MO	4141.489		32	

Table IIb.(3) CE II 4137.646 A
(con't)

Element	Wavelength (Angstrom)	Intensity Arc	Spark	Plasma
TB	4141.560	25		
CD II	4141.580		10	
TH	4141.641	8	10	
LA II	4141.740	200	200	
HF II	4141.830		8	
CR	4142.193	32	8	
W	4142.256	16	10	
CE	4142.343	32	6	
CE	4142.398	32	32	
TH	4142.482	10	10	

Table IIa. CL II 4132.480 A

	Element	Wavelength (Angstrom)	Intensity Arc	Spark	Plasma
	GD	4131.482	50	50	
	ER	4131.502	10		
	NB	4131.527	2	10	
----	AR II	4131.730			79
	LA II	4131.740	3	3	
	U	4131.784	6	6	
	LU	4131.790	10		
	PR	4131.820	13	4	
	CE	4131.855	5	2	
	MO	4131.920	20	10	
---	CS II	4132.003			10
	V I	4132.017	13	10	
	FE I	4132.060	320	200	
	YB	4132.140	3		
	CO I	4132.155	16		
	W	4132.214	6	8	
	TB	4132.220	3		
--	MO	4132.230	20	20	
	PR	4132.230	16	8	
	EU	4132.240	4		
	GD	4132.281	25	20	
	RE I	4132.282	20		
	LI I	4132.290	400		
	CE	4132.313	8	2	
	BA	4132.431	10	4	
	CR II	4132.441		2	
	CE	4132.447	4		
	TB	4132.470	5		
	*CL II	4132.480			200
	CA I	4132.497		2	
	LA II	4132.501		10	
	ND	4132.554	4	5	
	LI I	4132.560			0.4 VAF
	LI I	4132.620			0.4 VAF
	CE II	4132.637	8	2	
--	TM	4132.690	13	16	
	MO	4132.750	13	10	
	TH	4132.751	10	13	
	SE	4132.760			200
	O II	4132.820			100

Table IIa. CL II 4132.480 A
(con't)

	Element	Wavelength (Angstrom)	Intensity Arc	Spark	Plasma
	TB	4132.830	10		
	DY	4132.850	10	2	
	V I	4132.877	2		
	FE I	4132.903	8		
---	IN	4132.904		16	
	MO	4133.002	16		
	SC I	4133.006	8	3	
	CE	4133.048	3		
	EU	4133.115	4		
	IR	4133.148	10		
	SM II	4133.171	4	4	
--	U	4133.201	8	8	
	I	4133.246			16
	LA	4133.334	16	25	
	ND	4133.361	16	10	
	DY	4133.370	10	4	
	RE I	4133.417	200		
	NB	4133.418	2	5	

Table IIb.(2) CL II 4132.480 A

Element	Wavelength (Angstrom)	Intensity Arc	Spark	Plasma
FE I	4127.612	100	79	
Y I	4128.304	160	32	
RH I	4128.870	320	160	
TA	4129.378	200	40	
EU II	4129.737	160	50	70 VAF
GD	4130.378	200	10	
BA II	4130.664	50	63	9 VAF
FE I	4132.060	320	200	
LI I	4132.290	400		
*CL II	4132.480			200
LI I	4132.560			0.4 VAF
LI I	4132.620			0.4 VAF
RE I	4133.417	200		
CE II	4133.800	32	8	6 VAF

Table IIb.(2) CL II 4132.480 A
(con't)

Element	Wavelength (Angstrom)	Intensity Arc	Spark	Plasma
FE I	4134.681	160	100	
RH I	4135.272	320	160	
OS	4135.784	200	50	
RE I	4136.446	160		
FE	4137.004	100	79	
NB I	4137.095	100	63	

Table IIa.   CO II 2388.915 A

Element	Wavelength (Angstrom)	Intensity Arc	Intensity Spark	Intensity Plasma
AR II	2387.960			40
OS	2387.990	13	5	
ZR I	2388.009	6		
RH	2388.020	4		
MO	2388.050		3	
V I	2388.081	25		
AU II	2388.100		2	
TH	2388.143	4	2	
-- CO I	2388.177	5		
RU	2388.189	20	8	
FE II	2388.202		3	
ZR	2388.210	2	2	
N II	2388.240			3
AL II	2388.250			2
V II	2388.260		4	
NB	2388.267	5	25	
PD II	2388.289	2	40	
ZN I	2388.300	2		
TA	2388.370	2	25	
CO	2388.376	3		
FE II	2388.387		2	
AU II	2388.390		3	
YB	2388.410		3	
IR	2388.410	10	2	
-- U	2388.420	2	3	
W	2388.550		8	
FE II	2388.627	25	32	20 VAF
BR	2388.630			5
-- MO II	2388.700		32	
PB	2388.774	40	20	
W	2388.800		5	
OS	2388.870	6		
*CO II	2388.915	10	32	50 VAF
IR	2388.916		3	
NI	2388.919	10	20	
V I	2388.920	25	2	
BR	2388.940			16
TM	2388.950	8	3	
MO	2388.960		10	
W	2389.072	13	3	

Table IIa. CO II 2388.915 A
(con't)

	Element	Wavelength (Angstrom)	Intensity Arc	Spark	Plasma
	AL II	2389.080			3
	MN	2389.080		6	
	OS	2389.080	16	3	
	AR	2389.110			5
	TA	2389.110	6	13	
	RE	2389.130	32	6	
--	V	2389.149		2	
	MO	2389.200	20	20	
	ZR	2389.209	25		
	RU	2389.258	4		
	W	2389.260	4	2	
	RH	2389.270		13	
	RE	2389.280	5		
	SE	2389.380			10
	FE	2389.402	2	2	
---	ZR	2389.405	2	4	
	CR I	2389.460	25		
	GE I	2389.470	2		
	ZR II	2389.530	2	6	
	PT	2389.533	25	20	
	IN I	2389.543	50		
	TM	2389.560		6	
	CO II	2389.565	13	20	
	MN	2389.590		6	
----	TA	2389.600	3	13	
	BR	2389.690			63
	V	2389.699	5	100	
	CR	2389.760		25	
	ER	2389.790		3	
	TE	2389.790			25
	W	2389.800		8	
	LA	2389.840		3	
	RH	2389.840		10	
	PD II	2389.860		2	

Table IIb.(3) CO II 2388.915 A

Element	Wavelength (Angstrom)	Intensity Arc	Spark	Plasma
RU	2383.963	13		
RE	2383.970	10	3	
MN	2384.050	40	3	
ZR	2384.166	25		
V I	2384.276	16		
TA	2384.280	6	20	
NI I	2384.380	16	4	
FE	2384.386	20	5	
PT II	2384.460		16	
OS	2384.620	32	5	
RH	2384.650	25		
IR	2384.810	4	79	
W	2384.817	13	13	
CU II	2384.840		16	
CO I	2384.860	10	4	
FE II	2385.007	8	16	
NI	2385.010	10	2	
PD II	2385.010		50	
PD	2385.170		16	
Y	2385.230	13		
RH	2385.440	3	25	
OS	2385.490	16	2	
RE	2385.500	13		
CR I	2385.740	25		
TE I	2385.780	630		17 VAF
CO	2385.816	10		
V	2385.816		100	
IR	2385.863	20	3	
OS	2386.040	2	20	
MO	2386.050		20	
RH	2386.140	79	8	
IR	2386.153	16		
CR I	2386.190	20		
AG II	2386.330		25	
CO II	2386.363	10	25	14 VAF
V I	2386.413	20		
PT II	2386.506	2	25	
IR	2386.580	25	20	
NI I	2386.580	20	5	
CO	2386.727	4	16	

Table IIb.(3) CO II 2388.915 A
(con't)

Element	Wavelength (Angstrom)	Intensity Arc	Spark	Plasma
OS	2386.750	10		
CR I	2386.780	20		
PT I	2386.808	25	5	
IR	2386.892	50	16	
V I	2386.956	25	10	
MO	2386.960		32	
MN	2387.020	2	32	
TA II	2387.060	20	50	80 VAF
MO	2387.180	13		
ZR II	2387.183	16	10	
OS	2387.290	40	16	
DY II	2387.360	8		6 VAF
CO I	2387.462	10	10	
RE	2387.480	20	6	
NB	2387.524	10	25	
NI I	2387.540	16	2	
AU I	2387.750	32	10	
RE	2387.750	16		
NI II	2387.762	6	32	
IR	2387.860		32	
RU	2387.900	63	3	
OS	2387.990	13	5	
V I	2388.081	25		
RU	2388.189	20	8	
NB	2388.267	5	25	
PD II	2388.289	2	40	
TA	2388.370	2	25	
IR	2388.410	10	2	
FE II	2388.627	25	32	20 VAF
MO II	2388.700		32	
PB	2388.774	40	20	
*CO II	2388.915	10	32	50 VAF
NI	2388.919	10	20	
V I	2388.920	25	2	
W	2389.072	13	3	
OS	2389.080	16	3	
RE	2389.130	32	6	
MO	2389.200	20	20	
ZR	2389.209	25		
CR I	2389.460	25		

Table IIb.(3) CO II 2388.915 A
(con't)

Element	Wavelength (Angstrom)	Intensity Arc	Spark	Plasma
PT	2389.533	25	20	
IN I	2389.543	50		
CO II	2389.565	13	20	
V	2389.699	5	100	
CR	2389.760		25	
ND	2389.920		20	
FE I	2389.971	16		
PT II	2390.064	16		
MO	2390.100	10	20	
RU	2390.120	10	3	
W II	2390.367	10	20	
RE	2390.440	20	6	
IR	2390.460		20	
V	2390.466		25	
AG II	2390.540		79	
IR	2390.617	40	6	
RH	2390.620	10	50	
YB	2390.730	10	20	
V I	2390.778	25	3	
MO	2390.780		25	
ZN II	2390.780		25	
OS	2390.950	32	3	
OS	2391.120	13		
IR	2391.178	50		
IR	2391.250		25	
V I	2391.263	25	3	
RE	2391.300	20	6	
CO	2391.369	10		
FE II	2391.474	8	20	
TH	2391.518		16	
MO	2391.710	3	20	
CU	2391.720		20	
ZR	2391.745	16		
OS	2391.780	5	20	
NB	2391.910		20	
NI II	2392.100	10	10	
LU	2392.190	32	100	
IR	2392.220		25	
OS	2392.220	13		
RH	2392.230		16	

Table IIb.(3) CO II 2388.915 A
(con't)

Element	Wavelength (Angstrom)	Intensity Arc	Spark	Plasma
MO	2392.330	2	20	
CR	2392.370	25	2	
RE	2392.380	20	6	
RH	2392.420		40	
RU	2392.424	79	6	
OS	2392.560	10		
CO	2392.600	10	16	
CU I	2392.627	1000	50	
RE	2392.710	10		
CR I	2392.890	40	2	
V I	2392.897	25		
W II	2392.927	8	16	
AG II	2392.960		25	
NI I	2392.960	10	3	
IR	2393.030	3	25	
NI I	2393.110	10		
RE	2393.120	16		
HF II	2393.183	20	40	
TA	2393.230		20	
RU	2393.253	79		
HF II	2393.362	50	79	140 VAF
W	2393.420	10	2	
V	2393.575		500	
OS	2393.590	10	2	
RE	2393.670	20		
PB I	2393.794	2500	1000	6 VAF
ZN	2393.810	16		
HF II	2393.828	100	160	140 VAF
OS	2393.860	32	5	
CO II	2393.900	8	16	
RE	2393.910	20	20	

Table IIa. CO II 2389.565 A

Element	Wavelength (Angstrom)	Intensity Arc	Spark	Plasma
FE II	2388.627	25	32	20 VAF
BR	2388.630			5
MO II	2388.700		32	
PB	2388.774	40	20	
W	2388.800		5	
-- OS	2388.870	6		
CO II	2388.915	10	32	50 VAF
IR	2388.916		3	
NI	2388.919	10	20	
V I	2388.920	25	2	
BR	2388.940			16
TM	2388.950	8	3	
MO	2388.960		10	
-- W	2389.072	13	3	
AL II	2389.080			3
MN	2389.080		6	
OS	2389.080	16	3	
AR	2389.110			5
TA	2389.110	6	13	
RE	2389.130	32	6	
V	2389.149		2	
MO	2389.200	20	20	
ZR	2389.209	25		
RU	2389.258	4		
W	2389.260	4	2	
RH	2389.270		13	
RE	2389.280	5		
-- SE	2389.380			10
FE	2389.402	2	2	
ZR	2389.405	2	4	
CR I	2389.460	25		
GE I	2389.470	2		
ZR II	2389.530	2	6	
PT	2389.533	25	20	
IN I	2389.543	50		
TM	2389.560		6	
*CO II	2389.565	13	20	
MN	2389.590		6	
TA	2389.600	3	13	
BR	2389.690			63

Table IIa. CO II 2389.565 A
(con't)

	Element	Wavelength (Angstrom)	Intensity Arc	Spark	Plasma
	V	2389.699	5	100	
	CR	2389.760		25	
	ER	2389.790		3	
	TE	2389.790			25
--	W	2389.800		8	
	LA	2389.840		3	
	RH	2389.840		10	
	PD II	2389.860		2	
	ND	2389.920		20	
	FE I	2389.971	16		
	CO I	2389.982	8		
	ZN II	2390.040			5
---	PT II	2390.064	16		
	NB	2390.075	2		
	MO	2390.100	10	20	
	OS	2390.110	2		
	U	2390.110	2		
	RU	2390.120	10	3	
	PD	2390.130	2	4	
	PT	2390.140	3		
	ZR	2390.149	2		
	ER	2390.150	6	3	
	HG	2390.160			5
	TA	2390.210		4	
----	FE II	2390.223			
	W II	2390.367	10	20	
	CO	2390.420	4		
	MO	2390.430		8	
	RU	2390.430	5	4	
	RE	2390.440	20	6	
	IR	2390.460		20	
	V	2390.466		25	
	KR II	2390.500			4
	AG II	2390.540		79	

Table IIb.(3) CO II 2389.565 A

Element	Wavelength (Angstrom)	Intensity Arc	Spark	Plasma
OS	2384.620	32	5	
MO	2384.650		13	
RH	2384.650	25		
IR	2384.810	4	79	
W	2384.817	13	13	
CU II	2384.840		16	
ND	2384.850		10	
FE II	2385.007	8	16	
PD II	2385.010		50	
CU II	2385.070		10	
PD	2385.170		16	
Y	2385.230	13		
W	2385.260	4	13	
RH	2385.440	3	25	
OS	2385.490	16	2	
W II	2385.490	3	10	
RE	2385.500	13		
RH	2385.600		10	
V	2385.618		10	
CR I	2385.740	25		
TE I	2385.780	630		17 VAF
V	2385.816		100	
IR	2385.863	20	3	
OS	2386.040	2	20	
MO	2386.050		20	
RH	2386.140	79	8	
IR	2386.153	16		
CR I	2386.190	20		
AG II	2386.330		25	
CO II	2386.363	10	25	14 VAF
V I	2386.413	20		
PT II	2386.506	2	25	
IR	2386.580	25	20	
NI I	2386.580	20	5	
CO	2386.727	4	16	
CR I	2386.780	20		
PT I	2386.808	25	5	
IR	2386.892	50	16	
V I	2386.956	25	10	
MO	2386.960		32	

Table IIb.(3) CO II 2389.565 A
(con't)

Element	Wavelength (Angstrom)	Intensity Arc	Spark	Plasma
MN	2387.020	2	32	
TA II	2387.060	20	50	80 VAF
MO	2387.180	13		
ZR II	2387.183	16	10	
OS	2387.290	40	16	
PT	2387.358	4	10	
DY II	2387.360	8		6 VAF
CO I	2387.462	10	10	
RE	2387.480	20	6	
NB	2387.524	10	25	
NI I	2387.540	16	2	
AU I	2387.750	32	10	
RE	2387.750	16		
NI II	2387.762	6	32	
RH	2387.820		10	
IR	2387.860		32	
RU	2387.900	63	3	
OS	2387.990	13	5	
V I	2388.081	25		
RU	2388.189	20	8	
NB	2388.267	5	25	
PD II	2388.289	2	40	
TA	2388.370	2	25	
FE II	2388.627	25	32	20 VAF
MO II	2388.700		32	
PB	2388.774	40	20	
CO II	2388.915	10	32	50 VAF
NI	2388.919	10	20	
V I	2388.920	25	2	
MO	2388.960		10	
W	2389.072	13	3	
OS	2389.080	16	3	
TA	2389.110	6	13	
RE	2389.130	32	6	
MO	2389.200	20	20	
ZR	2389.209	25		
RH	2389.270		13	
CR I	2389.460	25		
PT	2389.533	25	20	
IN I	2389.543	50		

Table IIb.(3) CO II 2389.565 A
(con't)

Element	Wavelength (Angstrom)	Intensity Arc	Spark	Plasma
*CO II	2389.565	13	20	
TA	2389.600	3	13	
V	2389.699	5	100	
CR	2389.760		25	
RH	2389.840		10	
ND	2389.920		20	
FE I	2389.971	16		
PT II	2390.064	16		
MO	2390.100	10	20	
W II	2390.367	10	20	
RE	2390.440	20	6	
IR	2390.460		20	
V	2390.466		25	
AG II	2390.540		79	
IR	2390.617	40	6	
RH	2390.620	10	50	
YB	2390.730	10	20	
V I	2390.778	25	3	
MO	2390.780		25	
ZN II	2390.780		25	
PT II	2390.800		10	
PD II	2390.840		13	
W II	2390.890	4	10	
OS	2390.950	32	3	
OS	2391.120	13		
IR	2391.178	50		
IR	2391.250		25	
V I	2391.263	25	3	
RE	2391.300	20	6	
FE II	2391.474	8	20	
TH	2391.518		16	
MO	2391.710	3	20	
CU	2391.720		20	
ZR	2391.745	16		
OS	2391.780	5	20	
NB	2391.910		20	
NI II	2392.100	10	10	
LU	2392.190	32	100	
IR	2392.220		25	
OS	2392.220	13		

Table IIb.(3) CO II 2389.565 A
(con't)

Element	Wavelength (Angstrom)	Intensity Arc	Spark	Plasma
RH	2392.230		16	
MO	2392.330	2	20	
CR	2392.370	25	2	
RE	2392.380	20	6	
RH	2392.420		40	
RU	2392.424	79	6	
NI II	2392.580	4	10	
CO	2392.600	10	16	
CU I	2392.627	1000	50	
TA	2392.710		13	
CR I	2392.890	40	2	
V I	2392.897	25		
W II	2392.927	8	16	
AG II	2392.960		25	
IR	2393.030	3	25	
RE	2393.120	16		
HF II	2393.183	20	40	
MO	2393.200	5	10	
TA	2393.230		20	
RU	2393.253	79		
HF II	2393.362	50	79	140 VAF
V	2393.575		500	
RE	2393.670	20		
PB I	2393.794	2500	1000	6 VAF
ZN	2393.810	16		
HF II	2393.828	100	160	140 VAF
RU	2393.836		13	
OS	2393.860	32	5	
RH	2393.890		13	
CO II	2393.900	8	16	
RE	2393.910	20	20	
IR	2393.930		25	
RU	2393.971	50		
CR	2394.010		50	
ND	2394.200		20	
V I	2394.272	6	32	
OS	2394.290	50	10	
IR	2394.326	32		
NI II	2394.516	20	20	8 VAF

Table IIa.   CO I   2378.622 A

	Element	Wavelength (Angstrom)	Intensity Arc	Spark	Plasma
	NB II	2377.630		16	
	PR	2377.670		6	
	RH	2377.810		50	
	ER	2377.830	6	2	
	TH	2377.832	10	8	
	MO	2377.850		5	
	RB	2377.850			5
	FE	2377.870	2		
	U	2377.870	2	16	
--	PD II	2377.916		32	
	W	2377.920	4		
	IR	2377.938	4		
	PT	2377.958	10	4	
	NB	2377.977	3	10	
	IR	2377.983	32	16	
	I	2377.990			32
	HF	2377.993		6	
	XE	2378.040			2
	ND	2378.060		2	
	PT	2378.063		10	
	PR	2378.090		10	
	W II	2378.120	4	8	
--	OS	2378.140	32	5	
	IN I	2378.141	16		
	TI I	2378.147	10		
	U	2378.160		32	
	ZR I	2378.243	10		
	V	2378.263	2		
	IR	2378.297	2		
	TA	2378.310	5	20	
	HG	2378.330	20	20	
--	AL I	2378.408	40	20	
	RE	2378.420	4		
	TA I	2378.520	2		
	FE II	2378.527		3	
	OS	2378.540	25	5	
	CA	2378.590		2	
	U	2378.600		8	
	W II	2378.600	5	13	
	*CO II	2378.622	25	50	31 VAF

Table IIa. CO I 2378.622 A
(con't)

	Element	Wavelength (Angstrom)	Intensity Arc	Intensity Spark	Intensity Plasma
	MO	2378.650	6	3	
	PD	2378.721		10	
	OS	2378.740	25	5	
	HF II	2378.810		2	
--	ER	2378.850		6	
	CO	2378.908	5		
	RH I	2378.930	8		
	NB	2378.937	2		
	PR	2378.980		10	
	FE II	2378.980	2	2	
	IN I	2378.996	10		
	W	2379.040	2		
	RE	2379.110	5	13	
---	NB	2379.112		4	
	GE I	2379.140	3	4	
	V	2379.150	4	20	
	CO I	2379.154	4		
	FE II	2379.275	13	16	
----	CO I	2379.358	4		
	IR	2379.379	25	8	
	OS	2379.390	40	10	
	LA	2379.420		63	
	K	2379.500			20
	PD II	2379.500		2	
	CS	2379.540			2
	W	2379.560		8	
	RU	2379.570		10	

Table IIb.(2) CO I 2378.622 A

Element	Wavelength (Angstrom)	Intensity Arc	Spark	Plasma
SB	2373.623	79	25	
CR I	2373.730	63		
OS	2374.330	25	3	
ZR I	2374.422	20	10	
OS	2374.510	25		
AG	2375.060	320	320	
RE	2375.070	25	6	
RU	2375.271	79	5	
RU	2375.631	50	79	
RE	2375.830	20	6	
AU I	2376.240	25	3	
OS	2376.290	25	5	
OS	2377.030	50	32	
IR I	2377.275	25	5	
RE	2377.380	20	6	
OS	2377.610	50	16	
IR	2377.983	32	16	
OS	2378.140	32	5	
HG	2378.330	20	20	
AL I	2378.408	40	20	
OS	2378.540	25	5	
*CO II	2378.622	25	50	31 VAF
OS	2378.740	25	5	
IR	2379.379	25	8	
OS	2379.390	40	10	
TI I	2379.690	100	200	
TL I	2379.690			7 VAF
OS	2379.840	25	8	
HF II	2380.302	32	63	
TL	2380.340	20	63	
CO I	2380.485	20	10	
OS	2380.820	32	20	
HF II	2381.000	20	40	
PD	2381.024	25	10	
RE	2381.140	40	6	
AS I	2381.180	79	4	
IR I	2381.622	32	2	12 VAF
RU	2381.993	50	160	
BH C	2382.000	32		
FE II	2382.039	40	100	65 VAF

Table IIb.(2) CO I 2378.622 A
(con't)

Element	Wavelength (Angstrom)	Intensity Arc	Spark	Plasma
OS	2382.460	32	5	
V	2382.467		100	
RH	2382.890	50	5	
TE I	2383.250	500	320	11 VAF
CR I	2383.330	20		
RH	2383.400	50	10	
CO II	2383.460	16	32	21 VAF
RE	2383.480	25	5	

Table IIa. CR II 2677.159 A

	Element	Wavelength (Angstrom)	Intensity Arc	Spark	Plasma
	RU	2676.190	8	100	
	RH	2676.250		100	
	P	2676.280			10
	W	2676.310		6	
	MN	2676.331	16		
	V	2676.350		20	
	RU	2676.353	50	3	
	CE	2676.355	2		
---	U	2676.410	4	4	
	W	2676.412	10		
	CE	2676.467	3		
	RH I	2676.483	2	2	
	MO	2676.484	3	25	
	ZR I	2676.538	3		
	CR	2676.542		3	
	CE	2676.623	4		
	HF II	2676.630	10	13	
	V I	2676.634	6		
--	U	2676.687	6	4	
	MN	2676.747		10	
	CE	2676.755	4		
	RU	2676.762		20	
	IR I	2676.828	32	10	
	DY	2676.840	3		
	FE	2676.877	2	25	
--	CL II	2676.950			160
	RU	2676.972	13		
	W	2676.997		5	
	CS	2677.010			2
	PD II	2677.025		10	
	RE	2677.037	25		
	V I	2677.119	10	4	
	P I	2677.120	10		2
	HE I	2677.135			5
	PT I	2677.149	790	200	
	*CR II	2677.159	32	320	42 VAF
	TE	2677.160			5
	XE	2677.180			25
	KR II	2677.200			6
	MN	2677.246		16	

Table IIa. CR II 2677.159 A
(con't)

	Element	Wavelength (Angstrom)	Intensity Arc	Spark	Plasma
	LU	2677.250	10		
	W	2677.278	13	10	
	CE	2677.293	2		
	RU	2677.318	13		
	BR	2677.340			4
	NE	2677.360			16
--	YB	2677.370		40	
	HF II	2677.578	10		
	ER	2677.589	2		
	CE	2677.592	5		
---	CD I	2677.640	100	25	
	NB	2677.660	2	50	
	U	2677.756	3	2	
	LA I	2677.760	2		
	RE	2677.764	25		
	LU	2677.770		10	
	W II	2677.791	8	20	
	V	2677.802	63	320	
	MN	2677.846		40	
	PD II	2677.851		32	
	TA	2677.860	2		
----	AIR	2677.900		3	
	W	2677.910	8	2	
	HO	2677.950		20	
	CO	2678.040		3	
	RU	2678.045		6	
	FE	2678.050	20	10	
	U	2678.052	3	13	
	NA II	2678.086	6		40
	NB	2678.098		10	

Table IIb.(3) CR II 2677.159 A

Element	Wavelength (Angstrom)	Intensity Arc	Spark	Plasma
FE I	2673.213	32	16	
EU	2673.409	130	50	
RU	2673.477	50	3	
NB	2673.567	10	500	
RU	2673.603	50	3	
IR I	2673.607	40	10	
RE I	2674.337	100		
RH	2674.441		200	
PT	2674.574	200	10	
FE	2675.283	32	16	
BH B	2675.300	63		
TA II	2675.901	160	200	68 VAF
AU I	2675.950	250	100	96 VAF
RU	2676.353	50	3	
IR I	2676.828	32	10	
PT I	2677.149	790	200	
*CR II	2677.159	32	320	42 VAF
CD I	2677.640	100	25	
V	2677.802	63	320	
EU	2678.280	160	100	
V	2678.568	32	160	
ZR II	2678.632	79		20 VAF
RU II	2678.758	100	320	83 VAF
FE I	2679.062	200	200	
NI II	2679.241		500	
V	2679.322	63	320	
TM	2679.570	32	50	
CO	2679.756	79		
MO	2679.854	32	20	
RE	2679.916	32		
TI I	2639.926	100	13	
TA	2680.060	32	50	
NA I	2680.335	63	10	
NA I	2680.443	40		
FE I	2680.454	63	32	
RH I	2680.631	63	10	
TA	2680.665	32	40	
FE	2681.589	50	25	
TA	2681.869	50		

Table IIa. CR II 3578.687 A

	Element	Wavelength (Angstrom)	Intensity Arc	Spark	Plasma
	CO I	3577.688	2		
	NB	3577.718	10	16	
	FE	3577.750	2		
	SM II	3577.774	20	3	
	U	3577.783	4		
	CE	3577.798	2	2	
	PR	3577.846	6		
	V I	3577.874	50	40	
	MN	3577.880	50	25	
	U	3577.919	10		
----	RB II	3577.959			16
	DY	3577.989	160	50	
	TB	3578.000	8		
	CO II	3578.030		32	
	CO I	3578.075	20		
	PR	3578.106	10	5	
	EU	3578.110	6		
---	ZR II	3578.226	5	6	
	NB	3578.234		5	
	TI I	3578.266	8		
	U	3578.327	8		
	ER	3578.330	3		
	GD	3578.340	5	5	
	FE I	3578.380	40	5	
	TB	3578.400	8		
	PR	3578.424	10	3	
--	DY	3578.470	4	2	
	EU	3578.490	2		
	PR	3578.571	2		
	GD	3578.580	8		
	NB	3578.584	16		
	V	3578.639	32	79	
	*CR II	3578.687	500	400	13 VAF
	TI II	3578.687	25	5	
	TB	3578.700	8		
	HG	3578.747			40
	PR	3578.750	8	5	
	TA	3578.762	2		
	DY	3578.800	2		
	CE	3578.822	3		

Table IIa. CR II 3578.687 A
(con't)

	Element	Wavelength (Angstrom)	Intensity Arc	Spark	Plasma
	LA II	3578.890	2	3	
	CO I	3578.903	8	2	
--	SE	3578.930			13
	CE	3578.950	3		
	CO I	3579.029	5		
	MO	3579.073		25	
	TA	3579.076	16		
	EU	3579.080	6		
	V I	3579.091	6		
	HO	3579.120		4	
	DY	3579.125	3		
--	RE I	3579.127	50		
	TB	3579.200	50	50	
	TH	3579.343	4	8	
	EU	3579.360	4		
	U	3579.364	6	10	
--	DY	3579.420	2		
	ER	3579.440	16		
	TA	3579.448		32	
	CA	3579.450		2	
	SM	3579.499	4	2	
	GD	3579.550	5	5	
	FE	3579.558	2		
	MN	3579.661	3		
	SM	3579.668	4	3	
	BA I	3579.672	10	8	

Table IIb.(1) CR II 3578.687 A

Element	Wavelength (Angstrom)	Intensity Arc	Spark	Plasma
IR I	3573.724	8	100	
TI II	3573.737	20	40	
DY	3573.838	79	79	
MO	3573.882	20	25	
FE	3573.888	40	32	
CR I	3574.039	50	16	

Table IIb.(1) CR II 3578.687 A (con't)

Element	Wavelength (Angstrom)	Intensity Arc	Spark	Plasma
TM	3574.060	20	32	
DY	3574.160	200	100	
V	3574.346		40	
GD	3574.738	25	25	
HO	3574.780	10	20	
CO I	3574.964	200	25	
CO I	3575.361	200	25	
FE	3575.375	40	32	
ZR I	3575.792	100	5	
NB	3575.848	50	79	
FE I	3575.980	79	25	
DY II	3576.250	320		6 VAF
SC II	3576.340	20	50	82 VAF
FE	3576.760	79	40	
NI II	3576.762	2	40	
ZR II	3576.854	16	25	10 VAF
DY	3576.873	200	50	
YB	3577.050	6	20	
V	3577.225		20	
CE	3577.458	320	13	
V I	3577.874	50	40	
MN	3577.880	50	25	
DY	3577.989	160	50	
CO II	3578.030		32	
V	3578.639	32	79	
*CR II	3578.687	500	400	13 VAF
MO	3579.073		25	
RE I	3579.127	50		
TB	3579.200	50	50	
TA	3579.448		32	
DY	3580.043	79	32	
RE	3580.133	79		
NB	3580.273	100	320	
V I	3580.825	50	50	
SC II	3580.927	13	40	67 VAF
FE I	3581.195	1000	630	
LA II	3581.680		20	
FE	3582.201	32	32	
V I	3582.814	40	25	
RE I	3583.022	100		

Table IIb.(1) CR II 3578.687 A
(con't)

Element	Wavelength (Angstrom)	Intensity Arc	Spark	Plasma
RH I	3583.098	200	130	
FE	3583.341	50	16	

Table IIa. CR II 2055.520 A

	Element	Wavelength (Angstrom)	Intensity Arc	Spark	Plasma
	TI II	2054.530	4	16	
	OS	2054.600	3		
	HG	2054.610			20
	W II	2054.678	10	20	40 VAF
----	IR	2054.770	2	32	
	V	2054.840	10	16	
	OS	2054.850	10	2	
	SI	2054.850	3		
	HF	2054.860	2		
	K	2054.900			20
	CU II	2054.966	16	32	
	RH	2054.990	3	3	
---	RE	2055.270	4	10	
	HF	2055.300	4		
--	RH	2055.310	2	40	
	OS	2055.400	3	5	
	ZR	2055.420		2	
	IR	2055.440	10	4	
	PD	2055.490		13	
	TA	2055.490	3		
	NI I	2055.500	10	2	
	*CR II	2055.520	100	320	49 VAF
	OS	2055.530	5	20	
	V II	2055.560		2	
--	IR	2055.720	16	25	
	BE I	2056.000		6	1 VAF
---	W II	2056.020	5	10	
	PT II	2056.057		13	
	RE	2056.060	2	6	
----	IR	2056.170		20	
	RE	2056.330	8	2	
	IR	2056.400	2		
	BE I	2056.520	100		
	IR	2056.520	4		

Table IIb.(2) CR II 2055.520 A

Element	Wavelength (Angstrom)	Intensity Arc	Spark	Plasma
RH	2051.250	2	100	
IR I	2052.220	13	10	49 VAF
W II	2054.678	10	20	40 VAF
*CR II	2055.520	100	320	49 VAF
BE I	2056.000		6	1 VAF
BE I	2056.520	100		
SI I	4058.130	16		23 VAF
TA	2059.080	6	25	68 VAF
PB	2059.630	500		

Table IIa.   CR I   4254.346 A

	Element	Wavelength (Angstrom)	Intensity Arc	Spark	Plasma
	CE	4253.356	40	3	
	GD	4253.370	50	4	
	SR	4253.500	2		
	CL II	4253.510			79
	TH	4253.544	8	3	
	ZR I	4253.569	20		
	MO	4253.576	8	5	
	ER	4253.582	10	2	
----	GD	4253.619	50	50	
	DY	4253.650	3		
	NB	4253.695	25	40	
	U	4253.700	8	2	
	SM II	4253.715	5	2	
	O II	4253.740			50
	BE I	4253.760			16
	EU	4253.814	16	3	
	CE	4253.831	3		
---	U	4253.848	13	3	
	ND	4253.868	13	3	
	TH	4253.875	6	4	
	FE	4253.930	2		
	O II	4253.980			100
	TB	4254.000	5		
	CE	4254.004	3		
	GD	4254.030	3		
	W	4254.060	8	2	
--	BE I	4254.120			5
	GA	4254.130		2	
	BI I	4254.152	10	10	
	W	4254.288	4	3	
	ER	4254.320	6		
	*CR I	4254.346	5000	1000	
	CE	4254.370	8		
	NB	4254.392	10	16	
	PR	4254.420	32	20	
	V II	4254.425	3	5	
	MO	4254.429	10	5	
	HO	4254.430	100	20	
--	TH	4254.458	8	6	
	NB	4254.694	10	10	

Table IIa. CR I 4254.346 A
(con't)

Element	Wavelength (Angstrom)	Intensity Arc	Spark	Plasma
CA	4254.701	20		
N I	4254.750	16		
--- YB	4254.770	4	4	
KR II	4254.850			100
CE	4254.905	8		
FE	4254.938	2		
AR	4254.950			10
MO	4254.955	25	25	
-- PR	4255.000	4		
RE	4255.110	2		
CE	4255.178	2		
TB	4255.240	25		
EU	4255.270	16		
RE	4255.339	2		

Table IIb.(1) CR I 4254.346 A

Element	Wavelength (Angstrom)	Intensity Arc	Spark	Plasma
LA II	4249.991	100	50	
FE I	4250.130	250	160	
MO	4250.689	5	130	
FE I	4250.790	400	250	
SM II	4251.788	200	200	
MO	4251.873	63	63	
NB	4252.972	32	50	
GD	4253.619	50	50	
*CR I	4254.346	5000	1000	
NB	4255.442	32	50	
SM II	4256.396	160	160	

Table IIa.   CU I   3273.962 A

	Element	Wavelength (Angstrom)	Intensity Arc	Spark	Plasma
	MN	3273.015	20	20	
	U	3273.023	5		
	V I	3273.027	32	5	
	ZR II	3273.047	50	79	25 VAF
	PT II	3273.058		5	
	RU I	3273.078	63	20	
	ER	3273.080	25	16	
	TB	3273.120	16		
	TA	3273.132	63	3	
	ND	3273.173	8	2	
----	U	3273.225	2	2	
	SM II	3273.316	6	3	
	ER	3273.325	6		
	AR	3273.360			5
	OS	3273.384	16	5	
	I	3273.450			3
---	SM	3273.477	32	8	
	FE II	3273.499		6	
	NB	3273.509		20	
	CE	3273.516	4		
	O II	3273.540			32
	MO	3273.582		25	
	U	3273.596	5		
	SC I	3273.619	32	13	
	RU	3273.621	2	5	
	HF II	3273.655	20	10	
	AU II	3273.680		2	
--	TH	3273.884	10	16	
	NB	3273.886	20	100	
	CE	3273.926	5		
	CO	3273.931	10		
	CA	3273.958	2	4	
	*CU I	3273.962	320	1600	31 VAF
	CE	3273.964	5		
	SB	3273.970		4	
	IN II	3274.020		16	
	TI I	3274.047	6		
	CE	3274.064	10		
	CE	3274.112	8		
--	MO	3274.204		5	

Table IIa. CU I 3273.962 A
(con't)

Element	Wavelength (Angstrom)	Intensity Arc	Spark	Plasma
NA II	3274.220	16		40
TB	3274.240	63		
EU	3274.290	3		
TH	3274.399	6	3	
AG	3274.410		5	
FE	3274.453	79	63	
--- TA	3274.458	32		
IR I	3274.561	4		
PR	3274.576	8		
CE	3274.616	8		
MO	3274.628		32	
BE II	3274.640			50
CA I	3274.661	20		
--- RU I	3274.706	63	25	
ZR II	3274.713	25	32	10 VAF
ER	3274.747	10		
RH	3274.778	2		
NB	3274.788		10	
CE	3274.864	32	8	
NI II	3274.900		5	
XE	3274.940			2
TA	3274.947	200	32	

Table IIb.(2) CU I 3273.962 A

Elemgnt	Wavelength (Angstrom)	Intensity Arc	Spark	Plasma
OS	3269.209	200	20	
GE I	3269.494	320	320	
FE I	3271.002	320	320	
RH I	3271.612	200	63	
ZR II	3273.047	50	79	25 VAF
*CU I	3273.962	320	1600	31 VAF
ZR II	3274.713	25	32	10 VAF
TA	3274.947	200	32	
OS	3275.201	200	16	
V II	3276.124	50	200	19 VAF

Table IIa. CU I 3247.540 A

	Element	Wavelength (Angstrom)	Intensity Arc	Spark	Plasma
	TH	3246.581	5	5	
	CE II	3246.674	32	3	
	CE I	3246.674	32	3	
	NB	3246.689		5	
	NB	3246.780	5		
	EU	3246.780	4	2	
----	SM	3246.843	10	2	
	TA	3246.900	32		
	FE I	3246.962	100	63	
	CO I	3246.997	32		
	KR II	3247.000			13
	U	3247.034	8		
---	LA I	3247.040	8	2	
	CE	3247.118	3		
	SM	3247.170		10	
	FE II	3247.171		10	
	CO I	3247.179	79		
	TB	3247.180	16	3	
	U	3247.209	2		
	FE	3247.213	10	10	
	CE	3247.250	3		
	CR I	3247.274	20		
	FE I	3247.278	20	10	
--	EU	3247.300	4		
	SM	3247.366	4		
	FE II	3247.393		2	
	NB	3247.474	50	100	
	CS	3247.500			4
	ER	3247.520	20	2	
	EU	3247.530	50	5	
	*CU I	3247.540	5000	2000	56 VAF
	MN	3247.542	130		
	SB	3247.547	2	10	
	AR	3247.550			3
	AG	3247.550	16	16	
	CE	3247.552	16		
	IN	3247.610		16	
	MO	3247.621	32	20	
	IR I	3247.667	3		
	HF I	3247.670	400	79	

Table IIa. CU I 3247.540 A
(con't)

	Element	Wavelength (Angstrom)	Intensity Arc	Spark	Plasma
	U	3247.709	4	5	
	XE	3247.740			4
--	TB	3247.790	8		
	RN	3247.800			10
	CE	3247.898	3		
	V	3247.908		5	
	OS	3247.996		5	
---	KR II	3248.030			6
	CE	3248.065	5		
	SM	3248.140	10	4	
	NE II	3248.150			6
	FE I	3248.206	200	160	
---	CA	3248.280		3	
	U	3248.336	5	3	
	DY	3248.364	13	4	
	CE	3248.428	13		
	NI I	3248.457	160	2	
	PT I	3248.502	2		
	MN	3248.516	100	100	
	TA	3248.522	100	3	
	CE	3248.528	4		

Table IIb.(1) CU I 3247.540 A

Element	Wavelength (Angstrom)	Intensity Arc	Spark	Plasma
PD I	3242.703	2000	630	39 VAF
FE I	3244.190	320	200	
LA II	3245.120	400	320	
FE I	3245.984	200	160	
NB	3247.474	50	100	
*CU I	3247.540	5000	2000	56 VAF
FE I	3248.206	200	160	
MN	3248.516	100	100	
TI II	3248.602	25	200	
TI I	3248.602	25	200	
CD II	3250.170		100	

Table IIb.(1) CU I 3247.540 A
(con't)

Element	Wavelength (Angstrom)	Intensity Arc	Spark	Plasma
NB	3250.270	5	100	
MO	3250.747	2	100	
FE I	3251.235	320	160	
DY	3251.260	100	100	
PD I	3251.640	200	500	
TI II	3251.911	50	160	
CD I	3252.525	320	320	

Table IIa.   CU II 2246.995 A

	Element	Wavelength (Angstrom)	Intensity Arc	Spark	Plasma
	MO	2246.050		16	
	SN I	2246.053	100	100	25 VAF
	RH	2246.070		16	
	AG II	2246.130		4	
	CO	2246.150		3	
	NB	2246.175	10		
	U	2246.230		4	
---	K	2246.320			5
	V	2246.333		2	
	W	2246.360	4	6	
	RH	2246.380		50	
	AG II	2246.412	25	320	2 VAF
	BI	2246.418		2	
	NB	2246.423	3		
	AU	2246.430		16	
	ND	2246.470		13	
	RE	2246.470	25	5	
---	NB	2246.495		4	
	PT II	2246.505	2	10	
	W II	2246.630	6	13	
	AU II	2246.680		20	
--	NB	2246.761		10	
	BI	2246.770	8		
	CD	2246.800			2
	PB I	2246.888	32	100	9 VAF
	IR	2246.900	10	100	
	FE II	2246.907		20	
	CD II	2246.929		2	
	MO	2246.950		20	
	NB	2246.980		2	
	W	2246.980		5	
	SB	2246.990		4	
	*CU II	2246.995	32	500	39 VAF
	PD II	2247.010		8	
	GE	2247.040	2		
	U	2247.080		8	
	MO	2247.110		16	
--	NI II	2247.230	2	13	
	W	2247.320		4	
	PD II	2247.400			3

Table IIa. CU II 2246.995 A
(con't)

	Element	Wavelength (Angstrom)	Intensity Arc	Spark	Plasma
---	FE	2247.460	3		
	BH C	2247.500	13		
	RE	2247.510	4	6	
	V I	2247.516	5		
	PT II	2247.593		10	
	IR	2247.660		20	
	W	2247.670	3	13	
	OS	2247.680	40	6	
	FE II	2247.692		50	
	CR	2247.700		8	
----	HG	2247.700			5
	CO I	2247.869	4		
	CR	2247.920		13	
	NB	2247.993	10		

Table IIb.(2) CU II 2246.995 A

Element	Wavelength (Angstrom)	Intensity Arc	Spark	Plasma
RH	2242.050	25	25	
OS	2242.100	32	5	
RE	2242.410	25	10	
PB	2242.610	16	16	
CU II	2242.613	25	50	
IR II	2242.680	50	320	11 VAF
Y II	2243.060	25	32	33 VAF
HF II	2243.180	16	20	
RU	2243.220	20		
RE	2243.290	16	6	
RE	2244.120	16	4	
CU I	2244.265	25	10	
NI I	2244.480	16	3	
NI I	2244.530	40		
PT I	2244.970	25	10	
CO II	2245.127	16	32	
RE	2245.140	16	2	
OS	2245.250	20		

Table IIb.(2) CU II 2246.995 A
(con't)

Element	Wavelength (Angstrom)	Intensity Arc	Spark	Plasma
PT II	2245.518	25	32	36 VAF
IR	2245.760	10	160	
SN I	2246.053	100	100	25 VAF
AG II	2246.412	25	320	2 VAF
RE	2246.470	25	5	
PB I	2246.888	32	100	9 VAF
*CU II	2246.995	32	500	39 VAF
OS	2247.680	40	6	
ZR I	2248.050	40		
RH	2248.740	20	3	
AG II	2248.740	16	160	1 VAF
W II	2248.750	20	25	67 VAF
FE I	2248.857	32		
PT	2249.300	25	13	
BI I	2249.382	16		
RE	2249.460	20		
PT	2249.900	25	8	
MO	2249.930	20	20	
ZR	2250.748	32		
SN	2251.150	25	50	
RE	2251.290	16		

Table IIa. DY II 3531.712 A

	Element	Wavelength (Angstrom)	Intensity Arc	Spark	Plasma
	IR	3530.739	10	2	
	W	3530.756	8	6	
	V II	3530.773	40	100	
	MO	3530.809		3	
	NB	3530.824	3	16	
	PR	3530.837	10	2	
	ZR II	3530.851	5	5	
	V I	3530.869		20	
	HF	3530.874	5		
	DY	3530.880	4		
	CE	3530.946	10		
----	W	3531.021	6	6	
	CR	3531.083	16		
	CA	3531.090		2	
	U	3531.113	8	20	
	OS	3531.129	16	13	
	EU	3531.140	10	10	
---	AR II	3531.220			3
	HF	3531.227	5	3	
	YB	3531.250	2	5	
	ER	3531.264	13		
	MO	3531.303	3	3	
	CS	3531.376			4
	ND	3531.390	4		
	RU	3531.390	63	10	
	W	3531.440		10	
	FE I	3531.446	3		
	TH	3531.451	4		
--	TA	3531.584	32	3	
	CE	3531.593	20	2	
	RB II	3531.602			100
	PR	3531.623	4	2	
	U	3531.641	16		
	TB	3531.700	16	50	
	Y	3531.707	6		
	RN I	3531.710			5
	*DY II	3531.712	100	100	30 VAF
	ND	3531.712	6	4	
	HO	3531.740	10	20	
	RU	3531.794	8		

Table IIa. DY II 3531.712 A
(con't)

	Element	Wavelength (Angstrom)	Intensity Arc	Spark	Plasma
	IR I	3531.798	2		
	EU	3531.832	13	2	
	MN	3531.848	40	32	
	S	3531.900			5
--	TH	3531.935	4	3	
	MN	3531.998	50	8	
	SM	3532.008	5		
	U	3532.051		2	
	RU	3532.055	5		
	IR	3532.060	2		
	MN	3532.121	50	32	
--	TA	3532.208	16	2	
	EU	3532.230	5		
	IR	3532.267	13	3	
	V	3532.276		25	
	RN	3532.300			10
	CE	3532.390	4		
	DY	3532.440	6	2	
--	ZR	3532.460	8		
	NB	3532.530	3	5	
	SM	3532.541	20	4	
	MO	3532.550		10	
	FE	3532.576	5	2	
	CE II	3532.608	6		
	HG	3532.630			200
	N	3532.650			16
	TB	3532.700	32	8	

Table IIb.(3) DY II 3531.712 A

Element	Wavelength (Angstrom)	Intensity Arc	Spark	Plasma
CO I	3526.849	320	25	
FE I	3527.797	100	79	
NI I	3527.982	200	16	
RH I	3528.024	1000	160	24 VAF
NB	3528.476	3	50	
OS I	3528.602	400	50	
CO I	3529.033	200		
TI I	3529.430	1000	200	
TL I	3529.430			2 VAF
CO I	3529.813	1000	32	
FE	3529.820	130	79	
OS	3530.062	100	20	
V II	3530.773	40	100	
TB	3531.700	16	50	
*DY II	3531.712	100	100	30 VAF
OS I	3532.800	100	20	
FE	3533.010	50	79	
FE I	3533.202	50	50	
CO I	3533.358	200		
DY	3534.963	130		
NB	3535.301	320	500	
TI II	3535.412	16	130	
HF II	3535.545	16	50	
SC II	3535.729	16	32	24 VAF
DY II	3536.024	130	10	10 VAF
TM	3536.210	40	63	
FE I	3536.557	320	200	

Table IIa. DY II 4000.454 A

Element	Wavelength (Angstrom)	Intensity Arc	Spark	Plasma
HO	3999.560	6	2	
BR	3999.620			8
CR	3999.679	40	10	
-- NB	3999.709		5	
DY	3999.820	3	2	
MO	3999.864		5	
CR	3999.945	3		
U	3999.950	6	8	
-- N	3999.980			16
TB	4000.010	5		
V I	4000.078	8	2	
W	4000.093		8	
GD	4000.160	10		
PR	4000.190	50	25	
-- FE I	4000.266	8		
TH	4000.287	8	3	
MO	4000.386	6	5	
FE	4000.452	32	10	
*DY II	4000.454	400	320	8 VAF
TB	4000.460	16	2	
PR	4000.478	8	4	
ND	4000.493	10	5	
MO	4000.497	8	8	
XE	4000.550			3
ND	4000.562	8	3	
NB	4000.605	2	50	
HO	4000.620	4		
CR	4000.630	5		
CE	4000.675	5		
W	4000.694	13	10	
- EU	4000.698	3		
KR I	4000.720			2
U	4000.732	6		
CE	4000.799	4		
EU	4000.807	3		
YB	4000.890		6	
PR	4000.910	20	8	
- U	4000.910	5	6	
CE II	4001.055	20	2	
TH	4001.065	10	5	

Table IIa. DY II 4000.454 A
(con't)

	Element	Wavelength (Angstrom)	Intensity Arc	Spark	Plasma
	ZR I	4001.087	10		
	NB	4001.130	10	16	
	V	4001.180		2	
	MN	4001.183	13	5	
	RE I	4001.190	5		
----	K II	4001.200			40
	ZR I	4001.227	10		
	GD	4001.240	79	3	
	U	4001.246	10	3	
	TB	4001.280	5		
	EU	4001.310	6	2	
	W	4001.373	10	10	
	LA I	4001.380	3		
	CR	4001.444	200	79	
	BR	4001.450			4

Table IIb.(2) DY II 4000.454 A

Element	Wavelength (Angstrom)	Intensity Arc	Spark	Plasma
LA II	3995.750	630	320	22 VAF
GD	3996.325	100	100	
TM	3996.516	200	40	
DY	3996.699	200	79	
FE I	3997.396	320	160	
CO I	3997.906	200	20	
FE I	3998.055	160	100	
TI I	3998.640	160	100	
*DY II	4000.454	400	320	8 VAF
CR	4001.444	200	79	
FE I	4005.246	250	200	

Table IIa.   ER II 4007.967 A

Element	Wavelength (Angstrom)	Intensity Arc	Spark	Plasma
GD I	4006.969	3		
SM	4006.980	3		
NB	4006.995		5	
W	4007.003		6	
CE	4007.022	2		
TH	4007.027	20	20	
MN	4007.037	10	5	
DY	4007.137	6		
TI I	4007.192	16		
TA	4007.233	4	2	
FE I	4007.273	79	50	
BR	4007.330			10
HF II	4007.350	5	4	
YB	4007.360	5		
ND	4007.435	20	20	
MO	4007.450	4	5	
CE	4007.451	4		
U	4007.469	2	2	
PD I	4007.470	5		
SM II	4007.486	50	25	
RU	4007.535	20	10	
IN II	4007.543			16
CE	4007.588	16	4	
ZR I	4007.601	25		
IN II	4007.608			10
LA	4007.662	3	2	
EU	4007.687	5		
U	4007.689	2	4	
TB	4007.750	3		
DY	4007.770	13	2	
PR	4007.780	8	3	
S II	4007.780			5
SE II	4007.900			160
U	4007.934	8	3	
CO I	4007.943	3		
*ER	4007.967	32	6	
EU	4007.980	6		
MN	4008.020	16	5	
IR I	4008.052	13	2	
MO	4008.054	4	5	

Table IIa. ER II 4007.967 A
(con't)

	Element	Wavelength (Angstrom)	Intensity Arc	Spark	Plasma
	TI I	4008.062	50	6	
	KR II	4008.080			25
	SM II	4008.091	10	3	
	V II	4008.169	2	10	
	ER	4008.185	8		
--	TH	4008.216	10	8	
	RU I	4008.269	20	20	
	NB	4008.280	5	10	
	SM	4008.330	8	10	
	GD I	4008.331	16	10	
	ND	4008.416	5	3	
	CE	4008.446	6		
---	HF II	4008.460	5	8	
	KR II	4008.480			10
	DY	4008.490	5		
	CE	4008.664	8		
	MO	4008.667		20	
	U	4008.700	2		
----	PR II	4008.714	160	50	5 VAF
	W I	4008.753	50	50	
	ND	4008.754	13	10	
	BR	4008.760			20
	EU	4008.872	8	2	
	FE	4008.873	5		
	U	4008.918	8		
	GD	4008.922	20	3	
	TI I	4008.928	79	32	
	IR I	4008.967	5		

Table IIb.(3) ER II 4007.967 A

Element	Wavelength (Angstrom)	Intensity Arc	Spark	Plasma
MO	4002.970		20	
CU I	4003.038	40		
TH	4003.111	10	10	
U	4003.196	8	5	
MN	4003.255	20	5	
TH	4003.316	16	16	
CR	4003.330		20	
U	4003.405	6	10	
SM II	4003.454	32	20	
OS	4003.480	50	6	
TA	4003.704	16	5	
EU	4003.715	20	25	
FE	4003.767	32	79	
CE	4003.771	40	20	
TI I	4003.806	63	50	
W	4003.836		6	
GD	4003.855	8	8	
CR	4003.921	32	13	
ND	4004.010	20	16	
OS	4004.024	50	6	
U	4004.063	16	20	
ND	4004.264	16	10	
U	4004.617	3	5	
PR	4004.714	20	25	
W	4004.747		6	
FE	4004.843	10	6	
RE	4004.932	32		
MO	4005.120		20	
OS	4005.155	32	20	
FE I	4005.246	250	200	
W	4005.404	8	10	
TH	4005.549	20	32	
TB	4005.550	100	130	
CE	4005.639	20	6	
RU I	4005.640	25	32	
V II	4005.712	10	32	
W	4005.895	6	6	
NB	4005.927	2	5	
TI I	4005.971	32	3	
MO	4006.049	20	20	

Table IIb.(3) ER II 4007.967 A
(con't)

Element	Wavelength (Angstrom)	Intensity Arc	Spark	Plasma
FE	4006.315	63	32	
AS II	4006.340		50	
TH	4006.386	10	10	
U	4006.395	3	5	
SM II	4006.585	8	8	
RU I	4006.598	25	16	
FE	4006.631	20	16	
CD	4006.680		5	
MO	4006.700	5	5	
PR	4006.704	10	5	
SM II	4006.834	8	5	
TA	4006.835	32	20	
MO	4006.950		5	
NB	4006.995		5	
W	4007.003		6	
TH	4007.027	20	20	
MN	4007.037	10	5	
FE I	4007.273	79	50	
HF II	4007.350	5	4	
ND	4007.435	20	20	
MO	4007.450	4	5	
SM II	4007.486	50	25	
RU	4007.535	20	10	
CE	4007.588	16	4	
U	4007.689	2	4	
*ER	4007.967	32	6	
MN	4008.020	16	5	
MO	4008.054	4	5	
TI I	4008.062	50	6	
V II	4008.169	2	10	
TH	4008.216	10	8	
RU I	4008.269	20	20	
NB	4008.280	5	10	
SM	4008.330	8	10	
GD I	4008.331	16	10	
HF II	4008.460	5	8	
MO	4008.667		20	
PR II	4008.714	160	50	5 VAF
W I	4008.753	50	50	
ND	4008.754	13	10	

Table IIb.(3) ER II 4007.967 A
(con't)

Element	Wavelength (Angstrom)	Intensity Arc	Spark	Plasma
TI I	4008.928	79	32	
TH	4009.066	10	8	
U	4009.170	8	16	
GD	4009.210	50	2	
MO	4009.366	20	25	
ND	4009.367	8	5	
U	4009.410	4	8	
TI I	4009.663	63	25	
NB	4009.714	5	10	
FE I	4009.717	130	100	
W	4009.806		10	
C II	4009.900		10	
DY	4010.080	10	4	
MO	4010.136	5	5	
EU	4010.180	8	4	
MO	4010.300		10	
W	4010.375	10	6	
ND	4010.454	20	6	
PR	4010.643	8	5	
AS II	4010.660		10	
W	4011.024	4	5	
ND	4011.069	16	10	
DY	4011.295	13	8	
U	4011.450	8	10	
RE	4011.514	32		
MN	4011.531	16	16	
SM II	4011.719	8	4	
TH	4011.752	16	16	
U	4011.780	10	5	
MN	4011.905	13	10	
MO	4011.966	25	25	
W	4012.096		5	
U	4012.161	6	4	
NB	4012.168		100	
ND II	4012.250	79	40	58 VAF
CE II	4012.388	63	20	4 VAF
CE I	4012.388	63	20	
TI II	4012.391	32	50	
CR	4012.469	63	63	
TH	4012.497	16	16	

Table IIb.(3) ER II 4007.967 A
(con't)

Element	Wavelength (Angstrom)	Intensity Arc	Spark	Plasma
ND	4012.704	16	10	
EU	4012.816	20	25	

Table IIa. ER II 3372.750 A

	Element		Wavelength (Angstrom)	Intensity Arc	Spark	Plasma
	DY		3371.760	8	5	
	TH		3371.800	4	10	
	CE		3371.845	8		
	O	II	3371.850			5
	RU	I	3371.860	63	20	
	NE	II	3371.870			13
	S		3371.900			20
	NI	I	3371.993	400	10	
--	CA		3372.000	2		
	U		3372.010	8	4	
	MO		3372.065	5	2	
	FE	I	3372.080	40	6	
	OS		3372.083	40	8	
	NB		3372.090	5	3	
	CR		3372.130		32	
	SC	II	3372.151	6	160	68 VAF
	W		3372.193		13	
	TI	II	3372.208	10	16	
--	RH	I	3372.254	320	200	
	FE		3372.351			
	TB		3372.360	16	16	
	CE		3372.391	3		
	S		3372.480			3
--	AG	II	3372.505		3	
	PR		3372.510	20	3	
	OS		3372.524	3	2	
	RH	I	3372.526	10		
	CE		3372.539	8		
	NB		3372.562	10	200	
	U		3372.600	4		
	RB		3372.610			32
	CE		3372.648	3		
	V		3372.668		6	
	CA		3372.680		3	
	TH		3372.696		2	
	RN		3372.700			20
	P	II	3372.700			50
	TB		3372.720	50		
	*ER	II	3372.750	32	20	29 VAF
	HO		3372.790	13	16	

Table IIa. ER II 3372.750 A
(con't)

	Element	Wavelength (Angstrom)	Intensity Arc	Spark	Plasma
	PT	3372.791	2		
	TH	3372.794	3	4	
	PR	3372.797	10	4	
	TI II	3372.800	79	400	45 VAF
	FE	3372.861	2		
	AR I	3372.880			3
	MO	3372.920	5		
--	PT	3372.996	10		
	PD I	3373.001	790	500	
	OS	3373.038	16	5	
	ND	3373.095	4		
	MO	3373.126	5	25	
	S	3373.190			5
	OS	3373.204	6	8	
	EU	3373.228	3	2	
	CO I	3373.230	63		
---	W	3373.233	6	6	
	ND	3373.300	8		
	ZR II	3373.417	13	10	
	CE	3373.455	25	3	
	AR I	3373.481			320
	F II	3373.490			16
----	TH	3373.499	2		
	TB	3373.540	8		
	TI I	3373.581	4		
	CU II	3373.594		8	
	K	3373.600			32
	ER	3373.670	5		
	U	3373.727	10		
	CE II	3373.729	25	3	

Table IIb.(3) ER II 3372.750 A

Element	Wavelength (Angstrom)	Intensity Arc	Spark	Plasma
TH	3367.823	8	13	
NI I	3367.892	79		
MO	3367.969	2	100	
CR II	3368.054	32	130	
DY	3368.116	63	5	
TM	3368.330	10	32	
RH I	3368.375	320	50	
TA	3368.442		20	
RU I	3368.451	100	63	
IR I	3368.481	25	20	
SC II	3368.946	50	20	32 VAF
EU	3369.052	40	5	
NB	3369.157	5	50	
TI II	3369.212	6	25	
MO	3369.253	20	16	
RU	3369.282	13	63	
FE	3369.549	320	200	
NI I	3369.573	500	100	
TM	3369.640	10	20	
RU	3369.669	32	2	
MO	3369.939	5	20	
RU	3370.054	3	10	
NB	3370.158		79	
OS	3370.202	50	10	
CR	3370.231	32		
W II	3370.240	5	10	
CO I	3370.326	79	2	
TI I	3370.439	79	16	
CU	3370.454		16	
MO	3370.520	2	25	
OS I	3370.588	320	32	
NB	3370.611	3	32	
FE	3370.786	320	200	
CO II	3370.940		50	
V I	3371.114	32	20	
U	3371.292	8	16	
NB	3371.329	10	16	
W	3371.354	10	10	
TI I	3371.454	100	16	
TA	3371.539	63	20	

Table IIb.(3) ER II 3372.750 A
(con't)

Element	Wavelength (Angstrom)	Intensity Arc	Spark	Plasma
MO	3371.692		25	
TH	3371.800	4	10	
RU I	3371.860	63	20	
NI I	3371.993	400	10	
FE I	3372.080	40	6	
OS	3372.083	40	8	
CR	3372.130		32	
SC II	3372.151	6	160	68 VAF
W	3372.193		13	
TI II	3372.208	10	16	
RH I	3372.254	320	200	
TB	3372.360	16	16	
NB	3372.562	10	200	
TB	3372.720	50		
*ER II	3372.750	32	20	29 VAF
HO	3372.790	13	16	
TI II	3372.800	79	400	45 VAF
PD I	3373.001	790	500	
MO	3373.126	5	25	
CO I	3373.230	63		
ZR II	3373.417	13	10	
W	3373.754	10	10	
CO I	3373.972	63		
RU I	3373.978	63	4	
V I	3374.033	16	10	
NB	3374.090		32	
AIR	3374.100		10	
NI I	3374.221	400	6	
NB	3374.247	5	16	
CO I	3374.303	63		
TI II	3374.352	10	32	
TM	3374.510	63	32	
TH	3374.581	5	10	
RU I	3374.646	79	20	
ZR II	3374.726	20	32	
MO	3374.768	2	20	
W II	3374.870	6	13	
RU I	3374.903	32		
NB	3374.925	20	20	
CU II	3374.953		10	

Table IIb.(3) ER II 3372.750 A
(con't)

Element		Wavelength (Angstrom)	Intensity Arc	Spark	Plasma
TB		3375.030	32	16	
W		3375.116	10	10	
MO		3375.222	20	20	
GA		3375.300		13	
YB		3375.480	32	100	
MO		3375.648	20	10	
U		3375.780	13	10	
TA		3376.053	32	16	
V	I	3376.057	79	63	
W	II	3376.144	10	40	
LA	II	3376.329	100	50	
CR		3376.397	32	20	
ND		3376.490	20	10	
TA		3376.490	16	63	
LU		3376.500	100	10	
MO		3376.689		25	
NB		3376.730	8	16	
OS		3376.957	16	10	
CO	I	3377.063	100		
CE		3377.127	50	5	
RH		3377.138	40	10	
NB		3377.374		20	
V	I	3377.394	32	16	
TH		3377.430		20	
ZR	II	3377.455	25	10	
TI	I	3377.485	16	10	
TI	I	3377.585	20	16	
V	I	3377.625	63	32	

Table IIa.   EU II 3819.660 A

	Element	Wavelength (Angstrom)	Intensity Arc	Spark	Plasma
	MO	3818.663	8	16	
	TH	3818.672	8	5	
	CE II	3818.688	5	2	
	HO	3818.690		6	
	PT	3818.691	40	10	
	ER	3818.704	10		
	PR	3818.707	2	2	
	SE II	3818.750			5
	U	3818.757	8	3	
	DY	3818.760	5		
	GD	3818.760	10	16	
	CE	3818.810	2		
	ND	3818.837	4	6	
	NB	3818.858	8	320	
----	CE II	3819.024	20	6	
	RU I	3819.033	50	32	
	AR II	3819.040			5
	TH	3819.046		2	
	U	3819.050	5	4	
	EU	3819.068	2		
	PR	3819.068	3	3	
	NB	3819.147	13	10	
---	CE	3819.204	3	3	
	MO	3819.220		5	
	ER	3819.271	13	2	
	TH	3819.280	2	5	
	HF	3819.376	16	20	
--	FE	3819.500	2		
	CE	3819.534	2	3	
	CR I	3819.564	63	40	
	HE I	3819.606			50
	CS	3819.610			4
	*EU II	3819.660	500	500	110 VAF
	SM	3819.678	10	20	
	ND	3819.703	2	2	
	TA	3819.713	5	3	
	HE I	3819.761			4
	RU I	3819.767	13	6	
	MO	3819.778	6	6	
	MO	3819.873	6	6	

Table IIa. EU II 3819.660 A
(con't)

	Element	Wavelength (Angstrom)	Intensity Arc	Spark	Plasma
	PT	3819.880	3		
--	CO I	3819.910	20	3	
	HF	3819.925	13	10	
	V I	3819.963	63	32	
	ND	3819.965	16	4	
	CR I	3819.981	13	10	
	CE	3819.999	4	3	
	IR	3820.088	4		
	W	3820.109	8	10	
---	HG	3820.130			2
	CL II	3820.250			100
---	V I	3820.297	25	20	
	PR	3820.411	2	2	
	CE	3820.422	3	4	
	FE I	3820.429	790	630	
	ND	3820.431		40	
	CE	3820.545	3	2	

Table IIb.(1) EU II 3819.660 A

Element	Wavelength (Angstrom)	Intensity Arc	Spark	Plasma
RU I	3814.857	20	32	
V	3815.392		160	
NB	3815.511	20	32	
CE	3815.831	50	5	
FE I	3815.842	630	630	
AU	3816.130	20	40	
PR	3816.166	40	40	
BI	3816.173		25	
YB	3816.197	13	32	
CO I	3816.326	63	50	
CO I	3816.473	63		
MN	3816.753	63	50	
DY	3816.770	100	50	
CO I	3816.874	63	5	
IR I	3817.244	16	32	

Table IIb.(1) EU II 3819.660 A
(con't)

Element	Wavelength (Angstrom)	Intensity Arc	Spark	Plasma
RU	3817.269	50	63	
ND	3817.373	32	25	
TM	3817.400	63	16	
FE	3817.650	50	16	
ND	3817.675	40	40	
RH I	3818.187	50	25	
PR	3818.281	130	100	
Y II	3818.342	32	50	
CR I	3818.481	50	20	
NB	3818.858	8	320	
RU I	3819.033	50	32	
CR I	3819.564	63	40	
*EU II	3819.660	500	500	110 VAF
V I	3819.963	63	32	
FE I	3820.429	790	630	
ND	3820.431		40	
HF I	3820.728	1000	100	
FE	3821.178	100	100	
V I	3821.487	50	32	
ND	3821.770	50	40	
PR	3821.817	50	50	
FE I	3821.838	50	32	
V I	3822.009	63	40	
RU	3822.091	50	25	
RH I	3822.262	100	100	
V I	3822.888	40	25	
PR	3823.184	130	25	
MN	3823.513	79	79	
CR I	3823.522	40	32	
MN	3823.893	50	50	
FE I	3824.444	160	100	

Table IIa. EU II 4205.046 A

	Element	Wavelength (Angstrom)	Intensity Arc	Spark	Plasma
	CR I	4204.198	50	6	
	V II	4204.200	2	6	
---	CE	4204.296		3	
	KR	4204.310			3
	NB	4204.321	4	3	
	ND	4204.351	10	3	
	U	4204.371	16	10	
	TL	4204.400			2
	W	4204.409	20	10	
	CR	4204.471	79	32	
	RE	4204.530	25		
	SC I	4204.538	4	2	
	CL II	4204.540			20
---	OS	4204.560	13		
	PR	4204.580	5		
	MO	4204.609	16	10	
	CA	4204.623		2	
	Y II	4204.696	16	16	
	CE	4204.739	16		
--	MO	4204.809	25	20	
	SM II	4204.813	8	10	
	GD	4204.839	25		
	EU	4204.909	5		
	DY	4205.030	6	2	
	*EU II	4205.046	200	50	70 VAF
	CL II	4205.070			10
	IN II	4205.079			50
	V II	4205.086	5	20	
	BH CA	4205.100	6		
	IN II	4205.151			32
	CE	4205.161	6	2	
	SC I	4205.194	10		
	IN II	4205.217			16
	OS	4205.222	10		
	TB	4205.230	2		
--	ND	4205.255	5	4	
	NB	4205.311	16	16	
	SM	4205.361	8	4	
	TL	4205.400			2
	XE I	4205.404			10

Table IIa. EU II 4205.046 A
(con't)

	Element	Wavelength (Angstrom)	Intensity Arc	Spark	Plasma
---	FE I	4205.546	50	6	
	W	4205.559	6	13	
	ND	4205.595	20	16	
	U	4205.630	3		
	DY	4205.640	5	2	
	TB	4205.640	3		
	N	4205.650			5
	PR	4205.720	5		
	SM	4205.773	4	2	
----	CE	4205.792	4		
	MO	4205.809	2		
	TA	4205.876	100	32	
	CE	4205.892	4		
	ZR II	4205.919	10	3	
	ER	4205.920	6		
	TM	4206.000	20	5	
	RU I	4206.016	100	40	

Table IIb.(2) EU II 4205.046 A

Element	Wavelength (Angstrom)	Intensity Arc	Spark	Plasma
MO	4200.570	16	16	
FE I	4200.927	79	20	
CE	4201.300	4	20	
LA	4201.500	2	13	
PR	4201.529	32	13	
MN	4201.757	40	20	
RB I	4201.851	2000	500	1 VAF
FE I	4202.031	400	320	
OS	4202.062	100	4	
V II	4202.340	6	16	
CE II	4202.944	40	20	
SM II	4203.047	16	20	
CR I	4203.590	100	20	
TM	4203.730	250	25	
FE I	4203.987	200	130	

Table IIb.(2) EU II 4205.046 A
(con't)

Element	Wavelength (Angstrom)	Intensity Arc	Spark	Plasma
LA II	4204.038	200	25	
CR	4204.471	79	32	
Y II	4204.696	16	16	
MO	4204.809	25	20	
*EU II	4205.046	200	50	70 VAF
V II	4205.086	5	20	
NB	4205.311	16	16	
W	4205.559	6	13	
ND	4205.595	20	16	
TA	4205.876	100	32	
RU I	4206.016	100	40	
SM II	4206.124	20	20	
TA	4206.404	50	20	
FE I	4206.702	130	25	
PR	4206.739	50	50	
CR	4206.899	79	25	
W	4207.053	25	13	
FE I	4207.131	79	40	
AS II	4208.000		32	
PR	4208.305	20	13	
CR	4208.357	100	25	
TA	4208.440	2	32	
FE	4208.615	100	50	
TH	4208.893	16	16	
ZR II	4208.983	32	25	
CR	4209.368	100	40	
MO	4209.649	4	79	
V	4209.739		13	
CR	4209.756	79	20	
V II	4209.800		13	

Table IIa. FE II 2599.396 A

	Element	Wavelength (Angstrom)	Intensity Arc	Spark	Plasma
	XE	2598.420			4
	W	2598.421	10	5	
	HG	2598.450			5
	FE	2598.470			
	RU	2598.581	20		
	RE	2598.593	6		
----	AG II	2598.680	10	10	
	W II	2598.680	3	3	
	IN II	2598.692			10
	CS	2598.700			2
	W II	2598.738	13	20	
	TA	2598.746	16		
	RU	2598.766	4	50	
	IN II	2598.802			20
	CU II	2598.813		200	
	HF II	2598.851		2	
	FE	2598.854	20		
	U	2598.859	4	2	
	NB	2598.883		160	
---	MN	2598.902	5	100	
	DY	2598.970	3		
	MN II	2599.036	2	13	
	IR I	2599.040	25	10	
	TA	2599.090	2		
	OS	2599.129	5		
	FE	2599.131			
--	YB	2599.150	2	50	
	MO	2599.182		16	
	CO I	2599.207	5		
	HF II	2599.215	10	10	
	*FE II	2599.396	1000	1000	48 VAF
	TA	2599.397	100	32	
	IR	2599.400	40		
	NB	2599.521		5	
	FE I	2599.570	1000		
	W II	2599.642		5	
--	MO	2599.643	20		
	RU	2599.658	10	25	
	FE	2599.661	6		
	W	2599.764	2	13	

Table IIa. FE II 2599.396 A
(con't)

Element	Wavelength (Angstrom)	Arc	Intensity Spark	Plasma
U	2599.805	4	2	
RE	2599.859	79		
--- MN	2599.871	10		
OS	2599.907	13	3	
TI I	2599.915	63	10	
PT I	2599.916	5		
FE	2600.100	10		
MO	2600.120		10	
--- TA	2600.141	79		
TB	2600.150		20	
NB	2600.152	13	10	
DY	2600.170	5		
YB	2600.190		3	
FE	2600.202	10		
W	2600.218	10	2	
CU II	2600.266		200	
MN	2600.267		10	
U	2600.282	2	2	
BH C	2600.300	20		
CD	2600.320			2
LA II	2600.347		4	
CS	2600.360			20
MO	2600.366	20		

Table IIb.(1) FE II 2599.396 A

Element	Wavelength (Angstrom)	Arc	Intensity Spark	Plasma
SN	2594.423	63	79	
NB	2594.740	3	79	
V	2595.096		63	
TA	2595.586	40	50	
MN	2595.761	200	25	
RU	2595.796		100	
PD II	2595.969	3	160	
PT	2596.000	200	20	
TA	2596.450	79	160	

Table IIb.(1) FE II 2599.396 A
(con't)

Element	Wavelength (Angstrom)	Intensity Arc	Spark	Plasma
MN	2596.827		50	
RH	2597.070	3	160	
RU	2597.147		50	
SB I	2598.062	200	100	
SB I	2598.090			28 VAF
TA	2598.210	2	50	
FE II	2598.369	630	1000	24 VAF
RU	2598.766	4	50	
CU II	2598.813		200	
NB	2598.883		160	
MN	2598.902	5	100	
YB	2599.150	2	50	
*FE II	2599.396	1000	1000	48 VAF
TA	2599.397	100	32	
FE I	2599.570	1000		
CU II	2600.266		200	
NI II	2601.126		2000	
NB	2601.291	4	200	
RU	2602.249		100	
MN	2602.720		79	
PD II	2602.760	20	200	
MO	2602.798	25	100	
PT I	2603.142	320	20	
RH	2603.321		100	
TA II	2603.493	10	16	54 VAF
TA	2603.573	5	320	
MN	2603.719	5	50	
RE	2603.870	100		
RU	2604.120		79	

Table IIa. FE I 3719.935 A

	Element	Wavelength (Angstrom)	Intensity Arc	Spark	Plasma
	MO	3719.047		20	
	CE II	3719.081	3		
	EU	3719.167	32		
--	HF II	3719.279	16	32	
	U	3719.293	13		
	ER	3719.295	4		
	SM	3719.300	3		
	RU	3719.329	20	25	
	W	3719.405	13	10	
	TA	3719.418	3		
	CE	3719.430	8		
--	PR	3719.435	16	10	
	TH	3719.436	32	10	
	TB	3719.450	32	8	
	SM	3719.451	50	10	
	GD	3719.464	40	40	
	OS I	3719.522	40	13	
	MO	3719.553	5	3	
	ND	3719.595	10	8	
	NB	3719.635		50	
--	U	3719.690		8	
	MO	3719.692	3		
	SB II	3719.700		8	
	TM	3719.720	10	4	
	MO	3719.740		32	
	CE II	3719.797	16	5	
	BA I	3719.930	2		
	*FE I	3719.935	1000	630	
	CE	3719.946	2		
	TH	3719.969	2		
--	OS I	3720.132	79	40	
	PR	3720.222	16	6	
	MO	3720.254	10	40	
	TH	3720.309	16	10	
	AS II	3720.310		16	
	TB	3720.360	8		
	CE	3720.380	2	2	
	TI I	3720.384	40	10	
	U	3720.394	6	10	
	CA	3720.400		3	

Table IIa. FE I 3719.935 A
(con't)

	Element	Wavelength (Angstrom)	Intensity Arc	Spark	Plasma
---	AR	3720.430			10
	NB	3720.456	5	100	
	W	3720.513	8	10	
	ND	3720.540	8	8	
	SM	3720.572	4	4	
	CE	3720.595	2		
----	SM	3720.644	2		
	EU	3720.695	4		
	HO	3720.740	6	4	
	PT	3720.740	5		
	LA II	3720.750		3	
	CU I	3720.771	10		
	TH	3720.776	5		
	XE II	3720.800			20
	IR I	3720.821	6		
	PR	3720.830	10	8	
	BA I	3720.850	2		

Table IIb.(1) FE I 3719.935 A

Element	Wavelength (Angstrom)	Intensity Arc	Spark	Plasma
CR	3715.428		32	
V	3715.470	63	400	
LA II	3715.524	100	50	
FE I	3715.915	79	50	
GD	3716.369	160	130	
FE I	3716.448	160	100	
NB	3717.066	8	1000	
RE I	3717.290	160		
TI I	3717.396	79	50	
GD	3717.487	100	50	
TM	3717.920	100	10	
V II	3718.160	5	63	
FE I	3718.408	79	50	
SM II	3718.880	100	5	
PD I	3718.909	320	200	

Table IIb.(1) FE I 3719.935 A
(con't)

Element	Wavelength (Angstrom)	Intensity Arc	Spark	Plasma
MN	3718.930	79	100	
HF II	3719.279	16	32	
GD	3719.464	40	40	
NB	3719.635		50	
MO	3719.740		32	
*FE I	3719.935	1000	630	
OS I	3720.132	79	40	
MO	3720.254	10	40	
NB	3720.456	5	100	
TI II	3721.636	63	130	
TH	3721.831	40	32	
SM	3721.842	100	50	
GD	3722.035	50	32	
V I	3722.195	40	40	
YB	3722.293	6	63	
NI I	3722.484	200	20	
FE I	3722.564	500	400	
NB	3722.566		63	
TI I	3722.566	100	63	
SB II	3722.793	40	50	
NB	3723.444		32	
MO	3723.814	20	32	
YB	3724.211	16	50	
FE I	3724.380	200	160	
TI I	3724.574	100	50	

Table IIa. FE II 2611.872 A

	Element	Wavelength (Angstrom)	Intensity Arc	Spark	Plasma
	V I	2610.891	6	2	
	BR	2610.930			5
	KR	2610.980			10
	CE	2610.985	2		
	CR	2611.045		2	
	RU	2611.048	50	5	
	FE II	2611.072	20	79	
	W	2611.102	5		
----	U	2611.140	6	2	
	MO	2611.199	25	16	
	SC II	2611.209	5	10	
	V I	2611.256	20	10	
	W	2611.274		8	
	TI I	2611.285	79	16	
	IR I	2611.295	79	10	
	OS	2611.330	8		
	FE II	2611.332			
	TA	2611.340	100		
---	W	2611.396	6		
	TA	2611.459	16		
	TI I	2611.485	16	2	
	RU	2611.510	3	79	
	W	2611.580		6	
	RE	2611.604	32		
--	U	2611.625	5	2	
	MO	2611.627	16		
	CR	2611.628		3	
	NI II	2611.654		130	
	RH	2611.760		10	
	CD	2611.810			2
	NA II	2611.815			8
	*FE II	2611.872	500	500	24 VAF
	IN	2612.010		2	
	CR	2612.024	8		
	YB	2612.030		4	
	IR I	2612.040	10		
	RU	2612.068	100	32	
--	CS	2612.100			2
	ZR I	2612.181	4		
	W	2612.188	10	16	

Table IIa. FE II 2611.872 A
(con't)

Element	Wavelength (Angstrom)	Intensity Arc	Spark	Plasma
CD	2612.190			3
CR	2612.204	8		
IR I	2612.259	10	2	
NB	2612.260		5	
MO	2612.289	10	6	
-- SB	2612.301	50	63	
ER	2612.379	4		
NB	2612.382	8	2	
U	2612.456	10	4	
RU	2612.515		79	
CR	2612.555		2	
HF	2612.590	4	3	
TA	2612.609	50	40	
-- YB	2612.620		4	
MN	2612.627		13	
CO II	2612.630		20	
OS	2612.630	20	5	
W	2612.648	2	13	
RE	2612.724	20		
AU	2612.740		3	
FE I	2612.771	50	10	
MN	2612.855	16	2	
LU	2612.860	3		

Table IIb.(2) FE II 2611.872 A

Element	Wavelength (Angstrom)	Intensity Arc	Spark	Plasma
HF II	2607.032	250	2000	
FE II	2607.087	320	400	
TA	2607.840	20	160	
ZN I	2608.640	320	100	
RH	2609.170	3	200	
PD II	2609.862		200	
NI II	2610.092		1000	
LA II	2610.335	10	160	
NI II	2611.654		130	

Table IIb.(2) FE II 2611.872 A
(con't)

Element	Wavelength (Angstrom)	Intensity Arc	Spark	Plasma
*FE II	2611.872	500	500	24 VAF
HF II	2613.604	200	1600	
FE II	2613.823	400	400	
PB I	2614.178	200	79	23 VAF
AG II	2614.590		320	
NI II	2615.193		1000	
LU II	2615.420	100	250	15 VAF

Table IIa. FE II 2382.039 A

Element	Wavelength (Angstrom)	Intensity Arc	Intensity Spark	Intensity Plasma
W	2381.100	3	6	
NB	2381.128		5	
TA	2381.130	16	40	
RE	2381.140	40	6	
AR	2381.180			2
AS I	2381.180	79	4	
CO I	2381.251	4		
-- RB	2381.300			100
W II	2381.330	3	10	
PT	2381.360		13	
ER	2381.410		3	
MO	2381.470		13	
CR	2381.480	3	25	
TA	2381.520	8	40	
TH	2381.520		10	
-- ZR	2381.557	4	2	
W	2381.570	8		
IR I	2381.622	32	2	12 VAF
LU	2381.690		32	
ER	2381.720		5	
CO	2381.752	4	13	
PD II	2381.780			2
-- I	2381.790			20
W	2381.790		10	
IR	2381.820	8	50	
FE I	2381.830	3		
DY	2381.970	8		
RU	2381.993	50	160	
BH C	2382.000	32		
*FE II	2382.039	40	100	65 VAF
HG	2382.060			8
ZN I	2382.220	4		
NB	2382.246	2		
- PD	2382.250		2	
CO	2382.335	2	4	
W II	2382.340	3	10	
FE II	2382.355	3		
ZR	2382.360		2	
MO	2382.390		13	
AU II	2382.410		5	

Table IIa. FE II 2382.039 A
(con't)

	Element	Wavelength (Angstrom)	Intensity Arc	Spark	Plasma
	OS	2382.460	32	5	
---	V	2382.467		100	
	PD	2382.570		40	
	IN II	2382.593			40
	RH	2382.640		32	
	W	2382.670	3	13	
	IN II	2382.673			20
	RU	2382.741		4	
----	PD II	2382.756	2	2	
	ND	2382.820		10	
	DY	2382.850	2		
	RH	2382.890	50	5	
	FE II	2382.893		4	
	U	2382.930		2	
	W	2382.986	16	3	
	V	2383.001	8	79	

Table IIb.(3) FE II 2382.039 A

Element	Wavelength (Angstrom)	Intensity Arc	Spark	Plasma
OS	2377.610	50	16	
RH	2377.810		50	
AL I	2378.408	40	20	
CO II	2378.622	25	50	31 VAF
OS	2379.390	40	10	
LA	2379.420		63	
TI I	2379.690	100	200	
TL I	2379.690			7 VAF
CE	2380.160		50	
HF II	2380.302	32	63	
TL	2380.340	20	63	
V	2380.918	6	50	
RE	2381.140	40	6	
AS I	2381.180	79	4	
IR I	2381.622	32	2	12 VAF
IR	2381.820	8	50	

Table IIb.(3) FE II 2382.039 A
(con't)

Element	Wavelength (Angstrom)	Intensity Arc	Spark	Plasma
RU	2381.993	50	160	
*FE II	2382.039	40	100	65 VAF
V	2382.467		100	
RH	2382.890	50	5	
V	2383.001	8	79	
TE I	2383.250	500	320	11 VAF
PD II	2383.400		50	
RH	2383.400	50	10	
CO II	2383.460	16	32	21 VAF
SB	2383.630	79	20	
MN	2384.050	40	3	
IR	2384.810	4	79	
PD II	2385.010		50	
TE I	2385.760	630		17 VAF
V	2385.816		100	
RH	2386.140	79	8	
CO II	2386.363	10	25	14 VAF
IR	2386.892	50	16	

Table IIa. GA I 4172.056 A

	Element	Wavelength (Angstrom)	Arc	Spark	Plasma
			Intensity		
	AR I	4164.179			1000
	MO	4171.074	16	10	
	LA I	4171.132	8		
	W	4171.185	25	13	
	V I	4171.296	16	6	
----	TH	4171.349	10	13	
	CE II	4171.386	20		
	MO	4171.454	10	3	
	ZR I	4171.477	20		
	SM II	4171.554	6	13	
---	I I	4171.560			8
	SC I	4171.568	3		
	U	4171.591	32	32	
	N	4171.630			5
	CR	4171.675	63	8	
	FE	4171.700	8	2	
	ER	4171.708	16		
	GD	4171.710	25		
	CE	4171.769	2	3	
	TB	4171.800	8		
--	PR	4171.824	79	40	
	CU II	4171.854		5	
	MO	4171.900		25	
	TI II	4171.903	16	63	
	DY	4171.925	4	2	
	CE	4171.964	2		
	DY	4171.992	16		
	*GA I	4172.056	2000	1000	45 VAF
	FE	4172.127	79	50	
	CE II	4172.161	20		
	U	4172.180	3	3	
	HO	4172.230	2		
	YB	4172.230	2	2	
--	PR	4172.273	79	40	
	LA I	4172.316	8		
---	KR II	4172.510			20
	IR I	4172.559	160	13	
	OS	4172.570	63	3	
	TB	4172.590	13	2	
	TI I	4172.609	5		

Table IIa. GA I 4172.056 A
(con't)

Element	Wavelength (Angstrom)	Arc	Intensity Spark	Plasma
CE	4172.621	2		
FE I	4172.652	3		
FE I	4172.750	63	10	
MO	4172.751	3	2	
CR	4172.769	32	16	
P II	4172.790			5
-- EU	4172.800	13	3	
TB	4172.820	10		
KR I	4172.830			3
CE	4172.885	2	2	
U	4172.974	10	16	

Table IIb.(1) GA I 4172.056 A

Element	Wavelength (Angstrom)	Arc	Intensity Spark	Plasma
RU I	4167.512	100	160	
NB	4168.126	100	79	
AC	4168.400		100	
PD I	4169.842	200	50	
GD	4170.110	50	50	
TI II	4171.903	16	63	
*GA I	4172.056	2000	1000	45 VAF
FE	4172.127	79	50	
FE I	4175.640	100	79	
FE I	4176.572	100	50	

Table IIa. GA I 2943.637 A

	Element	Wavelength (Angstrom)	Intensity Arc	Spark	Plasma
	K I	2942.700	5		
	MN	2942.745	10		
	PT I	2942.755	20	3	
	YB	2942.820		5	
	OS	2942.848	32	8	
	MO	2942.850	10		
	U	2942.853	6	6	
	TH II	2942.862	10	10	31 VAF
	O	2942.880			5
----	AR II	2942.900			100
	O	2943.000			5
	TI II	2943.130		63	
	MN	2943.132		3	
---	RE I	2943.144	63		
	CO II	2943.150		100	
	IR I	2943.151	32	20	
	U	2943.184	5	8	
	V I	2943.196	32	25	
	CL	2943.200			4
	CE	2943.215	6		
	IR	2943.260	4		
	W	2943.326	6	6	
	TM	2943.360	6	16	
	MO	2943.380		25	
	RE I	2943.380	16		
--	U	2943.405	6	4	
	XE	2943.410			2
	RU I	2943.481	32		
	CO I	2943.484	32		
	SM	2943.492	8	3	
	ND	2943.500	3	5	
	LA II	2943.551	2	6	
	FE	2943.570	13	6	
	SN II	2943.620		2	
	IR	2943.628	3		
	V	2943.636	2	4	
	*GA I	2943.637	10	20	64 VAF
	CE	2943.673	8		
	IR	2943.725	3		
	TA	2943.769	10	5	

Table IIa. GA I 2943.637 A
(con't)

	Element	Wavelength (Angstrom)	Intensity Arc	Spark	Plasma
	SM	2943.786	8	2	
	V I	2943.827	6		
--	IR	2943.870	4	5	
	CD II	2943.890			5
	U	2943.895	10	25	
	MN	2943.908	2	2	
	NI I	2943.914	50	20	
	RU	2943.921	50	5	
	W	2943.959	5	4	
	CE	2943.987	6		
	MO	2943.989		3	
	ER	2944.071	13		
--	CS	2944.100			2
	GA I	2944.175	10	16	9 VAF
	RU	2944.184	13		
	U	2944.189	8	13	
	CL	2944.200			4
	ZR II	2944.200	2	3	
	MO	2944.213	25	2	
	BI	2944.287	5	4	
	RE	2944.321	10		
--	CE	2944.346	20		
	W	2944.395	32	20	
	FE II	2944.398	>3	630	
	YB	2944.460		3	
	HO	2944.500	10	20	
	FE	2944.510	2		
	DY	2944.564	2		
	V II	2944.571	50	320	
	XE	2944.610			2
	U	2944.637	2	8	

Table IIb.(3) GA I 2943.637 A

Element	Wavelength (Angstrom)	Intensity Arc	Spark	Plasma
IR I	2938.757	10		
PT	2938.813	16	2	
W II	2938.852		10	
W	2939.043	10	4	
FE I	2939.080	79	20	
RU I	2939.131	13		
IR I	2939.271	20	16	
TA	2939.280	200	40	
HO	2939.290		10	
MN II	2939.304	50		29 VAF
IN	2939.360		10	
CR	2939.450	6	20	
FE II	2939.508	3	32	
CE	2939.538	13		
RU	2939.692	13	5	
W II	2939.743	2	13	
MN	2939.904	13		
RU I	2939.944	32	3	
TB	2940.030	10	3	
TA	2940.060	100	40	
MO	2940.101	2	40	
W II	2940.199	4	20	
TA	2940.215	160	50	
CR	2940.225		32	
RU I	2940.358	50	3	
U	2940.367	6	13	
MN	2940.392	40		
EU	2940.460	16		
MN	2940.483	40		
YB	2940.510	3	25	
IR I	2940.542	16	10	
FE	2940.591	200	79	
HF I	2940.762	790	100	
CE	2940.785	16		
EU	2940.820	10		
CR	2940.978		10	
MO	2940.978	10		
MN	2941.041	25		
IR I	2941.079	20	5	
MO	2941.222	2	40	

Table IIb.(3) GA I 2943.637 A
(con't)

Element	Wavelength (Angstrom)	Intensity Arc	Spark	Plasma
FE I	2941.343	630	320	
U	2941.343	8	13	
V II	2941.369	40	320	
V II	2941.492	13	160	
NB	2941.543	50	320	
RE	2941.557	16		
TB	2941.700	3	10	
CR I	2941.880	13	25	
U II	2941.919	16	32	6 VAF
TI II	2941.995	100	160	
TI I	2941.995	100	160	
HO	2942.050		10	
MG I	2942.110	20	2	
U	2942.120	8	13	
TA	2942.137	160	40	
W II	2942.139	6	10	
RU	2942.250	32	100	
W II	2942.256	2	10	
PR	2942.302		10	
V I	2942.352	79	20	
W II	2942.610	2	10	
FE	2942.630	10	5	
MN	2942.745	10		
PT I	2942.755	20	3	
OS	2942.848	32	8	
MO	2942.850	10		
TH II	2942.862	10	10	31 VAF
TI II	2943.130		63	
RE I	2943.144	63		
CO II	2943.150		100	
IR I	2943.151	32	20	
V I	2943.196	32	25	
TM	2943.360	6	16	
MO	2943.380		25	
RE I	2943.380	16		
RU I	2943.481	32		
CO I	2943.484	32		
FE	2943.570	13	6	
*GA I	2943.637	10	20	64 VAF
TA	2943.769	10	5	

Table IIb.(3) GA I 2943.637 A
(con't)

Element	Wavelength (Angstrom)	Intensity Arc	Spark	Plasma
U	2943.895	10	25	
NI I	2943.914	50	20	
RU	2943.921	50	5	
ER	2944.071	13		
GA I	2944.175	10	16	9 VAF
RU	2944.184	13		
U	2944.189	8	13	
MO	2944.213	25	2	
RE	2944.321	10		
CE	2944.346	20		
W	2944.395	32	20	
FE II	2944.398	63	630	
HO	2944.500	10	20	
V II	2944.571	50	320	
HF	2944.711	20		
PT I	2944.754	16	2	
MO	2944.821	2	50	
TL I	2945.040	50	25	
FE	2945.055	100	32	
RU	2945.101	6	50	
ER	2945.280	16		
ZR II	2945.463	4	10	
TI II	2945.470		100	
MO	2945.665	20	2	
RU	2945.668	63	320	
FE	2945.700	10	5	
TB	2945.700	10	3	
HO	2945.830	3	63	
NB	2945.883	2	100	
U	2945.889	8	13	
YB	2945.900	10	63	
Y	2945.950	2	100	
MO	2946.009	20	40	
CE	2946.056	10		
NB	2946.116	4	20	
TA	2946.261	10	2	
YB	2946.290	3	16	
MO	2946.422	10	6	
W	2946.429		13	
V I	2946.527	10	5	

Table IIb.(3) GA I 2943.637 A
(con't)

Element	Wavelength (Angstrom)	Intensity Arc	Spark	Plasma
RE	2946.576	10		
MO	2946.690	3	25	
ND	2946.720		10	
TM	2946.840	10	32	
CR	2946.842	5	32	
NB	2946.898	3	32	
TA	2946.910	160	10	
IR I	2946.972	20	10	
W	2946.981	20	20	
RU I	2946.991	63	13	
HF II	2947.130	16	16	
MO	2947.285	2	25	
EU	2947.300	32		
FE	2947.363	32	20	
IR I	2947.367	13	2	
W	2947.384	13	10	
NI II	2947.453		10	
CR	2947.497		25	
FE II	2947.658	10	100	
TA	2947.800	10	2	
FE I	2947.876	630	200	
TM	2948.010	16	20	
V	2948.075	2	63	
U	2948.092	10	13	
TM	2948.150	16	10	
CD	2948.160		32	
OS	2948.229	13	5	
TI I	2948.255	100	32	
CE	2948.385	10		
Y I	2948.405	20	5	
FE	2948.433	79	63	

Table IIa. GD II 3422.466 A

Element	Wavelength (Angstrom)	Intensity Arc	Spark	Plasma
BA I	3421.480	2	2	
CE II	3421.542	5		
RE	3421.579	16		
CO I	3421.626	20	2	
AR	3421.640			10
CR	3421.640	2	6	
HO	3421.640	20	20	
EU	3421.667	5	5	
OS	3421.687	32	10	
U	3421.690	8	8	
CE	3421.714	3		
---- CR	3421.720	13		
PT	3421.720	2		
IR	3421.760	6		
TA	3421.800		20	
--- CE	3422.000	5		
SM	3422.063	5		
DY	3422.070	3	2	
FE	3422.140	2		
SM II	3422.189	3		
CE	3422.215	2		
-- PR	3422.260	5	3	
V	3422.260		16	
OS	3422.275	20	10	
RH I	3422.291	13		
MO	3422.309	10	10	
NI I	3422.332	10	5	
U	3422.352	20	16	
RN	3422.400			20
CE	3422.420	8		
W	3422.426	10	10	
TB	3422.440	3	16	
*GD II	3422.466	79	100	21 VAF
ER	3422.471	6		
SM	3422.476	2		
FE	3422.493	40	10	
CE	3422.507	20		
DY	3422.570	10		
ER	3422.620	4		
TH	3422.657	3		

Table IIa. GD II 3422.466 A
(con't)

	Element	Wavelength (Angstrom)	Intensity Arc	Spark	Plasma
	FE I	3422.660	100	50	
	TI II	3422.661	4	10	
	CE	3422.708	32	10	
--	GD	3422.713	8	6	
	CR II	3422.739	32	130	
	SM	3422.760	10	3	
	MO	3422.777		32	
	CO	3422.784	6		
	NB	3422.786	3	5	
	W	3422.800		6	
	NB	3422.849		32	
	ER	3422.870	3		
	CA	3422.878		2	
	DY	3422.878	16	5	
	NI I	3422.878	10	5	
	CO I	3422.896	20	4	
--	CD II	3422.964		3	
	TB	3423.050	16		
	U	3423.050	13	16	
	RA II	3423.100			5
	EU	3423.110	10	10	
	TH	3423.131	5	6	
--	TI I	3423.172	4		
	DY	3423.240	4	2	
	W	3423.303	6	6	

Table IIb.(2) GD II 3422.466 A

Element	Wavelength (Angstrom)	Intensity Arc	Spark	Plasma
RE I	3417.799	40		
FE I	3417.843	160	100	
SM II	3418.510	50	10	
FE I	3418.512	160	100	
GD	3418.735	50	25	
RE I	3419.400	79		
TA	3419.748	50	5	

Table IIb.(2) GD II 3422.466 A
(con't)

Element	Wavelength (Angstrom)	Intensity Arc	Spark	Plasma
RU	3420.078	63	8	
TB	3420.340	50	16	
NB	3420.631	5	50	
RE I	3420.758	40		
CO I	3420.792	79	2	
NB	3421.162	10	50	
CR II	3421.212	50	200	
PD I	3421.240	2000	1000	30 VAF
*GD II	3422.466	79	100	21 VAF
FE	3422.493	40	10	
FE I	3422.660	100	50	
CR II	3422.739	32	130	
MO	3422.777		32	
NB	3422.849		32	
NI I	3423.711	630	25	
V I	3423.864	25	25	
FE I	3424.288	200	160	
CO I	3424.506	79	2	
RE I	3424.604	320		
FE	3425.015	63	40	
EU	3425.020	50	40	
DY	3425.060	40	5	
TM II	3425.080	200	320	30 VAF
HO	3425.350	40	40	
NB	3425.424	32	320	
TM	3425.630	100	50	
NB	3425.848	50	32	
YB	3426.040	40	16	
FE I	3426.388	79	20	
NB	3426.570	5	200	
FE I	3426.636	79	63	
FE I	3427.121	50	50	
NB	3427.449	32	32	

Table IIa. GD II 3646.196 A

	Element	Wavelength (Angstrom)	Intensity Arc	Spark	Plasma
	FE	3645.222	10		
	CE II	3645.228	5		
	CU I	3645.232	20	5	
	SM II	3645.290	10	6	
	IR	3645.303	25	2	
	SC II	3645.311	50	50	35 VAF
	NB	3645.356	5	5	
	TB	3645.380	50	16	
	SM II	3645.387	8	6	
	ER	3645.399	25	13	
	Y	3645.403	4	4	
	LA II	3645.414	100	63	
	DY II	3645.416	320	100	13 VAF
	EU	3645.430	5		
--	CE II	3645.452	10	2	
	U	3645.461		2	
	FE I	3645.494	16	6	
	PR	3645.539	20	8	
	CR	3645.583	4	3	
	MO	3645.589	3	3	
	RE	3645.590	6		
	V	3645.596	16		
	W II	3645.599	4	20	
	TB	3645.600	8		
	ND	3645.626	16	6	
	GD	3645.634	20	20	
	PR	3645.660	32	20	
--	CE	3645.711	2		
	ND	3645.776	8		
	SM	3645.789	5		
	FE	3645.825	79	63	
	DY	3645.860	6	4	
	SM II	3645.898	4	2	
	HO	3645.900		4	
	V	3645.905		20	
	NB	3645.929		5	
	ER	3645.938	16	2	
-	PD I	3645.968	16		
	BH SR	3646.000	4		
	SM	3646.013	2		

Table IIa. GD II 3646.196 A
(con't)

	Element	Wavelength (Angstrom)	Intensity Arc	Spark	Plasma
	U	3646.030		4	
	CA	3646.053		2	
	RU	3646.114	2	8	
	CR	3646.161	20	8	
	*GD II	3646.196	200	160	10 VAF
	TI I	3646.200	63	25	
	U	3646.217	10		
	PR	3646.299	50	16	
--	RB II	3646.321			10
	TB	3646.460	8	3	
	U	3646.491	2	6	
	W II	3646.525	10	32	
	DY	3646.600	5	2	
	RE	3646.628	10		
	CE	3646.652	10	3	
---	EU	3646.660	10		
	EU	3646.750	20	5	
	ER	3646.782	5		
	U	3646.847	4	10	
	V	3646.847		20	
	DY	3646.850	4		
	MO	3646.869	3	5	
	PR	3646.878	4	2	
	TH	3646.886	5	4	
----	ZR	3646.892	3		
	CE	3646.965	16	5	
	EU	3647.030	2		
	TB	3647.060	32	8	
	CO I	3647.089	32	4	
	IR I	3647.105	3		
	W	3647.147	4	3	

Table IIb.(2) GD II 3646.196 A

Element	Wavelength (Angstrom)	Intensity Arc	Spark	Plasma
TI II	3641.331	63	160	
MN	3641.391	50	50	
W II	3641.408	13	40	
TA	3642.057	130	20	
TI I	3642.675	320	130	
SC II	3642.785	63	50	11 VAF
RE I	3642.987	100		
TM	3643.652	63	40	
HF II	3644.355	25	50	
CA I	3644.410	200	16	
V I	3644.706	79	50	
SC II	3645.311	50	50	35 VAF
LA II	3645.414	100	63	
DY II	3645.416	320	100	13 VAF
FE	3645.825	79	63	
*GD II	3646.196	200	160	10 VAF
CO I	3647.659	100	8	
LU	3647.770	100	5	
FE I	3647.844	500	400	
OS	3648.806	100	10	
V I	3648.966	79	50	
CO	3649.351	200	4	
SM II	3649.506	100	32	
FE I	3649.508	100	100	
HG I	3650.146	200	500	
LA II	3650.174	100	63	
FE I	3650.280	63	50	
CR	3650.344		40	
TB	3650.400	50	100	
MO	3651.107	2	50	
NB	3651.186	10	400	

Table IIa. GE I 2651.575 A

	Element	Wavelength (Angstrom)	Intensity Arc	Spark	Plasma
	CE	2650.584	4		
	TH	2650.588	10	5	
	RU	2650.591	13	5	
	MO	2650.682	10		
	CS	2650.700			20
	BE I	2650.702	10		
	W	2650.711	2	3	
	SI	2650.730			5
	YB	2650.740	2	4	
	BR	2650.760			2
	BE I	2650.760	20	16	
	BE I	2650.781	25		
----	PT I	2650.857	630	100	
	MN	2650.994		160	
	CE	2651.006	25	2	
	NE	2651.010			50
	W	2651.023		10	
---	PD II	2651.090		10	
	NB	2651.122	3	200	
	HF II	2651.165	100	320	
	GE I	2651.178	40	20	62 VAF
	TA	2651.221		79	
	FE II	2651.291		2	
	RU I	2651.292	63	5	
--	CE	2651.418	2		
	W	2651.441	10	4	
	SE	2651.480			10
	TA	2651.483	50		
	PT II	2651.503		4	
	RU	2651.507	20		
	V	2651.570		5	
	*GE I	2651.575	32	20	36 VAF
	FE I	2651.706	63	63	
	LA	2651.710		8	
	YB	2651.710	2	63	
	IR	2651.768		5	
	NB	2651.807	2	16	
--	CL	2651.820			5
	RU	2651.841	100	10	
	U	2651.844	3	16	

Table IIa. GE I 2651.575 A
(con't)

	Element	Wavelength (Angstrom)	Intensity Arc	Spark	Plasma
	MN	2651.861		13	
	W	2651.874	2	13	
	V I	2651.898	50	4	
	RE I	2651.902	100		
	AR	2651.950			2
	CE	2652.006	10		
	W	2652.012	8	2	
--	HG I	2652.042	100	63	1 VAF
	CR	2652.097		2	
	RU	2652.134	8	2	
	CE	2652.152	2		
	U	2652.188	6	4	
	YB	2652.230	2	63	
	W II	2652.297		2	
--	TA	2652.323	16		
	HF II	2652.330		3	
	CO II	2652.340	3	4	
	MO	2652.350		10	
	W II	2652.424		3	
	MN	2652.485	3	100	
	AL I	2652.489	160	63	
	CE	2652.531	2		
	FE II	2652.571		40	

Table IIb.(3) GE I 2651.575 A

Element	Wavelength (Angstrom)	Intensity Arc	Spark	Plasma
TI I	2646.637	20	16	
W	2646.735	13	10	
TA	2646.770	50	50	
PT I	2646.886	1000	100	
NI II	2647.056		500	
RE I	2647.124	100		
BA II	2647.289	10	40	
HF II	2647.294	200	630	
RU	2647.315	50	5	

Table IIb.(3) GE I 2651.575 A (con't)

Element	Wavelength (Angstrom)	Intensity Arc	Spark	Plasma
TA	2647.472	200	10	
FE I	2647.557	100	63	
MN	2647.615		16	
V I	2647.710	50	10	
W	2647.740	10	20	
MN	2648.044		25	
V	2648.470	2	63	
CO I	2648.635	5	40	
NI II	2648.719		79	
RU	2648.780	32	160	
MN	2648.938		50	
RE	2649.049	100		
U	2649.069	16	16	
HF II	2649.150	10	20	
MO I	2649.250	10	25	
V	2649.360	2	100	
MO	2649.460	32	10	
FE II	2649.464		63	
PD II	2649.470		200	
RU	2649.513	32	3	
W	2649.693		13	
YB	2649.780	4	10	
CO I	2649.940	50	5	
MO	2649.999		20	
RU	2650.253		10	
CO I	2650.270	50	25	
PB	2650.400	100	79	
RU	2650.405	50		
BE I	2650.470	100	16	6 VAF
FE II	2650.492		160	
BE I	2650.550	32		
BE I	2650.760	20	16	
PT I	2650.857	630	100	
MN	2650.994		160	
W	2651.023		10	
PD II	2651.090		10	
NB	2651.122	3	200	
HF II	2651.165	100	320	
GE I	2651.178	40	20	62 VAF
TA	2651.221		79	

Table IIb.(3) GE I 2651.575 A
(con't)

Element	Wavelength (Angstrom)	Intensity Arc	Spark	Plasma
RU I	2651.292	63	5	
TA	2651.483	50		
*GE I	2651.575	32	20	36 VAF
FE I	2651.706	63	63	
YB	2651.710	2	63	
NB	2651.807	2	16	
RU	2651.841	100	10	
U	2651.844	3	16	
MN	2651.861		13	
W	2651.874	2	13	
V I	2651.898	50	4	
RE I	2651.902	100		
HG I	2652.042	100	63	1 VAF
YB	2652.230	2	63	
MO	2652.350		10	
MN	2652.485	3	100	
AL I	2652.489	160	63	
FE II	2652.571		40	
SB	2652.606	50	79	
W	2652.609	10	13	
RH I	2652.660	100	25	
V	2652.780		40	
U	2652.828	13	10	
RE	2652.912	32		
W	2653.012	2	10	
RU	2653.076		10	
TA	2653.274	200	16	
MO	2653.349	25	160	
MN	2653.561		16	
W II	2653.567	10	16	
CR I	2653.586	13	32	
EU	2653.598	16	16	
HG I	2653.681	79	40	
CO II	2653.703	5	40	
HO	2653.720		10	
ER	2653.743	16	10	
YB II	2653.750	50	200	14 VAF
MO	2653.790		20	
V I	2653.827	20	16	
RU	2653.947		79	

Table IIb.(3) GE I 2651.575 A
(con't)

Element	Wavelength (Angstrom)	Intensity Arc	Spark	Plasma
RE	2654.117	50		
RH	2654.771		25	
MO	2655.027	50	8	
HG I	2655.121	79	40	
NI II	2655.470		400	
W	2655.549	4	10	
W II	2655.670	5	10	
V	2655.676	10	100	
MN	2655.792	5	10	
MO	2655.812		20	
NI II	2655.907		500	
MN	2655.914	5	20	
NB	2656.078	8	200	
FE	2656.149	63	40	
MN	2656.170		13	
V I	2656.225	40	4	
RU	2656.248	20	160	
RU	2656.562	32		

Table IIa. GE I 2094.230 A

Element	Wavelength (Angstrom)	Intensity Arc	Spark	Plasma
NB	2093.280		8	
K	2093.300			5
KR II	2093.370			3
MN	2093.390	32	20	
CO	2093.400	16		
V	2093.410		2	
NI II	2093.560	6	20	
CU II	2093.606	2	20	
CL	2093.640			10
W	2093.800	8	16	
IR	2093.970		10	
IR	2094.050		3	
PT	2094.080	2		
N II	2094.120			8
IR	2094.180	8	16	
SI I	2094.200	2		
*GE I	2094.230	8		75 VAF
CO II	2094.240		3	
AL II	2094.300			50
HF II	2094.330	3	2	
SN I	2094.350			
RE	2094.530	6		
I	2094.560			20
V	2094.662	10		
PT	2094.700		4	
W II	2094.749	10	20	64 VAF
CU II	2094.770		8	
AL II	2094.800			50
YB	2094.820		10	
IR	2094.850	3		
IR	2095.040		16	
V II	2095.040		6	
W	2095.040	2	5	
AS	2095.100	3	3	
NI I	2095.123	16	2	
AL II	2095.200			40

Table IIb.(3) GE I 2094.230 A

Element	Wavelength (Angstrom)	Intensity Arc	Spark	Plasma
B I	2089.590	160	20	30 VAF
HF II	2089.950	32	32	
ER	2089.960	8		
BH C	2090.000	32		
NI I	2090.389	16		
GA	2090.500		5	11 VAF
IR	2090.520	8	6	
OS	2090.710	13		
HF II	2090.830	40	40	
V I	2091.307	10	5	
TA	2091.330	8	20	
GA II	2091.340			11 VAF
IR	2091.590	8		
NI I	2091.710	10		
IR	2091.860	10	20	
MN	2092.130	32	20	
RE	2092.260	20		
RE II	2092.410	5	100	3 VAF
V I	2092.440	10		
MN	2092.480	16		
V	2092.495	10	5	
W	2092.540	10	3	
PD I	2092.620	13		
IR I	2092.630	20	20	28 VAF
RH	2092.640	10	25	
IR	2092.770	16	5	
MN	2093.390	32	20	
CO	2093.400	16		
W	2093.800	8	16	
IR	2094.180	8	16	
*GE I	2094.230	8		75 VAF
V	2094.662	10		
W II	2094.749	10	20	64 VAF
NI I	2095.123	16	2	
RE	2095.620	20	2	
NI I	2095.730	25	2	
V	2095.750	8	2	
CO	2095.770	16	2	
HF II	2096.180	100	160	
OS	2096.280	13	16	

Table IIb.(3) GE I 2094.230 A
(con't)

Element	Wavelength (Angstrom)	Intensity Arc	Spark	Plasma
OS	2096.570	10		
RE	2097.160	40	10	
PT	2097.270	10		
V I	2097.330	10		
PT II	2097.440	10	32	
CO	2097.511	20		
MN	2097.540	20	32	
OS	2097.600	25	8	
RE	2097.640	13		
TA	2097.960	8	16	
W	2098.250	10	20	
RU	2098.460	8		
SB I	2098.470	32	32	9 VAF
W II	2098.600	10	20	55 VAF
PT	2098.740	13	3	
RU	2098.820	8		
W	2098.870	10		
CO I	2098.942	13	3	
MN	2098.960	10		
RH	2098.990	16	320	

Table IIa.   H I   6562.849 A

	Element	Wavelength (Angstrom)	Intensity Arc	Spark	Plasma
----	SM	6562.640	2		
	H	6562.725			1000
	*H	6562.849			2000
	HF II	6562.860	2	10	
--	SM	6562.940	50		
	XE II	6563.190			16
	W	6563.224	2		
---	SN	6563.240		50	
	CO I	6563.421	200	5	
----	SM	6563.520	20		
	GD	6563.660	6		
	TA	6563.701	2		

Table IIb.(1) H I   6562.849 A

Element	Wavelength (Angstrom)	Intensity Arc	Spark	Plasma
HF II	6557.910	10	100	
RB	6560.837		160	

Table IIa.   HF II 3399.795 A

	Element	Wavelength (Angstrom)	Intensity Arc	Spark	Plasma
	CO	3398.811	20		
	ND	3398.881	10	2	
	CE	3398.925	16		
	W	3398.927	4	13	
	ER	3398.940	40	6	
	HO II	3398.980	40	63	23 VAF
	U	3398.995	16		
	SN	3399.040		25	
---	CE	3399.074	2		
	TB	3399.100	16	8	
	ND	3399.193	8	2	
	FE I	3399.234	5		
	F II	3399.290			6
---	RE I	3399.299	200		
	FE I	3399.336	200	200	
	DY	3399.340	3		
	ZR II	3399.349	100	40	
	RU	3399.375	63	3	
	NB	3399.400	20	32	
	ND	3399.410	10	4	
	U	3399.528	3		
	CR	3399.530		63	
--	ER	3399.601	5		
	CD II	3399.670		5	
	SM	3399.688	5		
	RH I	3399.696	500	63	
	NB	3399.711	16	16	
	*HF II	3399.795	63	100	
	U	3399.810	5	2	
	SM	3399.840	10	3	
	SB II	3399.930		2	
	TM	3399.950	50	40	
	U	3399.958	6	3	
	NB	3399.963	10	2	
	TB	3399.970	16	3	
	RU I	3399.974	4		
	GD	3399.993	16	16	
--	CS I	3400.000	32		
	XE I	3400.070			2
	MO	3400.100		10	

Table IIa. HF II 3399.795 A
(con't)

	Element	Wavelength (Angstrom)	Intensity Arc	Spark	Plasma
	NA II	3400.110			5
	OS	3400.117	20	10	
	TI I	3400.162	10		
	RB	3400.180			5
	HF	3400.211	16	2	
---	CE	3400.254	8		
	U	3400.327	2	2	
	ND	3400.351	10	4	
	V I	3400.395	100	8	
	U	3400.467	8	2	
	CO I	3400.469	5		
----	TB	3400.530	16	8	
	W	3400.550	5	3	
	HO	3400.570		4	
	YB	3400.600	6		
	RU	3400.604	10		
	U	3400.750	6	3	
	RU I	3400.754	32		
	ZR	3400.765	2		

Table IIb.(3) HF II 3399.795 A

Element	Wavelength (Angstrom)	Intensity Arc	Spark	Plasma
NB	3394.975	2	50	
HF II	3394.983	20	50	
CO I	3395.375	400	50	
CR	3395.610	2	100	
NB	3396.372	2	160	
EU	3396.580	100	10	
FE I	3396.978	130	25	
LU II	3397.070	50	20	14 VAF
BI I	3397.213	100	50	
NB	3397.320		50	
TM	3397.503	100	50	
CS I	3398.140	63		
HO II	3398.980	40	63	23 VAF

Table IIb.(3) HF II 3399.795 A
(con't)

Element	Wavelength (Angstrom)	Intensity Arc	Spark	Plasma
RE I	3399.299	200		
FE I	3399.336	200	200	
ZR II	3399.349	100	40	
RU	3399.375	63	3	
CR	3399.530		63	
RH I	3399.696	500	63	
*HF II	3399.795	63	100	
V I	3400.395	100	8	
FE I	3401.521	160	100	
RU I	3401.739	100	50	
OS I	3401.859	200	20	
NB	3402.017		50	
FE	3402.262	160	160	
CR II	3402.399	25	79	
TI II	3402.422	16	100	
OS I	3402.511	200	16	
V I	3402.571	63	25	
MO	3402.812	5	100	
NB	3403.015	20	79	
CR II	3403.322	32	200	
CD I	3403.653	790	500	
FE I	3404.359	100	50	
V	3404.425		50	
PD I	3404.580	2000	1000	68 VAF
RE I	3404.724	100		

Table IIa. HF II 2773.360 A

	Element	Wavelength (Angstrom)	Intensity Arc	Spark	Plasma
	RE	2772.427	3		
	RU	2772.453		160	
	IR I	2772.456	16	10	
	W	2772.483	10	63	
	FE	2772.511	32	20	
	CO I	2772.552	16		
	LU	2772.580	5	160	
	U	2772.593	8	13	
	TA	2772.596	16		
	KR II	2772.600			10
----	DY	2772.610	3		
	RU	2772.612	50		
	BR	2772.620			3
	TE	2772.640			5
	CO	2772.696	32		
	PB II	2772.710		2	
	SM	2772.810		20	
	FE	2772.827	16	4	
	PT I	2772.829	16	2	
---	EU	2772.900	3		
	CE	2772.927	2		
	RU	2772.963	32	8	
	HF	2773.017	20	3	
	CE	2773.022	2		
	MN	2773.025	10		
	IN	2773.040		3	
	OS	2773.069	13	5	
	U	2773.084	5	4	
--	RH	2773.110		16	
	RE	2773.115	32		
	NB	2773.203	16	10	
	FE	2773.236	100	40	
	PT I	2773.241	50	5	
	RH I	2773.310	3	2	
	CR	2773.312		40	
	*HF II	2773.360	320	2000	190 VAF
	HF II	2773.500	10	10	
	XE	2773.550			3
	IR I	2773.594	4		
	PT I	2773.597	6		

Table IIa. HF II 2773.360 A
(con't)

	Element	Wavelength (Angstrom)	Intensity Arc	Spark	Plasma
--	U	2773.607	6	10	
	MN	2773.664	16		
	CO I	2773.670	3		
	V I	2773.679	32	6	
	FE II	2773.680			
	W	2773.698	13	8	
	U	2773.770	3	2	
	MO	2773.784	16	25	
	TM	2773.800	10		
---	HO	2773.840		10	
	W II	2773.870	4	10	
	RB	2773.900			2
	FE	2773.900	10	5	
	IR I	2773.901	6	5	
	CE	2773.947	2		
	PT I	2773.996	50	2	
	W	2774.002	13	2	
	V I	2774.005	10		
	HF II	2774.016	25	50	
	OS	2774.016	13	6	
	ZR I	2774.038	10		
	CE	2774.059	2		
	NB	2774.060	2	5	
	TH	2774.072	6	6	
--	W	2774.091		10	
	FE	2774.140			
	OS	2774.151	5	2	
	ZR II	2774.157	10	10	
	U	2774.169		4	
	RH	2774.170		40	
	RU	2774.181		6	
	PT	2774.205	4		
	V	2774.276	25	100	
	YB	2774.320		3	

Table IIb.(2) HF II 2773.360 A

Element	Wavelength (Angstrom)	Intensity Arc	Spark	Plasma
CR I	2769.915	400	40	
ZN I	2770.865	320	25	
ZN I	2770.984	320	160	
PT I	2771.666	500	16	
FE I	2772.109	320	320	
*HF II	2773.360	320	2000	190 VAF
MO II	2775.400	79	100	12 VAF
TA	2775.877	200	32	

Table IIa. HG I 2536.519 A

	Element	Wavelength (Angstrom)	Intensity Arc	Spark	Plasma
	I	2535.570			32
	W	2535.576		10	
	U	2535.580		8	
	RU	2535.592		100	
	TA	2535.598	50		
	FE I	2535.604	1000		
	MN	2535.645		79	
	P I	2535.650	100		11 VAF
---	TH	2535.870	5	2	
	TI II	2535.871	20	63	
	U	2535.900	5	2	
	PD	2535.940			3
	TA	2535.960	4		
	CO I	2535.964	10	40	
	NI	2535.967	25		
	PT	2535.967	25	5	
	W	2535.988	4	13	
	YB	2536.010		6	
--	AR II	2536.040			20
	MN	2536.080		13	
	IR	2536.127	2		
	RU	2536.216	13	6	
	FE	2536.224	3		
	TA	2536.227	100		
	U	2536.238	4	2	
--	PT I	2536.487	100	10	
	CO I	2536.493		2	
	*HG I	2536.519	2000	1000	49 VAF
	TH	2536.558	5	10	
	BI	2536.560	5	2	
	U	2536.600	3	2	
	W	2536.605		13	
	IR	2536.665	3		
	IN II	2536.669			10
	TA	2536.670	2		
	FE II	2536.673		5	
	RH	2536.706	16	5	
	PR	2536.710		8	
	TB	2536.750	3	10	
--	LA II	2536.760		3	

Table IIa. HG I 2536.519 A
(con't)

	Element	Wavelength (Angstrom)	Intensity Arc	Spark	Plasma
	U	2536.794	4	2	
	CO	2536.802		2	
	FE II	2536.817	10	4	
	MN	2536.830		5	
	MO	2536.849	25	4	
	V	2536.926	10		
---	LU	2536.950	10	20	
	BR	2537.020			4
	RH	2537.039	16	100	
	FE II	2537.141		2	
	W	2537.145		16	
	PD II	2537.169		100	
	FE I	2537.173	3		
	TE	2537.190			5
----	IR I	2537.225	32	10	
	U	2537.296	10	2	
	HF II	2537.337	50	400	
	MN	2537.353		2	
	FE	2537.458	3		
	CO	2537.460		10	

Table IIb.(1) HG I 2536.519 A

Element	Wavelength (Angstrom)	Intensity Arc	Spark	Plasma
TL	2531.600		100	
TA	2532.125	79	100	
CO I	2532.175		79	
NB	2533.188	2	79	
FE II	2533.627	8	50	
CO	2533.810	5	63	
RU	2534.001	4	79	
P I	2534.010	50		3 VAF
FE II	2534.416	6	50	
V II	2534.520	10	79	
RH	2534.570	2	100	
PD II	2534.599		100	

Table IIb.(1) HG I 2536.519 A
(con't)

Element	Wavelength (Angstrom)	Intensity Arc	Spark	Plasma
TI II	2534.621	25	79	
RU	2535.592		100	
FE I	2535.604	1000		
MN	2535.645		79	
P I	2535.650	100		11 VAF
TI II	2535.871	20	63	
*HG I	2536.519	2000	1000	49 VAF
RH	2537.039	16	100	
PD II	2537.169		100	
HF II	2537.337	50	400	
RH	2537.729	4	50	
PD II	2537.974		100	
MO	2538.455	32	130	
NI II	2539.100		250	
PT I	2539.205	400	20	
PD II	2539.358		50	
RU	2539.719	13	100	
TB	2540.120	3	50	
RU	2540.300	10	100	
NB	2540.615	2	160	
MN	2541.109		79	
NB	2541.424	3	79	

Table IIa. HO II 3456.000 A

	Element	Wavelength (Angstrom)	Intensity Arc	Spark	Plasma
	PB II	3455.000			10
	BI II	3455.010			10
	W	3455.021		20	
	CE	3455.023	5		
	OS	3455.031	50	16	
	IR	3455.038	3		
	TE	3455.070			16
	DY	3455.080	3		
	RN	3455.100			2
	BE I	3455.200	20		
	V I	3455.212	5	3	
	RH I	3455.219	320	13	
	CO I	3455.234	2000	10	
	FE	3455.234	2		
----	BI II	3455.275		100	
	TH	3455.275	5	5	
	CR	3455.281	32	10	
	TB	3455.350	16	3	
	RU	3455.386	20		
	TI I	3455.394	10		
	RH I	3455.422	50	2	
	TI I	3455.425	8		
	CE	3455.471	2		
	CS	3455.480			4
	PB I	3455.490		63	
---	CR	3455.602	50	32	
	HO	3455.700	6	6	
	RU	3455.732	13		
	U	3455.742	5	4	
--	TI I	3455.755	6		
	ND	3455.767	2		
	MO	3455.860		10	
	SC I	3455.898	3	2	
	ZR I	3455.908	13		
	U	3455.933	3	2	
	TH	3455.949	4	4	
	PR	3455.972	32	3	
	TB	3455.990	8		
	*HO II	3456.000	63	63	53 VAF
	ND	3456.004	4		

Table IIa. HO II 3456.000 A
(con't)

	Element	Wavelength (Angstrom)	Intensity Arc	Spark	Plasma
	DY	3456.010	40		
	CE	3456.031	2		
	SM	3456.114	4		
	OS	3456.139	20	10	
	MO	3456.152	5	6	
	HG	3456.170			10
--	RA II	3456.200			5
	U	3456.297	3	2	
	CE II	3456.340	10		
	MO	3456.387	16	10	
	TI II	3456.390	25	130	
	CO I	3456.437	5		
--	TH	3456.441	5	4	
	NE	3456.520			6
	SR I	3456.520	3		
	MO	3456.524	3	4	
	NB	3456.538	16	10	
	ER	3456.561	13	6	
	DY	3456.566	50	32	
	HO	3456.600		4	
	U	3456.605	2	2	
	RU I	3456.620	63	8	
	TI I	3456.661	2		
	CE	3456.674	20		
-	NE II	3456.680			13
	CE II	3456.772	8		
	TB	3456.830	32		
	TE	3456.840			10
	KR I	3456.870			3
	V I	3456.914	20	2	
	CO I	3456.926	32	2	
	FE II	3456.934	3	4	
	TH	3456.936	3	3	
	NB	3456.988	5		

Table IIb.(3) HO II 3456.000 A

Element	Wavelength (Angstrom)	Intensity Arc	Spark	Plasma
V	3451.045		63	
GD	3451.241	50	40	
PD II	3451.346		400	
B II	3451.410		100	
NB	3451.636		32	
RE I	3451.808	100		
FE I	3451.918	100	63	
LA II	3452.184	50	40	
FE I	3452.277	160	8	
B II	3452.280	5	32	
NB	3452.346	5	200	
TI II	3452.470	13	100	
NI I	3452.890	630	50	
RU I	3452.903	63	6	
V	3453.084		63	
HO II	3453.130	32	20	10 VAF
LA II	3453.168	50	40	
CR I	3453.328	32	32	
CO I	3453.505	3200	200	
TM II	3453.660	160	79	19 VAF
NB	3453.970		100	
TB II	3454.060	79	32	27 VAF
YB	3454.070	40	250	
DY	3454.326	100	10	
CE	3454.475	10	40	
NB	3454.706	3	50	
NB	3454.914	3	79	
CR	3454.990		100	
RH I	3455.219	320	13	
CO I	3455.234	2000	10	
BI II	3455.275		100	
PB I	3455.490		63	
CR	3455.602	50	32	
*HO II	3456.000	63	63	53 VAF
TI II	3456.390	25	130	
DY	3456.566	50	32	
RU I	3456.620	63	8	
RH I	3457.071	100	4	
V	3457.152	2	160	
CR	3457.630	4	130	

Table IIb.(3) HO II 3456.000 A
(con't)

Element	Wavelength (Angstrom)	Intensity Arc	Spark	Plasma
RH I	3457.926	130	10	
CO I	3458.028	63		
YB	3458.280	13	100	
FE	3458.309	63	25	
OS	3458.385	200	13	
NI I	3458.474	790	50	
OS	3459.018	100	10	
CR	3459.290		32	
FE I	3459.918	79	50	
MN II	3460.328	63	500	
CR I	3460.430	40	32	
RE I	3460.470	1000		3 VAF
PD I	3460.774	320	630	16 VAF
DY	3460.971	100	3	

Table IIa. HO II 3398.980 A

	Element	Wavelength (Angstrom)	Intensity Arc	Spark	Plasma
	DY	3397.990	3	2	
	TM	3398.020	32	40	
	W	3398.097	10	10	
	CS I	3398.140	63		
	FE	3398.220	2		
----	NB	3398.254	16	40	
	U	3398.256	13	6	
	V I	3398.265	13	10	
	ER	3398.276	13	2	
	LA II	3398.290	2	3	
	ND	3398.327	13	4	
	TA	3398.327	32	20	
	W	3398.330		4	
	CE	3398.339	10		
	DY	3398.340	4	2	
	TB	3398.350	32	8	
	SM	3398.440		3	
	EU	3398.470		2	
---	U	3398.498	6		
	TH	3398.548	5	2	
	OS	3398.575	20	10	
	TI I	3398.634	20	6	
	CE	3398.639	4		
	YB	3398.650	10		
	U	3398.670	2		
	CE	3398.722	10		
--	RE	3398.790	16		
	CO	3398.811	20		
	ND	3398.881	10	2	
	CE	3398.925	16		
	W	3398.927	4	13	
	ER	3398.940	40	6	
	*HO II	3398.980	40	63	23 VAF
	U	3398.995	16		
	SN	3399.040		25	
	CE	3399.074	2		
	TB	3399.100	16	8	
--	ND	3399.193	8	2	
	FE I	3399.234	5		
	F II	3399.290			6

Table IIa. HO II 3398.980 A
(con't)

Element	Wavelength (Angstrom)	Intensity Arc	Spark	Plasma
RE I	3399.299	200		
FE I	3399.336	200	200	
DY	3399.340	3		
ZR II	3399.349	100	40	
RU	3399.375	63	3	
NB	3399.400	20	32	
-- ND	3399.410	10	4	
U	3399.528	3		
CR	3399.530		63	
ER	3399.601	5		
CD II	3399.670		5	
SM	3399.688	5		
RH I	3399.696	500	63	
-- NB	3399.711	16	16	
HF II	3399.795	63	100	
U	3399.810	5	2	
SM	3399.840	10	3	
SB II	3399.930		2	
TM	3399.950	50	40	
U	3399.958	6	3	
NB	3399.963	10	2	
TB	3399.970	16	3	
RU I	3399.974	4		

Table IIb.(3) HO II 3398.980 A

Element	Wavelength (Angstrom)	Intensity Arc	Spark	Plasma
CR	3394.295	16	160	
TI II	3394.575	63	200	
FE I	3394.589	160	79	
NB	3394.975	2	50	
HF II	3394.983	20	50	
CO I	3395.375	400	50	
CR	3395.610	2	100	
OS	3395.719	40	13	
NB	3395.725		32	

Table IIb.(3) HO II 3398.980 A
(con't)

Element	Wavelength (Angstrom)	Intensity Arc	Spark	Plasma
NB	3396.372	2	160	
EU	3396.580	100	10	
FE I	3396.978	130	25	
LU II	3397.070	50	20	14 VAF
BI I	3397.213	100	50	
NB	3397.320		50	
TM	3397.503	100	50	
TM	3398.020	32	40	
CS I	3398.140	63		
NB	3398.254	16	40	
ER	3398.940	40	6	
*HO II	3398.980	40	63	23 VAF
RE I	3399.299	200		
FE I	3399.336	200	200	
ZR II	3399.349	100	40	
RU	3399.375	63	3	
NB	3399.400	20	32	
CR	3399.530		63	
RH I	3399.696	500	63	
HF II	3399.795	63	100	
TM	3399.950	50	40	
V I	3400.395	100	8	
YB	3401.010	13	32	
NI I	3401.166	40		
OS	3401.174	40	16	
NB	3401.226	2	40	
FE I	3401.521	160	100	
RU I	3401.739	100	50	
OS I	3401.859	200	20	
W II	3401.890	6	40	
NB	3402.017		50	
CU I	3402.244	50	10	
FE	3402.262	160	160	
CR II	3402.399	25	79	
TI II	3402.422	16	100	
SM II	3402.463	50	10	
OS I	3402.511	200	16	
V I	3402.571	63	25	
MO	3402.812	5	100	
NB	3403.015	20	79	

Table IIb.(3) HO II 3398.980 A
(con't)

Element	Wavelength (Angstrom)	Intensity Arc	Spark	Plasma
CR II	3403.322	32	200	
NI I	3403.432	40		
CD I	3403.653	790	500	

Table IIa.   IN I   4511.323 A

	Element	Wavelength (Angstrom)	Intensity Arc	Spark	Plasma
	GD	4510.386	10	10	
	TH	4510.535	32	20	
----	AR I	4510.733			1000
	CE	4510.761	3		
	HO	4510.810	2		
---	AL	4510.840		6	
	ER	4510.917	2		
	CE	4510.921	6		
	TA	4510.982	200	50	
--	NB	4511.089	5	16	
	PR	4511.091	3		
	U	4511.158	4	8	
	TI	4511.170	40	10	
	ZR I	4511.170	5		
	RU I	4511.197	25		
	MN	4511.238	2		
	PT I	4511.257	2		
	NE II	4511.290			16
	ND	4511.290	25		
	IN I	4511.300	200		
	SM	4511.307	40		
	*IN I	4511.323	5000	4000	16 VAF
	CD	4511.340	5		
	PR	4511.349	4		
	NE II	4511.370			50
	V I	4511.432	2	2	
	PR	4511.455	5		
	TA	4511.503	320	40	
	NE I	4511.509			20
	TB	4511.520	40		
--	EU	4511.530	3	3	
	CE	4511.635	10		
	ER	4511.715	4		
	U	4511.746	5	8	
	PR	4511.815	3		
---	ND	4511.823	50	10	
	SM II	4511.834	100	100	
	CR	4511.903	79	100	
----	IR	4512.028	3		
	NB	4512.129	2	3	

Table IIa. IN I 4511.323 A
(con't)

Element	Wavelength (Angstrom)	Intensity Arc	Spark	Plasma
EU	4512.140	2		
MO	4512.148	25	25	
U	4512.180	2	4	
ER	4512.205	3		
PR	4512.272	8		
CA I	4512.282	10		
ND	4512.289	3		
SM	4512.305	2		

Table IIb.(1) IN I 4511.323 A

Element	Wavelength (Angstrom)	Intensity Arc	Spark	Plasma
*IN I	4511.323	5000	4000	16 VAF

Table IIa. IN II 2306.062 A

	Element	Wavelength (Angstrom)	Intensity Arc	Spark	Plasma
	PR	2305.100		10	
	CO I	2305.180	16	2	
	W	2305.223		3	
	NI II	2305.244	2	20	
	MO	2305.250		8	
----	YB	2305.330	8	100	
	HF II	2305.340	6	8	
	NB	2305.432		2	
	IR I	2305.466	32	5	
	RU	2305.518	10	20	
---	RE	2305.570	16		
	RU	2305.626		20	
	PT	2305.637	10	6	
	U	2305.645		13	
	MO	2305.658	5	20	
	TI I	2305.670	16	2	
	PD	2305.710		5	
	RE	2305.760	16	2	
--	OS	2305.840	3	6	
	MO	2305.870		10	
	W	2305.936		4	
	RH	2305.950		100	
	IN II	2305.994			20
	OS	2306.046	6	32	
	MO	2306.060	4		
	*IN II	2306.062			47 VAF
	CO	2306.100		4	
	IN II	2306.118			10
	FE I	2306.171	10		
	HE II	2306.220			20
--	OS	2306.300	10	2	
	FE	2306.380	10	3	
	NI	2306.400	10	4	
	OS	2306.480	5		
	SB	2306.500	32	32	
	MO	2306.506		20	
---	RE	2306.560	20	8	
	K	2306.580			10
	W	2306.600	13		
	CD	2306.609	20	32	

Table IIa. IN II 2306.062 A
(con't)

	Element	Wavelength (Angstrom)	Intensity Arc	Spark	Plasma
	RE	2306.630	16		
	PD	2306.750	3	2	
	IR	2306.770		32	
	CO	2306.780		5	
--	RH	2306.790	2	4	
	CR	2306.840	2	10	
	OS	2306.870	13	2	
	TA	2306.870		6	
	IN I	2306.879	25	32	
	U	2306.910		25	
	RU	2306.919	20		
	W II	2306.920	4	13	
	RH	2306.950		16	
	MO	2306.970	8	25	
	PD	2306.990			2
	CO	2307.010		6	

Table IIb.(3) IN II 2306.062 A

Element	Wavelength (Angstrom)	Intensity Arc	Spark	Plasma
ND	2301.160		5	
TH	2301.220		32	
CO II	2301.400	10	25	
TA	2301.466	2	32	
W	2301.650	8	10	
IR	2301.675		25	
RE	2301.823		40	
RH	2301.944		25	
PD II	2302.012		63	
NB	2302.085	16	32	
HG I	2302.090	16	16	
HF II	2302.110	5	6	
C	2302.120		16	
MO	2302.130		8	
TA	2302.235	6	32	
PD II	2302.319		16	

Table IIb.(3) IN II 2306.062 A
(con't)

Element	Wavelength (Angstrom)	Intensity Arc	Spark	Plasma
PT	2302.420		20	
NI II	2302.472	2	20	
NB	2302.698	8	8	
RU	2302.806		8	
MO	2302.842		20	
TA	2302.928	8	32	
NI I	2302.997	10	32	
NI II	2302.997	10	32	13 VAF
SI I	2303.020	10	6	
CU I	2303.116	32	20	
PT I	2303.200	25	8	
V	2303.242		6	
W	2303.278	4	13	
FE II	2303.351		16	
RH	2303.468		20	
TA	2303.491	6	20	
PR	2303.820		10	
W II	2303.826	10	16	
NI II	2303.857		16	
U	2303.930		8	
CO I	2304.185	10	6	
BA II	2304.235	63	79	73 VAF
MO	2304.251	6	50	
PT	2304.260		10	
FE I	2304.729	10	20	
NB	2304.770		20	
AU II	2304.810		32	
RU	2304.827	13	6	
OS	2304.830		10	
MN	2304.997	40	63	
W	2305.053	6	5	
RH	2305.060	2	10	
PR	2305.100		10	
NI II	2305.244	2	20	
MO	2305.250		8	
YB	2305.330	8	100	
HF II	2305.340	6	8	
IR I	2305.466	32	5	
RU	2305.518	10	20	
RU	2305.626		20	

Table IIb.(3) IN II 2306.062 A
(con't)

Element	Wavelength (Angstrom)	Intensity Arc	Spark	Plasma
PT	2305.637	10	6	
U	2305.645		13	
MO	2305.658	5	20	
PD	2305.710		5	
OS	2305.840	3	6	
MO	2305.870		10	
RH	2305.950		100	
OS	2306.046	6	32	
*IN II	2306.062			47 VAF
SB	2306.500	32	32	
MO	2306.506		20	
RE	2306.560	20	8	
CD	2306.609	20	32	
IR	2306.770		32	
CO	2306.780		5	
CR	2306.840	2	10	
TA	2306.870		6	
IN I	2306.879	25	32	
U	2306.910		25	
W II	2306.920	4	13	
RH	2306.950		16	
MO	2306.970	8	25	
CO	2307.010		6	
ND	2307.100		5	
CR	2307.208	10	20	
FE	2307.314		25	
YB	2307.410		6	
CO	2307.490		6	
PD II	2307.500		50	
OS	2307.610	5	5	
W	2307.660	2	5	
NI II	2307.785	2	16	
PR	2307.790		10	
CO II	2307.857	25	50	31 VAF
W II	2307.930		5	
MO	2307.980	5	20	
PT I	2308.040	32	20	
ZR	2308.090		5	
TA	2308.160		5	
MN	2308.180		25	

Table IIb.(3) IN II 2306.062 A
(con't)

Element	Wavelength (Angstrom)	Intensity Arc	Spark	Plasma
OS	2308.310	20	6	
U	2308.350		6	
TA	2308.460	2	8	
NI II	2308.519		20	
PD II	2308.604		32	
RE	2308.620	10	10	
FE	2308.765		16	
IR	2308.934	20	5	
FE I	2308.997	20	5	
CO I	2309.020	10	10	
RU	2309.155	5	8	
NB	2309.234	10	32	
YB	2309.260		20	
AU	2309.430		13	
MO	2309.450		20	
PD	2309.520		6	
IR	2309.570		5	
CU II	2309.610		20	
AG	2309.644	160	200	
OS	2309.810		5	
RH I	2309.823	25	10	
W	2309.840	3	6	
V	2309.842		130	
MO	2309.850		32	
NB	2309.940		16	
CR	2310.040		6	
FE	2310.097		5	
OS	2310.160	10	6	
W	2310.250		5	
NB	2310.318	2	8	
U	2310.360		10	
RH	2310.530		25	
IR	2310.533	2	16	
U	2310.670		10	
ND	2310.820		5	
W	2310.820	4	6	
PT II	2310.958	16	25	
NI I	2310.965	50	10	

Table IIa.   IR II 2242.680 A

Element	Wavelength (Angstrom)	Intensity Arc	Spark	Plasma
ND	2241.750		13	
AG II	2241.807		10	
CR	2241.840		32	
FE	2241.840	6	2	
V I	2241.842	25		
XE II	2241.860			4
RH	2242.050	25	25	
W	2242.060	13	10	
OS	2242.100	32	5	
CU II	2242.140			6
NI II	2242.150		4	
PR	2242.170		10	
MO	2242.210		25	
IR	2242.240		16	
NB	2242.288	5	3	
CE	2242.330		63	
FE II	2242.338		3	
IR	2242.351	8		
RE	2242.410	25	10	
CD	2242.430			2
GE I	2242.470	2		
P I	2242.492			2
PD	2242.540			6
NB	2242.579	5	50	
PB	2242.610	16	16	
CU II	2242.613	25	50	
V	2242.614	5		
*IR II	2242.680	50	320	110 VAF
U	2242.680		2	
AU	2242.710		32	
W	2242.710		8	
TA	2242.750	4		
NI	2242.760		8	
NI I	2242.880	3		
CO	2242.900		4	
W	2242.960	3	8	
NB	2242.962	4		
AL II	2243.050			32
Y II	2243.060	25	32	33 VAF
CU II	2243.100			6

Table IIa. IR II 2242.680 A
(con't)

	Element	Wavelength (Angstrom)	Intensity Arc	Spark	Plasma
	FE	2243.150		10	
	PD	2243.170			2
---	HF II	2243.180	16	20	
	P I	2243.189			2
	NI I	2243.210	3		
	RU	2243.220	20		
	CO I	2243.257	10		
	V	2243.267	3		
	RE	2243.290	16	6	
	CR	2243.310		25	
	SN	2243.311		8	
	OS	2243.320	2	10	
	U	2243.370		2	
----	K II	2243.400			5
	AG II	2243.440		16	
	V	2243.465		2	
	W	2243.500		10	
	OS	2243.540	10		
	PD	2243.541		20	
	FE II	2243.587		2	
	CR	2243.620		32	
	MO	2243.623		10	
	U	2243.650		2	
	TL	2243.660		2	
	MN	2243.670		5	

Table IIb.(2) IR II 2242.680 A

Element	Wavelength (Angstrom)	Intensity Arc	Spark	Plasma
RH	2237.710		100	
TL I	2237.850	63	3	2 VAF
RU	2238.330	50		
CU I	2238.454	40		
RE	2238.640	40	4	
IR I	2238.818	32	5	
TA II	2239.480	20	32	69 VAF

Table IIb.(2) IR II 2242.680 A
(con't)

Element	Wavelength (Angstrom)	Intensity Arc	Spark	Plasma
CD	2239.860	79	32	
V	2240.622		100	
RU	2241.070	63		
V	2241.535		200	
OS	2241.620	25	2	
V I	2241.842	25		
RH	2242.050	25	25	
OS	2242.100	32	5	
RE	2242.410	25	10	
CU II	2242.613	25	50	
*IR II	2242.680	50	320	110 VAF
Y II	2243.060	25	32	33 VAF
CU I	2244.265	25	10	
NI I	2244.530	40		
PT I	2244.970	25	10	
PT II	2245.518	25	32	36 VAF
IR	2245.760	10	160	
SN I	2246.053	100	100	25 VAF
AG II	2246.412	25	320	2 VAF
RE	2246.470	25	5	
PB I	2246.888	32	100	9 VAF
IR	2246.900	10	100	
CU II	2246.995	32	500	39 VAF
OS	2247.680	40	6	

Table IIa.   IR II 2126.810 A

	Element	Wavelength (Angstrom)	Intensity Arc	Spark	Plasma
	V	2125.830	4		
	CO II	2125.860		2	
	NI II	2125.900	4	25	
	MO	2125.910	2	32	
	CO	2125.949	5		
	IR	2125.970	5		
	CU II	2126.028	16	32	
----	RH	2126.060	2	5	
	TI	2126.090	5		
	CO I	2126.199	5		
	IR	2126.240	4		
---	TA	2126.340		8	
	RE	2126.470	10	2	
	ZR	2126.535		2	
	NB	2126.543	16	50	
--	V	2126.583		16	
	HF	2126.610	6	8	
	AU I	2126.644	6		
	PD II	2126.652		20	
	TA	2126.670	2	10	
	YB II	2126.720	40	200	32 VAF
	RE II	2126.730	3	4	
	TA	2126.780	2	5	
	NI II	2126.800	2	25	
	*IR II	2126.810	25	200	100 VAF
	S	2126.810			16
	RE	2126.850	32	6	
	V II	2126.935	3	10	
--	TM	2127.030		8	
	CD	2127.130			2
	CO I	2127.147	10		
	K	2127.250			20
	IR	2127.260	5	8	
---	MO	2127.260	3	16	
	V II	2127.370		5	
	SB I	2127.390	25	16	
	PT II	2127.417	2	25	
	LU	2127.430		4	
	W	2127.430	2	13	
	HF	2127.500	5	5	

Table IIa. IR II 2126.810 A
(con't)

Element	Wavelength (Angstrom)	Intensity Arc	Spark	Plasma
IR	2127.520	16		
--- TA	2127.560	8		
NI II	2127.770		20	

Table IIb.(2) IR II 2126.810 A

Element	Wavelength (Angstrom)	Intensity Arc	Spark	Plasma
IR	2121.880	13	4	
PT	2122.560	16	10	
OS	2122.830	13	3	
HF II	2122.940	16	16	
CU II	2122.966	16	50	
HF II	2123.680	40	40	
OS	2123.840	40	25	
NI	2124.110	32		
SI I	2124.120	200	50	180 VAF
RH	2124.130	40	10	
OS	2124.390	25	5	
HF II	2124.590	50	79	
NI I	2124.810	25	3	
FE	2125.010	32		
NB	2125.206	16	40	
OS	2125.340	25	4	
IR	2125.440	25	10	
NI I	2125.620	16	2	
CU II	2126.028	16	32	
NB	2126.543	16	50	
YB II	2126.720	40	200	32 VAF
*IR II	2126.810	25	200	100 VAF
RE	2126.850	32	6	
SB I	2127.390	25	16	
IR	2127.520	16		
NI	2127.910	16		
IR	2127.940	20	16	
OS	2127.970	5	50	
ER	2128.070		250	

Table IIb.(2) IR II 2126.810 A
(con't)

Element	Wavelength (Angstrom)	Intensity Arc	Spark	Plasma
NI I	2128.410	16		
PT I	2128.610	32	25	24 VAF
HF II	2129.100	63	100	
NI I	2129.950	16	3	
IR	2130.450	5	79	
CU I	2130.761	25	3	
FE	2130.962	20		
NB	2131.180	13	40	
IR	2131.660		50	
TA	2131.790	13	32	

Table IIa. IR I 2052.220 A

	Element	Wavelength (Angstrom)	Intensity Arc	Spark	Plasma
	RH	2051.250	2	100	
	V	2051.290		2	
---	HF	2051.700		2	
	V	2051.780		10	
---	W	2051.890	2		
	NI I	2052.060	13	6	
--	W	2052.150		8	
	*IR I	2052.220	13	10	49 VAF
	V	2052.380		3	
--	OS	2052.400	10	2	
	NI I	2052.428	6		
	O	2052.530			16
	RH	2052.570		20	
---	XE	2052.700			2
	OS	2052.780	3	5	
	RH	2052.840		10	
	PR	2052.870		13	
---	HG II	2052.929			100
	K	2053.100			2
	SB	2053.100			8
	W II	2053.130	8	8	
	SR	2053.200			3
	RH	2053.200	2	32	

Table IIb.(3) IR I 2052.220 A

Element	Wavelength (Angstrom)	Intensity Arc	Spark	Plasma
NI I	2047.333	13	6	
TA	2047.410	6	20	
BH C	2047.500	32		
AS I	2047.590	25	3	
CU II	2047.650		10	
TA	2047.730	2	5	
IR	2047.960	3	10	
RH	2048.040		10	
W II	2048.040	6	13	

Table IIb.(3) IR I 2052.220 A
(con't)

Element	Wavelength (Angstrom)	Intensity Arc	Spark	Plasma
OS	2048.280	25	5	
IR	2048.550		10	
RH	2048.650		200	
OS	2048.800	6	25	
RH	2048.950		5	
W	2049.020	6	6	
RE II	2049.110	32	6	4 VAF
PT II	2049.168		25	
CO II	2049.170	3	10	
PT I	2049.373	20	25	42 VAF
OS	2049.420	16	5	
IR	2049.560	2	32	
SB I	2049.570	160		15 VAF
W	2049.630	8	10	
BI	2049.690	25	20	
PT I	2049.830	10	10	
NB	2049.870	2	40	
IR	2050.220		25	
W	2050.400	2	10	
NI I	2050.827	16	2	
IR	2051.160	10	50	
RH	2051.250	2	100	
V	2051.780		10	
NI I	2052.060	13	6	
W	2052.150		8	
*IR I	2052.220	13	10	49 VAF
RH	2052.570		20	
OS	2052.780	3	5	
RH	2052.840		10	
PR	2052.870		13	
W II	2053.130	8	8	
RH	2053.200	2	32	
NI II	2053.290	3	25	
PB I	2053.320	13		
RH	2053.610		32	
OS	2053.840	6	5	
IR	2053.870	2	25	
SB	2054.040	32	5	
HF	2054.300		6	
CU II	2054.300		8	

Table IIb.(3) IR I 2052.220 A
(con't)

Element	Wavelength (Angstrom)	Intensity Arc	Spark	Plasma
NI II	2054.320		25	
RH	2054.370	3	25	
CU II	2054.410	3	20	
PD	2054.430		32	
TI II	2054.530	4	16	
W II	2054.678	10	20	40 VAF
IR	2054.770	2	32	
V	2054.840	10	16	
CU II	2054.966	16	32	
RE	2055.270	4	10	
RH	2055.310	2	40	
OS	2055.400	3	5	
PD	2055.490		13	
CR II	2055.520	100	320	49 VAF
OS	2055.530	5	20	
IR	2055.720	16	25	
BE I	2056.000		6	1 VAF
W II	2056.020	5	10	
PT II	2056.057		13	
RE	2056.060	2	6	
IR	2056.170		20	
BE I	2056.520	100		
V	2056.530		5	
PT II	2057.003	8	32	
RH	2057.040		10	
NB	2057.050	4	32	
RH	2057.180		16	

Table IIa.   K I   7664.907 A

	Element	Wavelength (Angstrom)	Intensity Arc	Intensity Spark	Intensity Plasma
	XE I	7664.020			10
----	FE I	7664.302	16		
	LA I	7664.340	8	3	
---	XE I	7664.560			32
	CU II	7664.700	5	63	
--	CD	7664.740		3	
	*K I	7664.907	10000	400	
---	SC I	7665.720	5		

Table IIb.(1) K I   7664.907 A

Element	Wavelength (Angstrom)	Intensity Arc	Intensity Spark	Intensity Plasma
HF II	7663.090	2	32	
CU II	7664.700	5	63	
*K I	7664.907	10000	400	

Table IIa. K I 4047.201 A

Element	Wavelength (Angstrom)	Intensity Arc	Spark	Plasma
V	4046.264		16	
NB	4046.270	3		
CE I	4046.340	32	10	
CE II	4046.340	32	10	
U	4046.402	3	5	
PT II	4046.450		20	
--- SC I	4046.486	10		
U	4046.521	5	5	
IR	4046.539	8		
HG I	4046.561	200	320	
PR	4046.635	8	3	
--- W	4046.701	10	13	
ND	4046.702	8	2	
CE	4046.734	2		
CR I	4046.760	32	3	
NI I	4046.761	2		
GD	4046.842	10	10	
CE	4046.853	3		
MO	4046.886	3	20	
-- ER	4046.960	8		
U	4047.050	6	8	
GD	4047.093	6	5	
PR	4047.098	20	13	
ND	4047.158	13	10	
TB	4047.160	10		
TE	4047.180			16
CS II	4047.184			20
*K I	4047.201	400	200	1 VAF
MO	4047.204	4	3	
CE	4047.275	20	2	
FE I	4047.310	3		
IR	4047.329	4		
CA	4047.351	2	3	
SM II	4047.363	8	6	
YB	4047.390		4	
CE	4047.392	4		
MO	4047.398	4	4	
-- RB	4047.400			4
HO	4047.500	3	3	
AR II	4047.510			2

Table IIa. K I 4047.201 A
(con't)

	Element	Wavelength (Angstrom)	Intensity Arc	Spark	Plasma
	MO	4047.563	4	4	
	U	4047.610	20	3	
	CE	4047.620	3		
	Y I	4047.632	50	10	
---	CE	4047.699	3		
	EU	4047.740	2		
	DY	4047.740	4	2	
	SE II	4047.770			8
	SC I	4047.792	25	10	
	GD	4047.846	160	50	
	ER	4047.850	5		
	CL	4047.870			6
	CE	4047.875	5		
----	W	4047.935	10	8	
	HF	4047.960	8	25	
	OS	4048.054	20	2	
	TH	4048.057	8	8	
	U	4048.065	8	2	
	PR	4048.140	8	4	

Table IIb.(2) K I 4047.201 A

Element	Wavelength (Angstrom)	Intensity Arc	Spark	Plasma
LA II	4042.911	400	320	19 VAF
K I	4044.140	790	400	
AR I	4044.418			31 VAF
CO I	4045.390	400		
HO	4045.430	200	79	
FE I	4045.815	400	320	
AR I	4045.966			3 VAF
HG I	4046.561	200	320	
*K I	4047.201	400	200	1 VAF
GD	4047.846	160	50	
MN	4048.755	63	63	
CR	4048.780	79	50	
GD	4049.900	100	63	

Table IIb.(2) K I 4047.201 A
(con't)

Element	Wavelength (Angstrom)	Intensity Arc	Spark	Plasma
LA II	4050.079	63	63	
RU I	4051.400	130	200	
RU I	4052.195	25	50	

Table IIa. LA II 4086.714 A

	Element	Wavelength (Angstrom)	Intensity Arc	Spark	Plasma
	CE	4085.748	8		
	TA	4085.797	5	79	
	ND	4085.815	10	6	
	AG	4085.871		25	
	SB II	4085.907		6	
	IR	4085.931	16	16	
----	CR	4085.987	8		
	CE	4085.998	3		
	FE	4086.000	2		
	MO	4086.025	16	16	
	SC I	4086.030	3		
	CE	4086.072	3		
	U	4086.145	5	6	
	NI	4086.151	4		
---	PR	4086.240	25	2	
	CO I	4086.307	400	16	
	U	4086.383	2	2	
	EU	4086.422	4		
	CE	4086.424	8	2	
--	TH	4086.520	13	13	
	TB	4086.600	16		
	NB	4086.632	3	2	
	SC I	4086.673	10	8	
	U	4086.674	8		
	NE II	4086.690			5
	*LA II	4086.714	500	500	30 VAF
	CE	4086.768	3		
	KR I	4086.900			2
	TB	4086.930	3		
	IR	4086.957	2		
--	W	4086.957		4	
	CE	4087.103	3		
	FE I	4087.103	50	5	
	EU	4087.106	4		
	O II	4087.140			40
	SC I	4087.163	10		
	PR	4087.206	50	3	
---	DY	4087.213	16	4	
	CE II	4087.303	8		
	GD	4087.332	5		

Table IIa. LA II 4086.714 A
(con't)

Element	Wavelength (Angstrom)	Intensity Arc	Spark	Plasma
PD I	4087.343	500	100	
SR I	4087.344	13		
HO	4087.350	4		
CE II	4087.360	6		
BA	4087.377	3		
DY	4087.388	6		
U	4087.390	6	13	
W	4087.394	6	6	
--- MO	4087.463	3	3	
ND	4087.471	10	2	
RH I	4087.520	3		
CE	4087.566	8	2	
EU	4087.609	2		
ER	4087.638	20		
HO	4087.650	4	2	
ZR I	4087.689	13		
TB	4087.700	8		
GD	4087.709	79	100	

Table IIb.(1) LA II 4086.714 A

Element	Wavelength (Angstrom)	Intensity Arc	Spark	Plasma
PR	4081.901	79	32	
TI I	4082.462	63	25	
SM II	4082.589	10	25	
CO I	4082.602	50		
RH I	4082.780	100	50	
MN	4082.944	79	63	
MN	4083.628	79	63	
Y I	4083.713	50	10	
MO	4084.383	40	40	
SM II	4084.403	79		
FE I	4084.499	130	79	
FE I	4085.008	79	32	
FE I	4085.324	100	63	
RU I	4085.429	40	50	

Table IIb.(1) LA II 4086.714 A
(con't)

Element	Wavelength (Angstrom)	Intensity Arc	Spark	Plasma
GD	4085.648	100	79	
TA	4085.797	5	79	
AG	4085.871		25	
CO I	4086.307	400	16	
*LA II	4086.714	500	500	30 VAF
FE I	4087.103	50	5	
PR	4087.206	50	3	
PD I	4087.343	500	100	
GD	4087.709	79	100	
CO I	4088.299	50		
AC	4088.370		100	
OS	4088.442	100	3	
YB	4089.685	50	6	
MN	4089.938	79	20	
U II	4090.135	25	40	9 VAF
GD I	4090.422	100	20	
V I	4090.579	63	25	

Table IIa. LI I 6707.844 A

	Element	Wavelength (Angstrom)	Intensity Arc	Spark	Plasma
	SM	6706.850	5		
---	SM	6707.100	2		
	SM II	6707.450	50		
---	RU	6707.524	5		
	*LI I	6707.844	3200	200	
	LI	6707.850	320		
	V I	6708.170	2		
	W	6708.180	20		
	F I	6708.284			40
	HF	6708.330		3	
---	TE	6708.340			32
	N	6708.810			50

Table IIb.(1) LI I 6707.844 A

Element	Wavelength (Angstrom)	Intensity Arc	Spark	Plasma
FE I	6705.117	13	13	
*LI I	6707.844	3200	200	
LI	6707.850	320		

Table IIa. LI I 4602.863 A

	Element	Wavelength (Angstrom)	Intensity Arc	Spark	Plasma
	P II	4601.960			320
	FE I	4602.010	20	2	
	SM	4602.025	2		
	LA I	4602.046	20		
	ER	4602.060	2		
	O II	4602.080			10
----	TA	4602.192	100	2	
	ND	4602.242	10		
---	TE	4602.370			790
	TB	4602.503	4		
	PR	4602.562	10		
	ZR I	4602.573	13		
	LU	4602.600	3		
--	EU	4602.630	16		
	BH LA	4602.700	3		
	HF	4602.713	6	3	
	AS II	4602.730		200	
	CE	4602.752	4		
	RU I	4602.808	16		
	NB	4602.860	2	5	
	*LI I	4602.863	790		4 VAF
	TH	4602.884	5	3	
	FE I	4602.944	320	100	
	GD	4602.944	10	2	
	V I	4602.946	6	6	
	TB	4602.950	8		
--	XE II	4603.028			320
	SM	4603.116	5		
---	TM	4603.210	10		
	TM	4603.420	32	25	
	PR	4603.478	4		
----	MO	4603.560	6	6	
	U	4603.665	25	40	
	BH YT	4603.700	5		
	ER	4603.730	3		
	CS II	4603.755			63
	NB	4603.801	2	3	
	PR	4603.810	10	2	
	ND	4603.823	25		
	IR	4603.836	2		

Table IIa. LI I 4602.863 A
(con't)

Element	Wavelength (Angstrom)	Intensity Arc	Spark	Plasma
I	4603.850			8

Table IIb.(1) LI I 4602.863 A

Element	Wavelength (Angstrom)	Intensity Arc	Spark	Plasma
TM	4599.000	79		
RU I	4599.085	100		
NI I	4600.372	200		
CR I	4600.752	160	160	
TA	4602.192	100	2	
*LI I	4602.863	790		4 VAF
FE I	4602.944	320	100	
TA	4604.852	200		
NI I	4604.994	320	10	
MN	4605.363	160	16	
LA II	4605.782	100	100	
NI I	4606.231	100		
SR I	4607.331	1000	50	4 VAF

Table IIa. LU II 2615.420 A

	Element	Wavelength (Angstrom)	Intensity Arc	Spark	Plasma
	W	2614.445	6	5	
	FE I	2614.490	40	5	
	ER	2614.555	3	2	
	RE	2614.561	63		
	CE	2614.566	5		
	RU	2614.586	50	4	
	AG II	2614.590		320	
	HF	2614.590		3	
	PT I	2614.610	10		
	CS	2614.620			8
	CR	2614.630		6	
	CL	2614.640			3
	SB	2614.666	4	3	
----	MG I	2614.670	5	2	
	I	2614.700			32
	NB	2614.763		32	
	RU	2614.865		32	
	FE II	2614.872		10	
---	CD	2614.960			2
	IR I	2614.984	25	5	
	I	2615.010			13
	RU	2615.093	63	100	
	PD	2615.120			25
	U	2615.120	2	2	
	W	2615.122	10	3	
	CL II	2615.130			10
--	NI II	2615.193		1000	
	TA	2615.250	40		
	YB	2615.260	3	13	
	CO I	2615.335	10	2	
	MO	2615.391	25	3	
	V	2615.400		50	
	*LU II	2615.420	100	250	150 VAF
	ER	2615.420	16	10	
	FE	2615.422	25	10	
	W II	2615.445	6	20	
	CE	2615.453	5		
	TA	2615.465	50		
	ZR II	2615.590		4	
--	TA	2615.656	50		

Table IIa. LU II 2615.420 A
(con't)

	Element	Wavelength (Angstrom)	Intensity Arc	Spark	Plasma
	RE	2615.681	10		
	W	2615.695		8	
	PR	2615.750		3	
	FE	2615.870	3		
	CE	2615.877	4		
--	IR I	2615.881	8	2	
	U	2615.950	6	2	
	OS	2615.962	5		
	TM	2615.990	5	6	
	IR I	2615.998	10	2	
	RN	2616.000			13
	CD	2616.020			2
	W	2616.046		3	
	U	2616.068	13	6	
--	RH	2616.073	2	3	
	BR	2616.210			3
	NB	2616.218	2	32	
	CR	2616.230		6	
	V	2616.248	4	63	
	CO I	2616.261	40	2	
	CS	2616.270			8
	TM	2616.310	8	16	
	LA II	2616.314		6	
	FE	2616.320	3		
	RU	2616.329		20	
	W	2616.339	6	3	
	EU	2616.360		2	

Table IIb.(3) LU II 2615.420 A

Element	Wavelength (Angstrom)	Intensity Arc	Spark	Plasma
TA	2611.340	100		
NI II	2611.654		130	
FE II	2611.872	500	500	24 VAF
RU	2612.068	100	32	
HF II	2613.604	200	1600	
FE II	2613.823	400	400	
TA	2614.173	200		
PB I	2614.178	200	79	23 VAF
AG II	2614.590		320	
NI II	2615.193		1000	
*LU II	2615.420	100	250	150 VAF
FE II	2617.616	320	400	
FE I	2618.017	160	63	
CU I	2618.366	500	100	
FE II	2619.076	5	160	
MN	2619.509	130	10	
PT I	2619.567	320	5	

Table IIa. LU II 2911.390 A

	Element	Wavelength (Angstrom)	Intensity Arc	Spark	Plasma
	P	2910.390			5
	NE	2910.440			4
	PT I	2910.448	3		
	W	2910.477	13	10	
	IN	2910.520		5	
	GD	2910.526	8	4	
	NB	2910.587	10	100	
	TH	2910.597	13	10	
	U	2910.638	4	4	
--	CR	2910.653		50	
	FE	2910.670	4		
	NB	2910.686	2	5	
	PD	2910.708		2	
	FE II	2910.759		2	
	GA II	2910.770			3
	TI II	2910.770		6	
	CD	2910.800			32
	C II	2910.820		2	
	CS	2910.820			2
	U	2910.824	13	8	
	CE	2910.852	2		
-	CR I	2910.902	63	8	
	FE	2910.920	10	4	
	MO	2910.933		10	
	BR	2910.940			5
	W	2911.001	10	8	
	FE	2911.010	3		
	PD II	2911.030		2	
	V II	2911.064	32	200	
	ER	2911.069	4		
	SE	2911.070			5
	FE	2911.080	3		
	U	2911.104	6	4	
	CE	2911.120	2		
-	CR I	2911.145	40	8	
	O II	2911.200			5
	CU I	2911.215	2		
	RE I	2911.231	8		
	SM	2911.272	16		
	U	2911.287	3	2	

Table IIa. LU II 2911.390 A
(con't)

	Element	Wavelength (Angstrom)	Intensity Arc	Intensity Spark	Intensity Plasma
	TH	2911.324	8	6	
	OS	2911.341	5	5	
	TE	2911.350			10
	*LU II	2911.390	100	320	24 VAF
	ER	2911.417	32	16	
	IN	2911.450		3	
	NE I	2911.461			25
	CE	2911.514	8		
	YB	2911.520	5	40	
	U	2911.548	13	6	
--	CD II	2911.640			16
	CA	2911.650		2	
	V	2911.655		10	
	CR	2911.682		40	
	PR	2911.700		16	
	SM	2911.712	6		
	NB	2911.745	8	100	
	U	2911.764	3	2	
	MO	2911.765	5		
	TB	2911.810	5	3	
	O II	2911.850			6
	SB	2911.860			5
---	TM	2911.870	4	6	
	MO	2911.915	32	50	
	TH	2912.012	8	6	
	P II	2912.040			5
	CL II	2912.060			16
	TI I	2912.082	32	16	
----	EU	2912.120	5		
	FE I	2912.158	160	160	
	W	2912.243	6	3	
	PT I	2912.256	320	25	
	FE I	2912.257	8	4	
	OS	2912.334	50	50	
	EU	2912.340	3	5	

Table IIb.(2) LU II 2911.390 A

Element	Wavelength (Angstrom)	Intensity Arc	Spark	Plasma
FE	2906.421	63	25	
V II	2906.457	40	160	
EU	2906.676	320	320	
TI	2906.680		100	
RH I	2907.209	100	32	
MN	2907.216	50		
V II	2907.471	40	160	
FE I	2907.520	100	79	
NB	2908.243	20	200	
U II	2908.275	13	32	6 VAF
V II	2908.817	63	400	34 VAF
FE	2908.859	79	40	
RU I	2908.883	160	8	
TA	2908.910	160	10	
CR I	2909.052	63	13	
OS I	2909.061	500	400	
FE	2909.503	63	32	
RU	2909.740		160	
HF II	2909.912	200	320	
V II	2910.019	32	160	
RH	2910.171	50	13	
ER II	2910.357	20	6	11 VAF
V II	2910.389	32	160	
NB	2910.587	10	100	
CR I	2910.902	63	8	
V II	2911.064	32	200	
*LU II	2911.390	100	320	24 VAF
NB	2911.745	8	100	
FE I	2912.158	160	160	
PT I	2912.256	320	25	
OS	2912.334	50	50	
RH I	2912.616	50	20	
RU	2913.170	50	3	
PT I	2913.542	320	25	
SN I	2913.542	100	100	
CR	2913.727	63	5	
TA	2914.125	200	32	
RU	2914.299	50		
FE I	2914.306	50	25	
MN	2914.603	160		

Table IIb.(2) LU II 2911.390 A
(con't)

Element	Wavelength (Angstrom)	Intensity Arc	Spark	Plasma
V I	2914.926	63	50	
TA	2915.337	160	50	
RH I	2915.419	79	40	
TA	2915.492	160	40	
RU I	2916.255	100	25	

Table IIa.   LU I   4518.570 A

	Element	Wavelength (Angstrom)	Intensity Arc	Spark	Plasma
	PR	4517.595	40	16	
	NE I	4517.736			100
	NE II	4517.736			100
	EU	4517.760	20		
	ND	4517.779	2		
	RU I	4517.818	63		
---	TM	4517.870	16		
	ER	4517.942	2		
	CE	4518.017	13		
	TI I	4518.032	100	63	
---	U	4518.078		4	
	TB	4518.208	2		
	CE	4518.280	4		
	HF	4518.294	10	4	
--	MO	4518.440	5	5	
	DY	4518.538	6	4	
	*LU	4518.570	320	40	
	U	4518.590		3	
	CR	4518.628	6		
	ER	4518.640	40		
	TH	4518.647	2	2	
	MO	4518.668	4	4	
	EU	4518.680	8	5	
	PR	4518.687	2		
--	TI I	4518.697	32	10	
	OS	4518.889	16		
	DY	4518.968	2		
--	HF	4519.031		8	
	RB	4519.071		10	
	BH CA	4519.100	3		
	PR	4519.115	3		
	W	4519.171	13	3	
	CL II	4519.190			20
	U	4519.223	2	3	
	TH	4519.273	6		
	CO I	4519.290	40		
--	BH ZR	4519.300	20		
	ER	4519.440	5		
	SM II	4519.527	2		
	EU	4519.556	3		

Table IIb.(2) LU I 4518.570 A

Element	Wavelength (Angstrom)	Intensity Arc	Spark	Plasma
V I	4514.193	25	20	
MO	4514.407	8	10	
GD	4514.517	63	10	
AIR	4514.800		10	
YB	4515.148	50	100	
MO	4515.185	16	16	
U	4515.280	25	40	
FE II	4515.345	10	10	
CR	4515.440	25	10	
AS II	4516.140		16	
CO I	4517.109	320	6	
MO	4517.132	32	32	
U	4517.230	3	10	
IN	4517.420		10	
PR	4517.595	40	16	
TI I	4518.032	100	63	
*LU	4518.570	320	40	
TI I	4518.697	32	10	
RB	4519.071		10	
MO	4519.586		40	
SM II	4519.635	160	79	
GD	4519.657	160	100	
V I	4520.170	25	20	
FE II	4520.236	40	32	
PR	4520.778	40	16	
TM	4521.090	16	16	
TA	4521.094	200	10	
CR	4521.141	25	10	
MO	4521.149	8	16	
U	4521.591	4	16	
CE	4522.077	25	16	
MO	4522.190	25	32	
NB	4522.192		10	
LA II	4522.372	200	400	
TM	4522.570	200	320	
EU	4522.580	500		
EU I	4522.602	200	200	
FE II	4522.634	63	50	
TI I	4522.805	100	63	
SM II	4523.036	40	20	

Table IIb.(2) LU I 4518.570 A
(con't)

Element	Wavelength (Angstrom)	Intensity Arc	Spark	Plasma
CE	4523.077	32	25	
BA	4523.237	63	10	
NB	4523.410	32	32	
IN	4523.570		10	

Table IIa. MG II 2795.530 A

	Element	Wavelength (Angstrom)	Intensity Arc	Spark	Plasma
	IR I	2794.553	3		
	MO	2794.573	3	3	
	TL II	2794.590			2
	NE I	2794.592			5
	TM	2794.600	63	20	
	FE I	2794.704	50	32	
----	YB	2794.800		3	
	MN I	2794.817	1000	5	24 VAF
	FE I	2795.007	50	32	
---	YB	2795.070		3	
	NE I	2795.101			32
	ZR I	2795.129	5		
	NB	2795.138		16	
	RA II	2795.210			130
	U	2795.232	20	13	
--	AR	2795.330			2
	CU II	2795.331		2	
	RU	2795.353	32	8	19 VAF
	CE	2795.384	2		
	AR	2795.450			2
	AU II	2795.530		16	
	*MG II	2795.530	160	320	20 VAF
	FE I	2795.544	100	63	
	W	2795.550		10	
	CO	2795.600		10	
	YB	2795.630		8	
	RH I	2795.701	16		
--	FE II	2795.767		3	
	KR	2795.810			79
	CR	2795.818	32	3	
	CO I	2795.819	16		
	FE	2795.850	16	10	
	NB	2795.865	4	3	
	CE	2795.945	2		
---	NE I	2795.963			8
	RE	2796.081	10		
	HG II	2796.130			8
	W	2796.147	13	3	
	CE	2796.206	2		
	CO I	2796.231	50	5	

Table IIa. MG II 2795.530 A
(con't)

Element	Wavelength (Angstrom)	Intensity Arc	Spark	Plasma
U	2796.236	6	2	
--- KR II	2796.260			2
TA	2796.339	400	79	
LA II	2796.366	2	5	
IR I	2796.456	10	5	
XE	2796.490			3

Table IIb.(2) MG II 2795.530 A

Element	Wavelength (Angstrom)	Intensity Arc	Spark	Plasma
TA	2790.710	160	10	
MG II	2790.787	40	79	1 VAF
RH I	2791.158	100		
TA	2791.370	25	160	
TA	2791.674	100	10	
NB	2791.740	3	100	
RU	2792.332	10	100	
NB	2793.048	10	100	
PT I	2793.266	100	5	
FE II	2793.888	8	160	
U II	2793.937	25	32	6 VAF
PT II	2794.208	10	100	
MN I	2794.817	1000	5	24 VAF
RU	2795.353	32	8	19 VAF
*MG II	2795.530	160	320	20 VAF
FE I	2795.544	100	63	
TA	2796.339	400	79	
TA	2796.565	160		
LU	2796.630	25	100	
RH I	2796.632	100		
OS	2796.727	100	16	
MO	2796.777	100	10	
TM	2797.269	63	100	
NB	2797.693	10	200	
TA	2797.760	100	100	
FE I	2797.775	160	79	

Table IIb.(2) MG II 2795.530 A
(con't)

Element	Wavelength (Angstrom)	Intensity Arc	Spark	Plasma
MG II	2798.060	32	79	2 VAF
MN I	2798.271	790	79	18 VAF
TA	2798.404	160		
NI I	2798.653	130		
BI I	2798.685	200	25	
FE II	2799.286		100	
V	2799.451	25	100	
AG II	2799.660	20	100	

Table IIa.   MG II 2802.695 A

Element	Wavelength (Angstrom)	Intensity Arc	Spark	Plasma
CE	2801.747	2		
ZN II	2801.790			25
TL II	2801.810			25
RU	2801.865	32		
ND	2801.887		2	
AG II	2801.930		10	
--- W	2801.949	8	6	
PB I	2802.003	250	100	19 VAF
IN	2802.010		3	
TA	2802.071	320	79	
RN	2802.100			3
U	2802.157	6	4	
RU	2802.162	32	40	
MN	2802.168	5		
AU	2802.190		200	
-- S	2802.230			8
FE	2802.250	3		
RE	2802.251	10		
BR	2802.270			3
NI	2802.274	50	16	
CE	2802.282	2		
MO	2802.354	16	25	
MN	2802.414	13		
BI II	2802.440		2	
-- PD II	2802.473		10	
TA	2802.493	16	2	
TI I	2802.500	100	16	
ER	2802.528	13		
TE	2802.530			10
BI II	2802.550		3	
U	2802.559	16	32	
MN	2802.650	2		
CU I	2802.683	10	2	
*MG II	2802.695	160	320	100 VAF
CE	2802.699	20	5	
DY	2802.700	2		
TA	2802.702	10	4	
BI II	2802.707		10	
NB	2802.715	2	5	
HG	2802.760	5	16	

Table IIa. MG II 2802.695 A
(con't)

	Element	Wavelength (Angstrom)	Intensity Arc	Spark	Plasma
	V	2802.797	16	25	
	MN	2802.800	13		
	RU	2802.806	50	160	
	RE	2802.840	5		
	EU	2802.860	160		
	TE	2802.860			16
--	ER	2802.868	6		
	W	2802.953	13	10	
	XE	2803.020			3
	CE	2803.042	8		
	TM	2803.110	6	10	
	FE	2803.120	32	16	
	CE	2803.121	2		
	MO	2803.131	3		
	NI I	2803.148	5		
	FE I	2803.172	16		
---	IN	2803.180		2	
	KR II	2803.200			20
	RE	2803.236	8		
	PT I	2803.239	400	5	
	W	2803.300		10	
	CR	2803.357		32	
----	TH	2803.381	2	8	
	V	2803.467	32	25	
	HG I	2803.472	20	20	
	YB	2803.480		40	
	BI II	2803.482	2	32	
	RU	2803.498	50	4	
	NB	2803.540		3	
	ER	2803.541	6	8	
	BI II	2803.570	2	10	
	KR II	2803.600			4
	FE	2803.620	50	20	
	MN	2803.620	13		
	BI II	2803.653		16	
	W II	2803.663	6	13	

Table IIb.(2) MG II 2802.695 A

Element	Wavelength (Angstrom)	Intensity Arc	Spark	Plasma
TA	2797.760	100	100	
FE I	2797.775	160	79	
MG II	2798.060	32	79	2 VAF
MN I	2798.271	790	79	18 VAF
TA	2798.404	160		
NI I	2798.653	130		
BI I	2798.685	200	25	
FE II	2799.286		100	
V	2799.451	25	100	
AG II	2799.660	20	100	
TA	2800.572	160	40	
TI II	2800.610		160	
CR	2800.771	13	160	
ZN I	2800.869	400	320	0.4 VAF
ZN I	2801.056	100	20	0.4 VAF
MN I	2801.064	630	63	14 VAF
PB I	2802.003	250	100	19 VAF
TA	2802.071	320	79	
AU	2802.190		200	
TI I	2802.500	100	16	
*MG II	2802.695	160	320	100 VAF
RU	2802.806	50	160	
EU	2802.860	160		
PT I	2803.239	400	5	
CO I	2803.772	100	13	
FE I	2804.521	320	200	
TI II	2805.010		200	
NI II	2805.674		200	
TA	2806.301	320	50	
TA	2806.581	200	50	
RU	2806.742	50	100	
OS	2806.906	100		
FE I	2806.984	200	200	

Table IIa. MG I 2852.129 A

	Element	Wavelength (Angstrom)	Intensity Arc	Spark	Plasma
	MO	2851.179	16		
	HF II	2851.206	25	50	
	CS	2851.230			20
	V	2851.254	2	16	
	TH	2851.261	10	10	
	I	2851.290			13
	NB	2851.295		5	
	CR	2851.356	20	79	
----	IR	2851.410	8		
	TH	2851.445	5	5	
	NB	2851.447	5	3	
	FE	2851.500	5	2	
	CE	2851.598	3		
---	MG I	2851.650	25		
	V I	2851.748	32	4	
	FE I	2851.798	200	160	
--	ZR II	2851.967	13	20	
	NB	2851.977	4	5	
	HF II	2852.012	20	50	
	W	2852.100		20	
	DY	2852.129	5		
	*MG I	2852.129	320	100	19 VAF
	IR	2852.131	20		
	MO	2852.131	10	10	
	CE	2852.237	3		
	FE	2852.350	2		
--	TA	2852.355	5	100	
	XE	2852.390			2
	RE	2852.399	10		
	CS II	2852.415			2
	HG	2852.420			8
	U	2852.469	6	4	
	IR	2852.479	2		
	TH	2852.502	2	6	
	AG	2852.530		5	
	V	2852.536	6	32	
	AU II	2852.540		5	
---	EU	2852.560	5		
	U	2852.750	16	16	
	NA I	2852.828	100	20	1 VAF

Table IIa. MG I 2852.129 A
(con't)

Element	Wavelength (Angstrom)	Intensity Arc	Spark	Plasma
I	2852.846			4
RE	2852.857	32		
--- V I	2852.866	63	6	
W	2852.909	8	10	
U	2852.935	3	6	
TA	2852.950	5	16	
FE I	2852.958	6	2	
NA I	2853.031	79	16	
PT I	2853.107	16	2	

Table IIb.(3) MG I 2852.129 A

Element	Wavelength (Angstrom)	Intensity Arc	Spark	Plasma
YB II	2847.175	400	1000	
LU	2847.510	40	130	
V	2847.572	16	160	
HG I	2847.830	16	100	
FE II	2848.046		63	
MO II	2848.232	130	200	15 VAF
CO II	2848.370		63	
TA	2848.525	320	50	
V	2849.050	6	50	
HF II	2849.215	320	1000	
RU	2849.289	3	100	
NB	2849.557	2	100	
FE II	2849.606		50	
IR I	2849.725	40	20	10 VAF
TA	2849.823	16	50	
CR II	2849.838	79	160	21 VAF
TA	2850.491	200	100	
SN	2850.618	79	100	
TA	2850.985	400	160	
TI II	2851.102	20	79	
SB	2851.112	50	50	
YB	2851.120	10	50	
HF II	2851.206	25	50	

Table IIb.(3) MG I 2852.129 A
(con't)

Element	Wavelength (Angstrom)	Intensity Arc	Spark	Plasma
CR	2851.356	20	79	
FE I	2851.798	200	160	
HF II	2852.012	20	50	
*MG I	2852.129	320	100	19 VAF
TA	2852.355	5	100	
NA I	2852.828	100	20	1 VAF
CR II	2853.218	5	100	
MO	2853.229	25	100	
V	2854.336	20	100	
PD II	2854.581	4	500	
RU	2854.722	2	63	
CR	2855.073	4	100	
FE II	2855.670	2	200	
CR II	2855.676	63	200	16 VAF
LA II	2855.902	3	50	
TI II	2856.240		100	
RU	2856.552		50	
CR II	2856.766	20	63	

Table IIa.   MN II   2576.104 A

	Element	Wavelength (Angstrom)	Intensity Arc	Spark	Plasma
	U	2575.225	6	2	
	RU	2575.242	32		
	O II	2575.300			100
	FE	2575.344	3		
---	AG	2575.354	2		
	AL	2575.411	32	32	
	U	2575.422	6	2	
	W	2575.465	10		
	I	2575.470			20
	TA	2575.473	79		
	IR I	2575.481	3	2	
	PD II	2575.490		100	
	MN	2575.509	160		
	HF II	2575.511		8	
	CO II	2575.560		16	
	ZN	2575.600			10
---	TE	2575.700			10
	IR I	2575.743	10	2	
	AG	2575.744	10	3	
	FE	2575.744	79	10	
	RH I	2575.750	2	4	
	CR	2575.803		4	
--	W	2575.897	10		
	CE	2575.933	6		
	NB	2575.963		10	
	RU	2576.082		50	
	CO	2576.104	32		
	*MN II	2576.104	320	2000	220 VAF
	ZR I	2576.105	8		
	W	2576.165	2	20	
	BR	2576.170			16
	RH	2576.229	2	2	
	HG I	2576.295	20	16	
	RE	2576.319	16		
--	TH	2576.336	6	3	
	W II	2576.360	2	16	
	PD II	2576.399		100	
	TA	2576.428		3	
	V	2576.480		50	
	PB II	2576.550			100

Table IIa. MN II 2576.104 A
(con't)

	Element	Wavelength (Angstrom)	Intensity Arc	Spark	Plasma
	MO	2576.563	2	25	
---	NB	2576.597	3	2	
	TH	2576.690	16	5	
	FE I	2576.691	40	5	
----	HF II	2576.830	160	790	
	FE II	2576.865	2	63	
	W	2576.865		16	
	RU	2576.954	13	8	
	XE II	2576.970			8
	W	2577.025	16	4	
	PD II	2577.097	3	160	

Table IIb.(2) MN II 2576.104 A

Element	Wavelength (Angstrom)	Intensity Arc	Spark	Plasma
ZR II	2571.391	320	400	31 VAF
HF II	2571.678	320	2000	150 VAF
MN	2572.758	200	50	
HF II	2573.912	1300	1000	
AL I	2575.100	200	79	4 VAF
MN	2575.509	160		
*MN II	2576.104	320	2000	220 VAF
HF II	2576.830	160	790	
HF II	2578.155	160	630	
TL I	2580.140	100	79	2 VAF
TA	2580.156	160		

Table IIa. MN I 4030.755 A

	Element	Wavelength (Angstrom)	Intensity Arc	Spark	Plasma
	U	4029.798		2	
	Y I	4029.840	5		
	ND	4029.914	10	3	
	U	4029.920	6	4	
	TA	4029.940	50	5	
	MO	4029.941		32	
	W	4029.950	6	6	
	TB	4029.970	3		
	EU	4029.996	5		
--	ZR I	4030.039	32	2	
	SE II	4030.070			160
	RU	4030.143	6		
	CE	4030.155	5		
	FE I	4030.194	20	4	
	EU	4030.203	10		
--	BR	4030.290			13
	TH	4030.293	8	5	
	CA	4030.300	10	2	
	CE II	4030.344	20	4	
	SR I	4030.377	40		
	SM II	4030.425	10	3	
	ND	4030.470	20	16	
	FE I	4030.492	130	63	
--	TI I	4030.514	79	20	
	SC I	4030.657	10	2	
	EU	4030.660	5		
	TA	4030.668	10		
	CR	4030.681	40	32	
	*MN I	4030.755	500	20	7 VAF
	U	4030.758	5	6	
	ZR I	4030.759	20		
	Y	4030.830	2	2	
	CE	4030.853	2		
	TH	4030.855	10	8	
	LU	4030.860		5	
	AL II	4030.867			8
	GD	4030.881	8		
	MO	4030.915	3	5	
--	RU I	4030.997	16	13	
	DY	4031.081	6		

Table IIa. MN I 4030.755 A
(con't)

	Element	Wavelength (Angstrom)	Intensity Arc	Spark	Plasma
	PR	4031.090	13	8	
	TH	4031.099	5	5	
	CS	4031.100			10
	CR	4031.130	32	6	
	AL II	4031.135			2
	V I	4031.219	10	3	
---	FE I	4031.243	2		
	U	4031.306	8	8	
	NB	4031.314		5	
	TH	4031.327	5	5	
	PB I	4031.330		5	
	CE II	4031.336	40	8	
	ZR II	4031.355	3		
	EU	4031.380	6	3	
	SC I	4031.397	10	2	
----	AR	4031.410			2
	ND	4031.545	10	3	
	IR I	4031.558	3		
	AL II	4031.633			2
	RE	4031.640	2		
	TB	4031.640	50	3	
	CE	4031.669	10		
	W	4031.675	8	6	
	LA II	4031.692	400	320	19 VAF
	HO	4031.750	4		
	TI I	4031.754	32		
	PR	4031.755	50	32	

Table IIb.(3) MN I 4030.755 A

Element	Wavelength (Angstrom)	Intensity Arc	Spark	Plasma
LA II	4025.882	50	50	
MO	4025.990	32	32	
U	4026.019	25	25	
TH	4026.158	10	10	
CR	4026.166	100	32	
NB	4026.325		10	
MN	4026.435	50	40	
TI I	4026.541	63	10	
TA	4026.944	40	32	
CR	4027.103	8	32	
NB	4027.976	5	10	
TI II	4028.345	20	79	
AU	4028.480		10	
W	4028.790	13	10	
HF II	4029.173	10	13	
RB II	4029.562		16	
FE I	4029.639	79	25	
ZR II	4029.681	40	16	
PR	4029.730	16	13	
MO	4029.941		32	
ND	4030.470	20	16	
FE I	4030.492	130	63	
TI I	4030.514	79	20	
CR	4030.681	40	32	
*MN I	4030.755	500	20	7 VAF
RU I	4030.997	16	13	
LA II	4031.692	400	320	19 VAF
PR	4031.755	50	32	
MN	4031.790	8	10	
ND	4031.807	16	16	
FE	4031.965	79	50	
RU I	4032.205	20	20	
DY	4032.480	20	13	
PR	4032.492	20	13	
NB	4032.524	32	50	
FE I	4032.630	79	16	
PR	4032.974	16	10	
GA I	4032.982	1000	500	27 VAF
TA	4033.069	100	10	
MN I	4033.073	400	20	6 VAF

Table IIb.(3) MN I 4030.755 A
(con't)

Element	Wavelength (Angstrom)	Intensity Arc	Spark	Plasma
U	4033.427	13	10	
SB	4033.543	63	63	
U	4033.728	13	13	
IR	4033.762	100	25	
PR	4033.857	50	32	
W	4033.913		10	
TH	4034.256	10	10	
MN I	4034.490	250	20	
TM	4034.740	10	10	
W	4035.355	10	10	
V II	4035.626	40	79	
MO	4035.661	3	25	
MN	4035.728	50	63	

Table IIa.   MN II 2593.729 A

	Element	Wavelength (Angstrom)	Intensity Arc	Spark	Plasma
	FE	2592.779	20	100	
	MO	2592.792	3	16	
	CE	2592.801	2		
	RE	2592.868	50		
	LA II	2592.870		3	
	CO II	2592.882		2	
	MN	2592.944	160	3	
---	SN	2593.040		5	
	V	2593.052		50	
	TA	2593.085	160		
	IR	2593.134	3		
	MG I	2593.182	5	2	
---	PD I	2593.269	3	100	
	MN	2593.281		2	
	BR	2593.340			2
	MO	2593.378	3	20	
	W	2593.382	13	4	
	HG I	2593.410	5	3	
	I II	2593.466			160
--	FE	2593.518	25		
	U	2593.571	20	6	
	RH	2593.618		16	
	RU	2593.636	10		
	TI I	2593.642	20	2	
	TA	2593.660	79	100	
	RU	2593.700	20		
	MO	2593.705	20	40	
	SI	2593.710			2
	FE II	2593.726	16	63	
	*MN II	2593.729	200	1000	190 VAF
	BR	2593.760			20
	NB	2593.764	2	100	
	CE	2593.802	2		
	U	2593.820		4	
	NA I	2593.828	20		
--	NA I	2593.927	16		
	MO	2594.005	16		
	FE	2594.038	20	2	
	BI I	2594.039	13	4	
	OS	2594.143	10	3	

Table IIa. MN II 2593.729 A
(con't)

	Element	Wavelength (Angstrom)	Intensity Arc	Spark	Plasma
	FE I	2594.150	20	2	
	CO I	2594.158	10		
---	YB	2594.180		5	
	TA	2594.247	5	5	
	PD II	2594.273		25	
	U	2594.275	5	2	
	NB	2594.338	2	10	
	RU	2594.392		32	
	TB	2594.400		10	
	KR II	2594.400			4
	SN	2594.423	63	79	
----	BR	2594.450			4
	YB	2594.490	2	5	
	BH C	2594.500	2		
	TA	2594.536	6		
	W	2594.543	4	16	
	TI I	2594.629	5		
	MN	2594.720		13	

Table IIb.(2) MN II 2593.729 A

Element	Wavelength (Angstrom)	Intensity Arc	Spark	Plasma
CU II	2590.526		250	
NB	2590.944	16	790	
CR I	2591.848	100	13	
IR I	2592.056	100	20	
GE I	2592.537	20	16	29 VAF
MN	2592.944	160	3	
TA	2593.085	160		
*MN II	2593.729	200	1000	190 VAF
MN	2595.761	200	25	
PT	2596.000	200	20	
SB I	2598.062	200	100	
SB I	2598.090			28 VAF
FE II	2598.369	630	1000	24 VAF

Table IIa. MO I 3798.252 A

Element	Wavelength (Angstrom)	Intensity Arc	Spark	Plasma
RB II	3797.276			2
SM II	3797.278	16		
MO	3797.300	5	5	
CE	3797.322	2		
TA	3797.403	6	3	
SN	3797.420		3	
TH	3797.516	8	10	
FE	3797.517	320	200	
U	3797.520		3	
RE I	3797.591	40		
CR I	3797.716	100	20	
SM	3797.727	25	5	
DY	3797.760	4		
U	3797.773	10		
CU II	3797.832		2	
ND	3797.890	20	10	
CS	3797.908			4
TH	3797.910	2	5	
H I	3797.910			0.1 VAF
HF II	3797.923	25	25	
IR	3797.924	2		
TB	3797.930	16		
FE I	3797.949	4	2	
RU I	3798.052	32	40	
IR	3798.059	6		
CE	3798.080	2	2	
TH	3798.103	5		
NB	3798.121	50	79	
YB	3798.170	4		
LA II	3798.190		2	
CE	3798.238	2		
ER	3798.250	5		
HO	3798.250		4	
*MO I	3798.252	1000	1000	
U	3798.259	2	8	
IR	3798.272	8		
BR	3798.280			6
TI I	3798.314	10	6	
YB	3798.440	4		
F II	3798.460			6

Table IIa. MO I 3798.252 A
(con't)

	Element	Wavelength (Angstrom)	Intensity Arc	Spark	Plasma
	CE	3798.510	3	3	
	FE I	3798.513	400	320	
	TM	3798.550	16		
	TB	3798.590	16		
	CE	3798.624	3	3	
	DY	3798.650	2		
	ER	3798.650	4		
	V I	3798.661	6	5	
	HF	3798.662	5	3	
	IR I	3798.666	5		
---	EU	3798.710	2		
	TM	3798.760	20	10	
	CL II	3798.800			50
	U	3798.837	16		
	RU I	3798.901	63	100	
	W	3798.922	6	10	
	TB	3798.950	8	8	
----	ZN I	3799.002	5		
	EU II	3799.010	100		
	CE II	3799.038	4		
	IN II	3799.055			10
	HO	3799.060		4	
	CE	3799.097	3		
	IN II	3799.118			10
	TB	3799.150	8		
	PD I	3799.190	200	160	
	U	3799.202	6	13	
	IN II	3799.204			20
	S	3799.210			8
	TH	3799.218	10	5	
	ND	3799.236	10	10	

Table IIb.(1) MO I 3798.252 A

Element	Wavelength (Angstrom)	Intensity Arc	Spark	Plasma
OS I	3793.909	130	320	
FE	3794.340	79	50	
LA II	3794.773	400	200	30 VAF
V I	3794.964	50	50	
FE I	3795.004	500	400	
TM II	3795.765	250	160	26 VAF
TA	3796.208		50	
GD	3796.393	160	160	
HO II	3796.730	20	40	12 VAF
FE	3797.517	320	200	
CR I	3797.716	100	20	
H I	3797.910			0.1 VAF
NB	3798.121	50	79	
*MO I	3798.252	1000	1000	
FE I	3798.513	400	320	
RU I	3798.901	63	100	
EU II	3799.010	100		
PD I	3799.190	200	160	
MN	3799.259	50	50	
RH I	3799.311	100	25	
RU I	3799.347	63	100	
FE I	3799.549	400	320	
V I	3799.912	63	50	
IR	3800.123	160	100	
PR	3800.303	100	50	
MN	3800.552	63	63	
SN	3800.996	200	160	
CE II	3801.529	25	3	4 VAF
NB	3802.923	50	50	

Table IIa. MO I 3864.110 A

	Element	Wavelength (Angstrom)	Intensity Arc	Spark	Plasma
	LA II	3863.110	2	2	
	RE	3863.153	10		
	DY	3863.200	5		
	ND II	3863.327	10	4	23 VAF
	ZR I	3863.330	2		
----	NB	3863.384	16	10	
	TH	3863.391	20	20	
	U	3863.403	5	20	
	V I	3863.404	2		
	SM	3863.407	16	8	
	ND II	3863.409	20	20	23 VAF
	FE II	3863.413			
	ER	3863.460	5	3	
	YB	3863.460		5	
	W	3863.470	8	6	
	O II	3863.490			5
	CE	3863.595	2		
	CO I	3863.607	32		
---	EU	3863.655	3		
	BR	3863.710			3
	FE I	3863.742	63	32	
	CE II	3863.746	3		
	RE I	3863.750	16		
	NB	3863.780	2		
--	V II	3863.866	25	16	
	V I	3863.866	25	16	
	ZR I	3863.874	20	6	
	BI II	3863.900			100
	EU	3864.040	2		
	SM	3864.051	10	2	
	PR	3864.059	3	5	
	EU	3864.100	6	10	
	*MO I	3864.110	1000	500	
	BR	3864.120			2
	CU II	3864.121		2	
	BI	3864.200		160	
	CS	3864.249			6
	AR I	3864.260			10
	V I	3864.300	32		
	U	3864.305	8	10	

Table IIa. MO I 3864.110 A
(con't)

	Element	Wavelength (Angstrom)	Intensity Arc	Spark	Plasma
	FE	3864.306	2		
	ZR I	3864.335	50	20	
	W	3864.336	13	10	
--	NB	3864.358	3	5	
	CS	3864.367			4
	O II	3864.420			20
	CE	3864.464	2	3	
	U	3864.478	8	10	
	LA II	3864.488	100	160	
	TI	3864.497	16	6	
	PR	3864.534	2	4	
--	CL II	3864.600			16
	O II	3864.660			5
	TI	3864.703	4		
	U	3864.722	4	4	
	HF II	3864.750	2	20	
	RE	3864.764	3		
	FE	3864.767	2		
	EU	3864.780	2		
	ER	3864.810	6		
--	RU I	3864.857	5		
	V I	3864.862	100	50	
	CE	3864.869	2		
	ND	3864.896	4		
	NB	3865.024		200	
	TH	3865.036	8	3	
	NB	3865.039	10		
	OS	3865.045	10	5	
	CE	3865.105	2		

Table IIb.(1) MO I 3864.110 A

Element	Wavelength (Angstrom)	Intensity Arc	Spark	Plasma
FE I	3859.216	100	100	
U II	3859.580	20	32	12 VAF
ND	3859.670	25	25	
FE I	3859.913	1000	630	
W	3859.983	16	32	
CO I	3861.165	320	16	
EU	3861.180	32	32	
FE	3861.342	79	50	
RU	3862.646	2	63	
ND II	3863.327	10	4	23 VAF
ND II	3863.409	20	20	23 VAF
FE I	3863.742	63	32	
*MO I	3864.110	1000	500	
BI	3864.200		160	
LA II	3864.488	100	160	
V I	3864.862	100	50	
NB	3865.024		200	
PR	3865.458	200	130	
OS I	3865.469	130	200	
FE I	3865.526	630	400	
U	3865.923	20	25	
ND	3865.985	32	32	
V II	3866.736	5	50	
FE I	3867.219	160	100	
V I	3867.602	63	32	
RU I	3867.839	63	32	
W	3867.975	32	32	
AR	3868.530		50	
MO	3869.080	25	32	

Table IIa. MO II 2020.300 A

Element	Wavelength (Angstrom)	Intensity Arc	Spark	Plasma
RE	2019.300	3	13	
OS	2019.400	5	2	
V II	2019.490		3	
PD I	2019.540	16	10	
TA	2019.540	4	20	
-- W	2019.570	2	8	
GA	2019.600		5	
K	2019.600			2
RE	2019.670	6		
IR	2019.780		32	
PD I	2019.790	13	4	
-- NB	2019.800	4	16	
OS	2019.870	3	2	
OS	2020.040	10	2	
-- HF	2020.070	3	4	
V	2020.110		2	
W	2020.130	8	2	
OS	2020.260	25	10	
O	2020.270			5
*MO II	2020.300	20	50	38 VAF
CD	2020.330			2
TA	2020.410	2	3	
O II	2020.440			5
BI	2020.500	2		
-- FE	2020.520	2	100	
V II	2020.560		3	
TL	2020.600		2	
- TA	2020.730	2	6	
IR	2020.820		25	
V	2020.850		6	
HG	2020.900			2
PT	2020.920	2		
NI II	2020.970		32	
- XE	2021.040			4
RE	2021.110	4		
RH	2021.180		10	
BI I	2021.210	40	16	
NI I	2021.280	5		
OS	2021.280	8		

Table IIb.(3) MO II 2020.300 A

Element	Wavelength (Angstrom)	Intensity Arc	Spark	Plasma
IR	2015.300		25	
CU II	2015.575		50	
AG II	2015.900		25	
V II	2016.540		25	
NB	2017.280	3	25	
IR	2017.410		50	
RH	2017.450		79	
RE	2017.880	40	6	
IR	2017.990	2	40	
OS	2018.140	20	5	
PT II	2018.329	5	25	
NI II	2019.040	2	32	
IR	2019.780		32	
OS	2020.260	25	10	
*MO II	2020.300	20	50	38 VAF
FE	2020.520	2	100	
IR	2020.820		25	
NI II	2020.970		32	
BI I	2021.210	40	16	
IR	2021.510		50	
CO II	2022.340	20	79	
IR	2022.830	4	50	
P I	2023.471	16		4 VAF
V II	2023.560		40	
RE II	2023.660	10	50	3 VAF
RH	2023.740		63	
IR	2023.970	40		
BI	2023.990	50		
CU I	2024.330	50	6	
IR	2024.470	3	25	
RH	2024.930	3	25	
RH	2025.150	2	40	

Table IIa. MO II 2038.440 A

	Element	Wavelength (Angstrom)	Intensity Arc	Spark	Plasma
	V	2037.510		6	
	W	2037.584	3	20	
	RH	2037.600		50	
	HG	2037.620			3
	OS	2037.620	2	5	
	MO	2037.680	2	20	
--	RH	2037.850	4		
	V	2037.850		13	
	U	2037.870	3	2	
	W	2037.906	4	13	
--	NB	2037.940	3	25	
	RE II	2037.960	4	16	
	I II	2037.980			2
	CR	2038.000		3	
	MO	2038.010	2	20	
	TA	2038.060	3	13	
	IR	2038.110		32	
	OS	2038.160	4		
--	XE	2038.310			2
	*MO II	2038.440	20		24 VAF
	V	2038.490	4	13	
--	IR	2038.600		20	
	CO	2038.678		10	
	OS	2038.720	16	5	
	NI I	2038.750	2	3	
	V	2038.872	10	2	
--	PD	2038.930		20	
	IR	2038.990		25	
	GA	2039.000		5	
	RE	2039.010	20		
	V	2039.073		2	
	HF	2039.080	6	2	
	W	2039.110	4	6	
-	IR	2039.140	5	5	
	RE	2039.230	20		
	XE	2039.240			3
	CR	2039.310	25		
	V	2039.310		16	
	RH	2039.350		32	
	W	2039.380	4	3	

Table IIa. MO II 2038.440 A
(con't)

Element	Wavelength (Angstrom)	Intensity Arc	Spark	Plasma
IR	2039.430	25	13	

Table IIb.(2) MO II 2038.440 A

Element	Wavelength (Angstrom)	Intensity Arc	Spark	Plasma
P I	2033.489	16		7 VAF
OS	2033.530	10	25	
IR	2033.570	20	6	
RE I	2033.780	20	16	
V	2034.066	10	2	
NI I	2034.397	16	2	
OS	2034.440	32	5	
IR	2034.800	10	2	
BH C	2035.000	13		
W II	2035.030	6	13	38 VAF
NI I	2035.080	10		
IR	2035.580	10	4	
PT	2035.780	16	16	
CU II	2035.843	13	32	
OS	2035.930	16	4	
RE I	2035.960	10		
PT II	2036.463	16	40	54 VAF
OS	2036.490	10	3	
CU II	2037.117	13	32	
*MO II	2038.440	20		24 VAF
OS	2038.720	16	5	
V	2038.872	10	2	
RE	2039.010	20		
RE	2039.230	20		
CR	2039.310	25		
IR	2039.430	25	13	
SB II	2039.600	16	16	7 VAF
PT	2039.700	13	20	
TE	2039.790	320		
SE I	2039.851			40 VAF
PT	2040.330	16	16	

Table IIb.(2) MO II 2038.440 A
(con't)

Element	Wavelength (Angstrom)	Intensity Arc	Spark	Plasma
OS	2040.680	10	4	
NI I	2041.139	10		
SN	2041.200	10		
PT	2041.450	10		
GE I	2041.690	5	160	34 VAF
BI	2041.960	100	16	
IR	2042.130	13		
IR	2043.210	10	2	

Table IIa. NA I 5889.953 A

	Element	Wavelength (Angstrom)	Intensity Arc	Spark	Plasma
	TB	5889.060	10		
	XE I	5889.120			20
----	SM	5889.695	20		
	SE II	5889.740			16
---	S I	5889.750			5
	*NA I	5889.953	10000	1000	100 VAF
	C II	5889.970		63	
	MO	5889.978	50		
	CR	5889.989	13		
	HG	5890.160			40
	IN	5890.260		10	
	W	5890.333	6		
---	HF	5890.450	8		
	CO I	5890.484	6		
	ND	5890.503	5		
	SM	5890.626	4		

Table IIb.(1) NA I 5889.953 A

Element	Wavelength (Angstrom)	Intensity Arc	Spark	Plasma
MO	5888.326	160	100	
*NA I	5889.953	10000	1000	100 VAF
C II	5889.970		63	
GE II	5893.460		100	

Table IIa. NA I 5895.923 A

	Element	Wavelength (Angstrom)	Intensity Arc	Spark	Plasma
	XE I	5894.988			100
	SB II	5895.090			160
	SM	5895.154	6		
---	TA	5895.196	2		
	EU	5895.288	25		
---	FE	5895.497	4		
	ND	5895.578	2		
	XE I	5895.620			2
	TM	5895.626	79	20	
--	PB	5895.700	20	2	
	*NA I	5895.923	5000	500	43 VAF
	IN	5896.020		5	
--	SM	5896.278	5		
	HF	5896.610	2		
	YB	5896.610	5		
--	TE	5896.650			25
	BH LA	5896.700	79		
	SM	5896.872	3		

Table IIb.(1) NA I 5895.923 A

Element	Wavelength (Angstrom)	Intensity Arc	Spark	Plasma
C II	5891.650		32	
GE II	5893.460		100	
*NA I	5895.923	5000	500	43 VAF
YB	5897.220	6	100	
V	5897.544		32	
CU II	5897.986		25	
YB	5898.800	3	50	
TI I	5899.322	160	160	
NB	5900.616	200	200	

Table IIa.   NB II 3094.183 A

	Element	Wavelength (Angstrom)	Intensity Arc	Intensity Spark	Intensity Plasma
	CE	3093.243	6		
	SI	3093.280		6	
	AU II	3093.300		5	
	ZR I	3093.321	3		
	CE II	3093.339	13		
	FE	3093.357	63	40	
	U	3093.370	4	3	
	AR II	3093.410			50
----	YB	3093.440		2	
	RU	3093.447	4		
	CA	3093.460	2	3	
	RH I	3093.480	2	3	
	CR	3093.488		100	
	W	3093.512	13	10	
	BI I	3093.580	10	8	
	OS	3093.587	130	16	
	CE	3093.614	20		
	RE I	3093.652	63		
	MO	3093.680	10		
---	CD II	3093.740	3	2	
	ER	3093.750	6		
	Y II	3093.770	10	16	
	V I	3093.792	32		
	FE I	3093.806	50	40	
	TI I	3093.813	5		
	DY	3093.820	10		
	U	3093.868	3	4	
	TA	3093.869	50	16	
	YB	3093.879	3	16	
	FE	3093.883	40	32	
	ND	3093.883	6		
	RU	3093.901	32	100	
--	CR	3093.947		25	
	CE	3093.950	4		
	CU I	3093.992	160	50	
	IR I	3094.009	20	10	
	W	3094.032	8	10	
	OS	3094.074	32	8	
	NE II	3094.080			13
	*NB II	3094.183	100	1000	83 VAF

Table IIa. NB II 3094.183 A
(con't)

	Element	Wavelength (Angstrom)	Intensity Arc	Spark	Plasma
	EU	3094.184	10		
	V	3094.199	20	130	
	FE	3094.328	2	2	
	NB	3094.359	2		
	IR I	3094.366	5		
--	RU	3094.393	50	3	
	U	3094.460	3	4	
	XE II	3094.530			13
	RU	3094.558	6	50	
	IR	3094.617	3		
	FE	3094.622	3	4	
	SC	3094.625	4		
	EU	3094.629	3		
--	MO	3094.664	160	25	
	SN II	3094.690			6
	V I	3094.692	40		
	TH	3094.740	10	6	
	LA	3094.760	2	3	
	ZR I	3094.799	5		
	CS	3094.820			4
	U	3094.830	10	10	
	FE I	3094.900	32	16	
	YB	3094.920		2	
--	CR	3094.927		16	
	AR	3094.980			3
	U	3095.041	13	13	
	RE	3095.061	40		
	ZR II	3095.071	8	5	
	CE II	3095.099	16		
	CE II	3095.099	16		
	KR II	3095.140			32

Table IIb.(2) NB II 3094.183 A

Element	Wavelength (Angstrom)	Intensity Arc	Spark	Plasma
CO I	3089.595	100		
RU I	3089.801	63	5	
OS	3090.085	100	16	
RU	3090.229	50	6	
CO I	3090.254	79		
OS	3090.296	100	13	
OS	3090.492	79	16	
MG I	3091.077	79	10	
FE I	3091.578	320	200	
RU	3091.873	50	5	
TA	3092.444	50	16	
AL I	3092.713	1000	1000	13 VAF
V	3092.720	100	50	
NA II	3092.729	50		200
FE	3092.778	50	32	
AL I	3092.842	50	20	13 VAF
MG I	3092.991	130	20	
V II	3093.108	100	400	60 VAF
FE	3093.357	63	40	
OS	3093.587	130	16	
RE I	3093.652	63		
FE I	3093.806	50	40	
TA	3093.869	50	16	
CU I	3093.992	160	50	
*NB II	3094.183	100	1000	83 VAF
RU	3094.393	50	3	
MO	3094.664	160	25	
TA	3095.393	63	20	
CO I	3095.718	63	2	
CR	3095.859	130	3	
CO I	3096.405	63	3	
RU	3096.568	63	63	
CO I	3096.700	63	3	
MG I	3096.899	160	25	
MN	3097.063	79	40	
NI I	3097.118	200	50	
EU	3097.456	200	5	
FE I	3098.192	63	63	
CO I	3098.196	100	5	
TM	3098.590	79	63	

Table IIb.(2) NB II 3094.183 A
(con't)

Element	Wavelength (Angstrom)	Intensity Arc	Spark	Plasma
NI I	3099.115	200	50	

Table IIa. NB II 3163.402 A

	Element	Wavelength (Angstrom)	Intensity Arc	Spark	Plasma
	TB	3162.420	16	3	
	CR	3162.436	25		
	TM	3162.440	16	40	
	TI II	3162.570	50	200	
	MO	3162.606	2	20	
	HF II	3162.611	40	32	
	ND	3162.632	6	2	
----	V	3162.713	5	32	
	TA	3162.724	63	6	
	IR	3162.739	2		
	FE II	3162.800		100	
	U	3162.819	3	4	
	DY	3162.824	79	63	
	IR	3162.824	2		
	CE	3162.835	4		
	TH	3162.836	10	10	
	PD II	3162.860		4	
---	TB	3162.930	16	8	
	XE II	3162.930			13
	U	3162.980	3	2	
	CE	3162.996	8		
	V	3163.021	5	32	
	FE II	3163.098		10	
	TA	3163.128	63	32	
	NB	3163.146		4	
--	RU	3163.184	13	100	
	BA	3163.272		2	
	MO	3163.288	4	3	
	CE	3163.345	16		
	SM II	3163.391	5	2	
	*NB II	3163.402	16	8	75 VAF
	W	3163.418	13	16	
--	TA	3163.545	16	3	
	U	3163.727	8	6	
	NA II	3163.731	32		63
	PR	3163.735	4		
	CR	3163.756	40	25	
	V	3163.760		25	
	YB	3163.800	2	20	
	TA	3163.825	40	5	

Table IIa. NB II 3163.402 A
(con't)

Element	Wavelength (Angstrom)	Intensity Arc	Spark	Plasma
RU	3163.828	10		
TB	3163.850	16	3	
IR	3163.870		2	
P	3163.870			50
--- FE	3163.874	40	25	
MO	3163.905	10	20	
V I	3163.905	10		
CR	3163.923		10	
HO	3164.030		4	
DY	3164.041	3		
TB	3164.100	8	3	
--- U	3164.151	5	13	
CE II	3164.154	40		
NI I	3164.166	4		
XE II	3164.230			4
FE I	3164.299	20	10	
ZR II	3164.310	10	8	

Table IIb.(3) NB II 3163.402 A

Element	Wavelength (Angstrom)	Intensity Arc	Spark	Plasma
CE	3158.434	20		
U	3158.526		4	
TH	3158.612	10	4	
CO I	3158.775	160		
W	3158.806	6	6	
CE	3158.812	16		
CA II	3158.869	100	200	
RU I	3158.889	63	13	
U	3158.890	3	4	
MO	3158.994	2	10	
TA	3159.049	25	3	
PT II	3159.080	3	16	
CR II	3159.105	4	25	
IR I	3159.149	50	2	
W	3159.184	10	10	

Table IIb.(3) NB II 3163.402 A
(con't)

Element	Wavelength (Angstrom)	Intensity Arc	Spark	Plasma
ND	3159.216	6	4	
MO	3159.230		25	
RH	3159.250	5	10	
RE I	3159.311	16		
MO	3159.343	20	5	
OS	3159.356	20	5	
V	3159.364		40	
TB	3159.390	16		
CR	3159.590	63	20	
CO I	3159.665	100	2	
U	3159.817	8	16	
HF	3159.823	20	2	
CR	3159.845		6	
NB	3159.853		10	
RU I	3159.923	63	25	
W II	3160.020	10	32	
CR	3160.097		10	
MO	3160.120		16	
TH	3160.179	8	4	
FE	3160.200	63	50	
OS	3160.286	32	10	
FE	3160.344	40	20	
ER	3160.350	20	2	
OS	3160.427	20	8	
FE I	3160.658	160	130	
V	3160.780		32	
FE	3160.917	5	4	
CE	3161.026	16		
DY	3161.028	10	5	
MN	3161.039	160	50	
TI II	3161.205	32	130	
V	3161.312		50	
ER	3161.370	20	3	
FE I	3161.371	79	63	
GD	3161.379	32	32	
OS	3161.445	79	13	
U	3161.528	8	6	
LU	3161.620		10	
CO I	3161.654	63		
TH	3161.693	10	8	

Table IIb.(3) NB II 3163.402 A
(con't)

Element	Wavelength (Angstrom)	Intensity Arc	Spark	Plasma
OS	3161.728	100	10	
TI II	3161.774	32	160	
PD II	3161.947		100	
FE I	3161.949	200	160	
ZR II	3162.214	16	16	
FE I	3162.335	63	50	
V	3162.357		20	
TB	3162.420	16	3	
CR	3162.436	25		
TM	3162.440	16	40	
TI II	3162.570	50	200	
MO	3162.606	2	20	
HF II	3162.611	40	32	
V	3162.713	5	32	
TA	3162.724	63	6	
FE II	3162.800		100	
U	3162.819	3	4	
DY	3162.824	79	63	
TH	3162.836	10	10	
PD II	3162.860		4	
TB	3162.930	16	8	
V	3163.021	5	32	
FE II	3163.098		10	
TA	3163.128	63	32	
NB	3163.146		4	
RU	3163.184	13	100	
CE	3163.345	16		
*NB II	3163.402	16	8	75 VAF
W	3163.418	13	16	
TA	3163.545	16	3	
U	3163.727	8	6	
NA II	3163.731	32		63
CR	3163.756	40	25	
V	3163.760		25	
YB	3163.800	2	20	
TA	3163.825	40	5	
TB	3163.850	16	3	
FE	3163.874	40	25	
MO	3163.905	10	20	
CR	3163.923		10	

Table IIb.(3) NB II 3163.402 A
(con't)

Element	Wavelength (Angstrom)	Intensity Arc	Spark	Plasma
HO	3164.030		4	
U	3164.151	5	13	
CE II	3164.154	40		
FE I	3164.299	20	10	
ZR II	3164.310	10	8	
TH	3164.437	10	10	
W	3164.442	10	8	
RE I	3164.520	32		
MO	3164.526	10	10	
OS	3164.612	63	13	
V II	3164.828	10	100	
TM	3164.870	20	10	
AU II	3164.900		8	
TI	3164.910		50	
RU	3164.948	3	63	
FE I	3165.006	100	63	
YB	3165.210	2	16	
U	3165.282	6	20	
W	3165.376	10	10	
ZR II	3165.446	6	4	
U	3165.505	10	8	
NI I	3165.508	16		
C II	3165.510		20	
TH	3165.579	10	4	
OS	3165.659	20	20	
HO	3165.670		4	
TB	3165.740	16	8	
SI	3165.780		6	
TH	3165.822	10	5	
FE I	3165.861	100	79	
V	3165.887		79	
ZR II	3165.974	10	13	11 VAF
TH	3166.054	10	6	
RU	3166.104	8	40	
CE II	3166.243	20		
ZR II	3166.256	5	5	
TA	3166.377	2	32	
FE	3166.438	100	79	
EU	3166.500	20		
OS	3166.512	200	20	

Table IIb.(3) NB II 3163.402 A
(con't)

Element	Wavelength (Angstrom)	Intensity Arc	Spark	Plasma
RU	3166.539	3	32	
CE	3166.608	20		
HO	3166.620	10	16	
SM	3166.930	4	4	
MO	3167.030		20	
U	3167.098	10	10	
ER	3167.103	16	3	
RE I	3167.157	20		
CE	3167.226	16		
CE II	3167.324	16		
LU	3167.330		16	
V	3167.436	5	160	
RU	3167.467	5	100	
TB	3167.520	16	8	
TA	3167.529	32	3	
W	3167.581	10	10	
MO	3167.725		20	
FE II	3167.859	2	100	
HO	3167.890		4	
FE	3167.922	100	32	
C II	3167.950		13	
CO I	3168.060	100		
V II	3168.138	13	40	
IR I	3168.178	20	2	
TA	3168.183		63	
TM	3168.190	25	63	
RU I	3168.239	40		
PR	3168.242	25	5	
OS	3168.281	100	16	
RE I	3168.384	160		
HF	3168.390	32	2	

Table IIa. NB I 3130.786 A

	Element	Wavelength (Angstrom)	Intensity Arc	Spark	Plasma
	RU I	3129.837	63	4	
	TH	3129.911	10	8	
	Y II	3129.933	8	50	
	TA	3129.946	50	8	
	SM	3129.950	5		
	GD	3129.965	2		
	OS	3130.002	32	8	
	AG	3130.009	25	16	
----	MO	3130.060		20	
	ZR I	3130.063	3		
	SE	3130.120			8
	W	3130.155	5	4	
	DY	3130.159	4	2	
	TI I	3130.175	2		
	CE	3130.197	16		
	SM	3130.235	5		
	LA II	3130.250	3	4	
	V II	3130.267	50	200	
	FE	3130.278	5	4	
	IR	3130.285	4		
---	TA	3130.290	16		
	CE	3130.334	32		
	U	3130.373	3		
	TI I	3130.376	2		
	HO	3130.380		4	
	P II	3130.380			32
	XE II	3130.400			2
	BE II	3130.416	200	200	11 VAF
	W	3130.456	10	8	
	SI	3130.480		5	
	CE	3130.516	5		
--	U	3130.564		16	
	CR	3130.565		13	
	FE II	3130.567	4	4	
	BA I	3130.570	2	3	
	IR	3130.578	3		
	TA	3130.578	100	32	
	CS	3130.700			4
	S	3130.710			16
	U	3130.732	10	6	

Table IIa. NB I 3130.786 A
(con't)

	Element	Wavelength (Angstrom)	Intensity Arc	Spark	Plasma
	EU	3130.740	100	100	
	*NB II	3130.786	1000	100	60 VAF
	RH I	3130.790	63	2	
	AR I	3130.800			20
	TI II	3130.800	25	100	
	GD	3130.814	3		
	CE	3130.872	32	2	
	CE	3130.919	5		
--	HO	3130.990	6	6	
	ER	3131.070	3		
	TH	3131.070	13	10	
	BE II	3131.072	200	160	41 VAF
	ZR I	3131.110	6		
	OS	3131.115	130	32	
	MO	3131.195		5	
	CR	3131.211	20	6	
	TA	3131.216	5	25	
---	TM II	3131.260	400	500	58 VAF
	U	3131.325	8	6	
	TB	3131.350	8	3	
	FE	3131.455	2		
	OS	3131.481	20	10	
	CE	3131.505	2		
---	U	3131.534	4	2	
	HG I	3131.546	400	320	
	U	3131.618		2	
	CE	3131.679	4		
	FE II	3131.718		32	
	EU	3131.750	20	5	
	SR II	3131.750		10	

Table IIb.(3) NB I 3130.786 A

Element	Wavelength (Angstrom)	Intensity Arc	Spark	Plasma
FE	3126.173	160	63	
V II	3126.215	63	100	
RU	3126.608	13	50	
CO I	3127.249	100		
NB	3127.526	10	50	
TA	3127.765	20	100	
RU	3127.913	10	100	
V	3128.284	2	63	
TI I	3128.640	13	63	
TI II	3128.640	13	63	
V	3128.686	3	63	
CR II	3128.699	32	160	
RE I	3128.952	100		
NI I	3129.314	130		
FE I	3129.330	100	63	
Y II	3129.933	8	50	
V II	3130.267	50	200	
BE II	3130.416	200	200	11 VAF
TA	3130.578	100	32	
EU	3130.740	100	100	
*NB II	3130.786	1000	100	60 VAF
TI II	3130.800	25	100	
BE II	3131.072	200	160	41 VAF
OS	3131.115	130	32	
TM II	3131.260	400	500	58 VAF
HG I	3131.546	400	320	
HG I	3131.833	200	100	
CR II	3132.058	25	130	
MO	3132.594	1000	320	
YB II	3132.628	16	160	
TA	3132.643	250	25	
CD I	3133.167	200	320	
V II	3133.328	50	200	
TM II	3133.890	200	200	22 VAF
NI I	3134.108	1000	160	
FE I	3134.111	200	130	
CR	3134.307	3	50	
HF	3134.718	79	130	
RU I	3134.802	10	100	
V	3134.931	32	160	

Table IIb.(3) NB I 3130.786 A
(con't)

Element	Wavelength (Angstrom)	Intensity Arc	Spark	Plasma
FE II	3135.364		100	
DY	3135.370	100	50	

Table IIa.   ND II 4012.250 A

	Element	Wavelength (Angstrom)	Intensity Arc	Spark	Plasma
	DY	4011.295	13	8	
	CE	4011.296	4		
	V I	4011.309	10	2	
	FE I	4011.412	5		
	U	4011.450	8	10	
----	RE	4011.514	32		
	MN	4011.531	16	16	
	PR	4011.532	5		
	TI I	4011.534	4		
	CE II	4011.560	16	3	
	TE	4011.680			32
	EU II	4011.683	25		
	SM II	4011.719	8	4	
	FE I	4011.728	2		
	RU	4011.729	6		
---	TH	4011.752	16	16	
	CE	4011.773	2		
	U	4011.780	10	5	
	W	4011.809	4	3	
	SE I	4011.880			200
	RB	4011.900			16
	MN	4011.905	13	10	
	HF	4011.944	3		
	MO	4011.966	25	25	
--	W	4012.096		5	
	K II	4012.100			20
	TA	4012.111	5	2	
	CE	4012.139	4		
	CO I	4012.160	2		
	U	4012.161	6	4	
	NB	4012.168		100	
	*ND II	4012.250	79	40	58 VAF
	ZR I	4012.252	20		
	RE	4012.260	25		
	MO	4012.270	3	3	
	CE II	4012.388	63	20	4 VAF
	CE I	4012.388	63	20	
	TI II	4012.391	32	50	
	TB	4012.450	5		
	CR	4012.469	63	63	

Table IIa. ND II 4012.250 A
(con't)

	Element	Wavelength (Angstrom)	Intensity Arc	Spark	Plasma
--	TH	4012.497	16	16	
	MO	4012.510	3	3	
	DY	4012.523	4		
	ER	4012.580	4	2	
	ND	4012.704	16	10	
--	U	4012.708	2	2	
	MO	4012.800	3	2	
	TI I	4012.806	13		
	EU	4012.816	20	25	
	DY	4012.820	5	2	
	TB	4012.870	6		
	PR	4012.910	8		
	IN II	4012.960			10
--	SE I	4012.960			160
	U	4013.029	8		
	W	4013.183	6	5	
	TA	4013.190	5		
	MO	4013.210		40	
	ND	4013.224	10	5	
	PR	4013.230	10	4	

Table IIb.(3) ND II 4012.250 A

Element	Wavelength (Angstrom)	Intensity Arc	Spark	Plasma
FE I	4007.273	79	50	
ND	4007.435	20	20	
SM II	4007.486	50	25	
RU I	4008.269	20	20	
MO	4008.667		20	
PR II	4008.714	160	50	5 VAF
W I	4008.753	50	50	
TI I	4008.928	79	32	
MO	4009.366	20	25	
TI I	4009.663	63	25	
FE I	4009.717	130	100	
MO	4011.966	25	25	

Table IIb.(3) ND II 4012.250 A
(con't)

Element	Wavelength (Angstrom)	Intensity Arc	Spark	Plasma
NB	4012.168		100	
*ND II	4012.250	79	40	58 VAF
CE II	4012.388	63	20	4 VAF
CE I	4012.388	63	20	
TI II	4012.391	32	50	
CR	4012.469	63	63	
EU	4012.816	20	25	
MO	4013.210		40	
FE I	4013.795	79	40	
FE	4013.824	200		
CO I	4013.945	320		
FE	4014.534	200	100	
W	4015.216	25	32	
PR	4015.389	50	32	
LA I	4015.393	100	2	
NB	4016.079	5	32	
PR	4016.748	25	20	
FE I	4017.153	79	50	

Table IIa.   ND II 4061.085 A

Element	Wavelength (Angstrom)	Intensity Arc	Spark	Plasma
U	4060.103	4	8	
CE	4060.169	6		
I II	4060.175			5
HF	4060.199	10	4	
U	4060.232	3		
W	4060.234	4	10	
TI I	4060.265	63	25	
HO	4060.300	3	2	
EU	4060.301	2		
NB	4060.313	5	5	
LA I	4060.321	79	5	
-- TB	4060.380	20		
CE	4060.471	10	2	
ND	4060.562	10	5	
ZR I	4060.579	10		
DY	4060.580	8	2	
-- AR II	4060.599			
O II	4060.600			32
W	4060.708	8	6	
CE	4060.716	8		
CR	4060.765	13	6	
U	4060.768	6		
NB	4060.787	10	10	
-- V I	4060.850	2	2	
TB	4060.860	25	3	
HG	4060.890			10
U	4060.907	4	5	
AU II	4060.990	3	4	
O II	4061.000			16
SM II	4061.049	8	8	
DY	4061.054	3	2	
SR I	4061.060	8	6	
XE II	4061.060			2
*ND II	4061.085	40	32	31 VAF
FE	4061.110	3		
GD	4061.164	3		
NB	4061.263	4	3	
- GD	4061.298	8	13	
PR	4061.336	4	2	
U	4061.349	13		

Table IIa. ND II 4061.085 A
(con't)

	Element	Wavelength (Angstrom)	Intensity Arc	Spark	Plasma
	TA	4061.399	50	32	
	CE II	4061.423	6		
	ZR I	4061.529	25		
	EU	4061.552	13	2	
	TB	4061.570	40	2	
---	DY	4061.574	3		
	PT II	4061.660		10	
	CE	4061.705	2		
	ND	4061.715	5	3	
	MN	4061.742	79	32	
	U	4061.744	2	5	
----	CE	4061.811	5		
	GD	4061.858	3		
	RE I	4061.862	20		
	FE	4061.955	3		
	NB	4061.977	2	10	
	SM	4062.052	6		
	SE II	4062.060			63
	MO	4062.076	79	79	
	P II	4062.080			16

Table IIb.(3) ND II 4061.085 A

Element	Wavelength (Angstrom)	Intensity Arc	Spark	Plasma
MO	4056.318	16	16	
PR	4056.543	100	63	
CO I	4057.199	100		
TI I	4057.624	40	6	
ZN II	4057.710	79		
PB I	4057.820	2000	320	
IN	4057.866	79	10	
MN	4057.950	79	20	
TI I	4058.144	50	6	
CO I	4058.190	100		
FE	4058.229	79	25	
GD I	4058.231	100	63	

Table IIb.(3) ND II 4061.085 A
(con't)

Element	Wavelength (Angstrom)	Intensity Arc	Spark	Plasma
CO I	4058.600	100		
FE I	4058.760	40	10	
CR	4058.772	79	50	
PR	4058.778	25	16	
SM II	4058.867	32	20	
MN	4058.930	79	63	
NB I	4058.938	1000	400	
MN	4059.392	20	16	
GD	4059.881	50	20	
TI I	4060.265	63	25	
LA I	4060.321	79	5	
*ND II	4061.085	40	32	31 VAF
TA	4061.399	50	32	
TB	4061.570	40	2	
MN	4061.742	79	32	
MO	4062.076	79	79	
PB	4062.144	20	20	
CE	4062.223	40	8	
FE I	4062.444	130	100	
U	4062.549	13	20	
CU I	4062.698	500	20	
PR II	4062.817	160	50	6 VAF
RU I	4062.988	13	32	
MN	4063.528	100	63	
FE I	4063.597	400	320	
RU	4064.105	16	25	
ZR I	4064.155	100	6	
TI I	4064.219	50	16	
RU I	4064.456	20	63	
SM II	4064.569	16	16	
TA	4064.630	40	5	
MO	4064.667	25	25	
LA I	4064.783	40	3	
AU I	4065.080	50	32	
V	4065.080	2	100	
MN	4065.082	20	20	
TI I	4065.101	79	32	
GD	4065.637	50	50	
CR	4065.716	79	32	

Table IIa.   NI I   3524.541 A

	Element	Wavelength (Angstrom)	Intensity Arc	Spark	Plasma
	W	3523.550		5	
	TH	3523.557	5		
	U	3523.565	4	16	
	CR	3523.590	10		
	CE	3523.613	2	2	
	ND	3523.625	16	8	
	OS	3523.636	160	32	
	ZR I	3523.636	5		
	TB	3523.660	32	50	
	EU	3523.670	2		
	CO I	3523.701	16	3	
	CE	3523.735	5		
----	MO	3523.890		10	
	HO	3523.970		4	
	DY II	3523.980			7 VAF
	ER	3523.983	25	8	
	CE	3524.008	8		
	DY	3524.030	16	20	
---	FE	3524.071	50	40	
	CD II	3524.072		8	
	CE	3524.073	8		
	RU	3524.152	6		
	HG II	3524.190			100
	GD	3524.198	25	16	
	MO	3524.228	8	3	
	CU I	3524.239	40	10	
	TI	3524.240	2		
	FE I	3524.241	63	50	
	ZR	3524.242	13		
	W	3524.246	10	8	
	HG I	3524.270			16
	I	3524.277			3
--	EU	3524.340	3		
	IR	3524.372	2		
	RU	3524.454		6	
	PR	3524.470	3	2	
	SM	3524.538	10		
	ZR I	3524.538	10	2	
	MN	3524.540	16		
	*NI I	3524.541	1000	100	7 VAF

Table IIa. NI I 3524.541 A
(con't)

	Element	Wavelength (Angstrom)	Intensity Arc	Spark	Plasma
	DY	3524.610	6	4	
	MO	3524.646	5	50	
	W	3524.681	6	5	
--	V II	3524.715	10	63	
	TI II	3524.870		5	
	RU I	3524.902	13	2	
	DY	3524.920	6		
	ER	3524.920	20	8	
	NB	3524.936	2	5	
	MO	3524.981	5	5	
---	BA I	3524.985	20	5	
	SM	3525.065	2		
	CO I	3525.089	2		
	TB	3525.130	8	8	
	U	3525.141	5	8	
	TH	3525.142	2		
	TI I	3525.161	10		
	NB	3525.232	16	32	
---	OS	3525.290	4	5	
	BR	3525.340			3
	BH SR	3525.400	8		
	PR	3525.490	4	2	
	SM II	3525.496	16	2	

Table IIb.(2) NI I 3524.541 A

Element	Wavelength (Angstrom)	Intensity Arc	Spark	Plasma
RU	3519.635	63	32	
NI I	3519.766	500	32	
CO	3519.818	2	25	
V II	3520.025	5	50	
RU	3520.130	63	40	
YB	3520.290	10	63	
FE I	3521.264	320	200	
CO I	3521.567	200	25	
V	3521.839	20	79	

Table IIb.(2) NI I 3524.541 A
(con't)

Element	Wavelength (Angstrom)	Intensity Arc	Spark	Plasma
FE I	3522.276	50	32	
NB	3522.363		50	
CO I	3523.434	320	25	
OS	3523.636	160	32	
TB	3523.660	32	50	
DY II	3523.980			7 VAF
FE	3524.071	50	40	
FE I	3524.241	63	50	
*NI I	3524.541	1000	100	7 VAF
MO	3524.646	5	50	
V II	3524.715	10	63	
NB	3525.232	16	32	
FE I	3526.042	79	50	
FE I	3526.169	50	25	
FE I	3526.676	79	50	
CO I	3526.849	320	25	
FE I	3527.797	100	79	
MO	3527.870		32	
RH I	3528.024	1000	160	24 VAF
NB	3528.476	3	50	
OS I	3528.602	400	50	
NB	3528.896	2	40	
TI I	3529.430	1000	200	
TL I	3529.430			2 VAF

Table IIa.   NI I   3414.765 A

Element	Wavelength (Angstrom)	Intensity Arc	Spark	Plasma
KR	3413.780			2
DY	3413.794	40	10	
U	3413.806	4	10	
BA I	3413.836		5	
I	3413.891			16
SM II	3413.895	6	2	
NI I	3413.939	320	10	
RU	3413.980	3		
--- EU	3414.020	4	2	
NB	3414.066	10	8	
TA	3414.140	20	100	
CE II	3414.168	8		
V I	3414.196	20	40	
OS	3414.245	10	8	
HO	3414.260		10	
--- RU	3414.282	13		
ND	3414.298	10	2	
DY	3414.300	2	2	
CR	3414.305	8		
CE	3414.313	8		
U	3414.360	8		
MO	3414.422		20	
AR II	3414.460			5
Y	3414.490	3	2	
TH	3414.513	6	8	
-- SM	3414.520		4	
AG I	3414.550	4		
CA	3414.600		2	
CE	3414.605	5		
U	3414.630	3		
RU I	3414.642	50	5	
ZR II	3414.661	20	16	
F II	3414.663			16
CO I	3414.736	200		
*NI I	3414.765	1000	50	6 VAF
CE	3414.766	5		
PR	3414.767	6		
EU	3414.773	40		
ER	3414.790	8		
KR II	3414.800			10

Table IIa. NI I 3414.765 A
(con't)

	Element	Wavelength (Angstrom)	Intensity Arc	Spark	Plasma
	NE II	3414.820			4
	DY	3414.830	32	5	
	V	3414.880		4	
	HO	3414.920	32	32	
	ZR I	3414.950	20		
--	SM II	3414.953	10	2	
	CE	3415.069	8		
	PR	3415.079	4	3	
	TB	3415.120	8		
	TH	3415.130	2		
	OS	3415.224	5	8	
---	IR	3415.241	10	2	
	TA	3415.270	3	32	
	MO	3415.273	5	5	
	CE	3415.307	3		
	U	3415.325	3		
	DY	3415.350	4		
	CA	3415.380	2	3	
	W	3415.419	3	4	
----	TB	3415.430	8		
	CO I	3415.530	20		
	FE I	3415.534	63	20	
	W	3415.550	6	6	
	ND	3415.553	16	6	
	CR	3415.570	8		
	U	3415.613		2	
	CE	3415.614	10		
	MO	3415.633	10	5	
	RB II	3415.648			32
	PR	3415.708	25	4	
	IR I	3415.744	6		
	CL	3415.750			4

Table IIb.(3) NI I 3414.765 A

Element	Wavelength (Angstrom)	Intensity Arc	Spark	Plasma
TI II	3409.809	20	40	
HF II	3410.171	25	63	
ZR II	3410.248	50	50	
MO	3410.620		25	
RU	3410.735	4	32	
FE	3411.356	79	32	
LA II	3411.760	5	25	
RH	3412.274	320	63	
CO I	3412.339	1000	100	
CO I	3412.633	1000	40	
NB	3412.935	5	160	
FE I	3413.135	400	320	
NB	3413.213		40	
TB	3413.760	50	32	
TA	3414.140	20	100	
V I	3414.196	20	40	
*NI I	3414.765	1000	50	6 VAF
HO	3414.920	32	32	
TA	3415.270	3	32	
NB	3415.974	50	50	
MO	3416.143	4	25	
HO II	3416.460	32	40	16 VAF
GD	3416.957	50	32	
TI II	3416.957	6	50	
V I	3417.064	32	25	
RU I	3417.353		63	
FE I	3417.843	160	100	
FE I	3418.512	160	100	
GD	3418.735	50	25	

Table IIa. NI II 2316.037 A

	Element	Wavelength (Angstrom)	Intensity Arc	Spark	Plasma
	IN I	2315.090	2		
	IR	2315.146	2	40	
	OS	2315.160	10	25	
	NB	2315.179	5	5	
	YB	2315.215	5	10	
	K	2315.220			20
	RE	2315.280	10		
----	AG II	2315.300		2	
	IR	2315.378	32		
	OS	2315.450	8	20	
	TA	2315.460	6	32	
	PT	2315.495	25	8	
	KR II	2315.520			8
---	OS	2315.580	10		
	ZR I	2315.580	3		
	MO	2315.620	3	16	
	V I	2315.635	25		
	PD	2315.650			5
	RH	2315.660	6		
	MN	2315.680		25	
	CO	2315.760	3	5	
--	AU	2315.850		6	
	PD II	2315.873	2	25	
	U	2315.880		10	
	SB	2315.883	20	10	
	RU	2315.898		3	
	RE	2315.970	32	10	
	TI I	2315.980	63		
	*NI II	2316.037	16	25	19 VAF
	OS	2316.120	5		
	O II	2316.120			10
	CO	2316.162	10		
	MO	2316.170		8	
	ZR	2316.200	4		
	W	2316.260		2	
--	CA	2316.270		2	
	AR	2316.310			100
	KR	2316.320			10
	OS	2316.380	10		
	ND	2316.410		4	

Table IIa. NI II 2316.037 A
(con't)

	Element	Wavelength (Angstrom)	Intensity Arc	Spark	Plasma
	MO	2316.450		13	
	N II	2316.460			8
	PD	2316.470	13		
	HF	2316.490	10	20	
--	CD	2316.510			2
	RE	2316.540	32	5	
	RH	2316.590	10		
	N II	2316.660			3
	ND	2316.690		3	
	W	2316.710		2	
	CO I	2316.735	5		
--	V I	2316.751	20		
	O II	2316.790			6
	XE II	2316.800			6
	CO I	2316.862	5		
	GA	2316.900		2	
	NB	2316.929	2	10	
	N II	2317.010			16
	AG II	2317.033	16	100	

Table IIb.(3) NI II 2316.037 A

Element	Wavelength (Angstrom)	Intensity Arc	Spark	Plasma
TM	2311.170		20	
TA	2311.180		16	
FE II	2311.224		20	
FE	2311.288		32	
V	2311.346		160	
NB	2311.455	2	16	
V I	2311.456	16		
SB I	2311.469	160	50	49 VAF
RH	2311.470		16	
CO II	2311.604	16	50	23 VAF
BH C	2312.000	32		
RE	2312.030	20		
FE II	2312.034		16	

Table IIb.(3) NI II 2316.037 A
(con't)

Element	Wavelength (Angstrom)	Intensity Arc	Spark	Plasma
NI I	2312.340	32	6	
AG II	2312.410	25	50	
CO II	2312.550	3	13	
U	2312.560	2	13	
CD	2312.570	25		
TA	2312.600	5	32	
RH	2312.650	2	100	
MN	2312.690		40	
RE	2312.820	16	5	
CD II	2312.840		200	1 VAF
NI II	2312.916		20	
RE	2312.980	16	3	
PT II	2313.037	3	20	
FE I	2313.102	25	3	
NB	2313.320		32	
RE	2313.360	20	8	
OS	2313.750	16	32	
RH	2313.880	5	20	
MN	2313.940	3	20	
LU	2313.980	3	16	
CO II	2314.054	25	32	18 VAF
IR	2314.111	2	40	
V	2314.187		100	
MO	2314.420		20	
ER	2314.430	3	40	
YB	2314.480	4	50	
AU	2314.640		16	
CR	2314.740	8	50	
OS	2314.840	20	2	
NB	2314.854	6	20	
IR	2314.900	10	40	
CO II	2314.980	25	32	13 VAF
W II	2315.020	13	13	
IR	2315.146	2	40	
OS	2315.160	10	25	
IR	2315.378	32		
OS	2315.450	8	20	
TA	2315.460	6	32	
PT	2315.495	25	8	
MO	2315.620	3	16	

Table IIb.(3) NI II 2316.037 A
(con't)

Element	Wavelength (Angstrom)	Intensity Arc	Spark	Plasma
V I	2315.635	25		
MN	2315.680		25	
PD II	2315.873	2	25	
SB	2315.883	20	10	
RE	2315.970	32	10	
TI I	2315.980	63		
*NI II	2316.037	16	25	19 VAF
MO	2316.450		13	
HF	2316.490	10	20	
RE	2316.540	32	5	
V I	2316.751	20		
AG II	2317.033	16	100	
CO	2317.055	2	20	
NI I	2317.162	32	13	
SN I	2317.228	100	100	
NB	2317.240		13	
ZR II	2317.248	20	10	
CE	2317.370		16	
IR	2317.442	2	25	
V	2317.568		32	
LA II	2317.820		20	
MO	2317.890		16	
RE	2317.890	2	16	
V	2318.070		250	
NB	2318.179		16	
PT	2318.294	25	10	
U	2318.470		25	
NI II	2318.480		20	
CE	2318.690		40	
MN	2318.920		32	
V	2318.999		160	
TA	2319.070	2	32	
RH I	2319.100	25	10	
RE	2319.220	32		
PD	2319.250	16		
IR	2319.313	25	8	
CR	2319.380	10	20	
PR	2319.420		13	
LA II	2319.441		20	
PD	2319.490	2	25	

Table IIb.(3) NI II 2316.037 A
(con't)

Element	Wavelength (Angstrom)	Intensity Arc	Spark	Plasma
TH	2319.525		13	
CU I	2319.561	20		
PT II	2319.888	4	25	
NI I	2320.030	32	5	20 VAF
CR	2320.080	10	32	
V I	2320.153	16	2	
OS	2320.180	25	13	
RU	2320.230	20		
AG II	2320.246	16	160	
FE I	2320.356	25	5	
MN	2320.450		63	
RU	2320.701	50		
YB	2320.828	20	5	

Table IIa. OS II 2255.847 A

Element	Wavelength (Angstrom)	Intensity Arc	Spark	Plasma
TA	2254.860	3	10	
SE	2254.890			10
FE II	2254.897		3	
IR	2254.910		10	
CS II	2254.940			20
NB	2254.949	5	10	
CU II	2254.971	3	10	
W	2254.987		10	
GA	2255.034	2		
CU	2255.080		32	
---- IR I	2255.101	25	10	
W	2255.103		6	
HF II	2255.150	40	63	
FE II	2255.150		13	
NB	2255.172		2	
RH	2255.260		10	
I	2255.270			32
GA	2255.290		3	
K II	2255.290			10
--- OS	2255.410	13		
RH	2255.418		20	
TE I	2255.480			3 VAF
W	2255.520	3		
RU	2255.521	79	3	
NB	2255.593	8	20	
IR I	2255.594	10		
-- IR I	2255.625	10		
CL	2255.640			4
CO	2255.650		4	
RH	2255.694		16	
W II	2255.720	2		
RE	2255.750	32	6	
FE II	2255.763		20	
CD	2255.770			2
TA	2255.770	3	20	
IN II	2255.788			10
IR I	2255.810	25	10	
*OS II	2255.847	130	2	83 VAF
FE I	2255.860	20		
NI I	2255.878	10		

Table IIa. OS II 2255.847 A
(con't)

Element	Wavelength (Angstrom)	Intensity Arc	Spark	Plasma
AU II	2255.910		5	
FE II	2255.980		3	
GE I	2256.000	2		
W	2256.018		10	
CR	2256.050		50	
-- NB	2256.070	2		
PT II	2256.104		20	
CD	2256.110			2
NI II	2256.146		10	
W II	2256.210	6	4	
RE	2256.220	40	6	
IR	2256.227	5		
--- HF II	2256.260		2	
IR	2256.430		10	
FE II	2256.433		10	
TA	2256.510	6	20	
XE II	2256.560			3
---- CO I	2256.573	10		
CR	2256.670		5	
CO	2256.740		32	
IR I	2256.752	2		
LA II	2256.760	2	32	

Table IIb.(3) OS II 2255.847 A

Element	Wavelength (Angstrom)	Intensity Arc	Spark	Plasma
ND	2250.870		16	
FE II	2250.930	5	20	
U	2251.090		3	
CO	2251.116		8	
V	2251.120		2	
W II	2251.130	10	10	
SN	2251.150	25	50	
MO	2251.330		32	
W II	2251.420	8	10	
NI I	2251.480	10	3	

Table IIb.(3) OS II 2255.847 A
(con't)

Element	Wavelength (Angstrom)	Intensity Arc	Spark	Plasma
PD II	2251.500	6	32	
CR	2251.520		20	
PT II	2251.521	3	32	
V	2251.555		8	
FE II	2251.559	4	4	
RE	2251.600	2	25	
HF II	2251.850	8	10	
FE I	2251.872	13	63	
CU II	2251.880		2	
PT II	2251.926		16	
W	2251.930		10	
MO	2252.030		5	
PD II	2252.030		32	
NB	2252.206	5	32	
OS	2252.220		4	
W	2252.280		3	
CO	2252.380		3	
ZR	2252.380		2	
MO	2252.406		20	
U	2252.450		8	
RE	2252.560	10	2	
ND	2252.590		10	
PD II	2252.618		10	
NB	2252.620		4	
CA	2252.650		2	
TA	2252.800		16	
CO	2252.820		2	
MO	2252.880		5	
NI	2252.886		5	
V	2252.950		8	
W	2253.020		3	
IR I	2253.040	10	2	
FE II	2253.124	13	32	
PT II	2253.130		25	
MO	2253.180		25	
IR	2253.230		25	
TI II	2253.250		13	
TA	2253.280	2	13	
AG II	2253.457		50	
AU II	2253.460		5	

Table IIb.(3) OS II 2255.847 A
(con't)

Element	Wavelength (Angstrom)	Intensity Arc	Spark	Plasma
IR	2253.487	10	3	
CO	2253.500		13	
ND	2253.500		25	
W	2253.510		3	
PD	2253.655	2	25	
NI II	2253.660		10	
NI II	2253.860	100	320	12 VAF
W	2253.910		10	
PB	2253.950	40	5	
HF II	2254.008	63	79	
RH	2254.070	3	25	
FE II	2254.078		10	
ZR II	2254.200		2	
RE	2254.220	20	10	
PD I	2254.260	25	8	
FE II	2254.390		2	
PD II	2254.448		10	
OS	2254.450		10	
IR	2254.470	2	10	
U	2254.520		4	
NB	2254.557	10	2	
W	2254.570	10	4	
OS	2254.690	13	3	
MO	2254.700		8	
BA II	2254.732	10	10	2 VAF
NI I	2254.800	10	3	
RE	2254.820	32	6	
CO I	2254.832		10	
TA	2254.860	3	10	
FE II	2254.897		3	
IR	2254.910		10	
NB	2254.949	5	10	
CU II	2254.971	3	10	
W	2254.987		10	
CU	2255.080		32	
IR I	2255.101	25	10	
W	2255.103		6	
HF II	2255.150	40	63	
FE II	2255.150		13	
NB	2255.172		2	

Table IIb.(3) OS II 2255.847 A
(con't)

Element	Wavelength (Angstrom)	Intensity Arc	Spark	Plasma
RH	2255.260		10	
GA	2255.290		3	
RH	2255.418		20	
TE I	2255.480			3 VAF
RU	2255.521	79	3	
NB	2255.593	8	20	
CO	2255.650		4	
RH	2255.694		16	
RE	2255.750	32	6	
FE II	2255.763		20	
TA	2255.770	3	20	
IR I	2255.810	25	10	
*OS II	2255.847	130	2	83 VAF
AU II	2255.910		5	
FE II	2255.980		3	
W	2256.018		10	
CR	2256.050		50	
PT II	2256.104		20	
NI II	2256.146		10	
W II	2256.210	6	4	
RE	2256.220	40	6	
HF II	2256.260		2	
IR	2256.430		10	
FE II	2256.433		10	
TA	2256.510	6	20	
CR	2256.670		5	
CO	2256.740		32	
LA II	2256.760	2	32	
W	2256.850	10	13	
FE II	2256.897		16	
MN	2256.930	3	13	
V I	2256.982	25	16	
MO	2256.983		13	
IR	2256.990		5	
YB	2257.010	4	40	
W	2257.100		10	
PT	2257.130	13	6	
MO	2257.190		13	
RH	2257.238		25	
ND	2257.260		16	

Table IIb.(3) OS II 2255.847 A
(con't)

Element	Wavelength (Angstrom)	Intensity Arc	Spark	Plasma
AG II	2257.375		5	
CR	2257.480		4	
IR	2257.500	2	50	
NB	2257.537	3	10	
MO	2257.705		8	
W	2257.741		6	
CR	2257.810		10	
CO	2257.860		3	
ZR	2257.880	2	3	
HF II	2257.890	10	16	
NI	2257.890		13	
U	2258.020		3	
MO	2258.060		4	
W	2258.128	10	8	
NI I	2258.146	16	2	
ND	2258.190		2	
MO	2258.220		5	
RH	2258.250		25	
OS	2258.332	13	32	
W	2258.360		3	
FE II	2258.372		2	
IR I	2258.508	16	50	
PD	2258.550		5	
CO	2258.590		2	
HF II	2258.680	16	20	
TA	2258.710		25	
PT II	2258.717		20	
HG I	2258.770	5	3	
V	2258.813	25	32	
IR I	2258.856	32	10	
OS	2258.975	3	25	
TE I	2259.040			17 VAF
W II	2259.068	3	10	
GA	2259.224	2	2	
IR	2259.270		32	
PT	2259.450		10	
TA	2259.550	4	13	
W	2259.555	13	6	
NI I	2259.571	320		
FE II	2259.589		20	

Table IIb.(3) OS II 2255.847 A
(con't)

Element	Wavelength (Angstrom)	Intensity Arc	Spark	Plasma
OS	2259.630	10	2	
U	2259.640		8	
W II	2259.690		5	
CO	2260.010	2	25	
W	2260.070	16	3	
FE II	2260.079	10	16	
OS	2260.110	25	50	
PD II	2260.130		25	
FE II	2260.228		5	
HF	2260.250	2	2	
TA	2260.250	2	3	
AU	2260.360		5	
LU	2260.400		4	
PT II	2260.497		25	
RE	2260.510	10	25	
MO	2260.520		5	
CU I	2260.528	25	6	
PD	2260.530		25	
ND	2260.540		10	
TA	2260.560		20	
W	2260.580		5	
IR	2260.650	3	40	
MO	2260.751		5	
LU	2260.790		2	
V	2260.830		4	

Table IIa. OS II 2282.260 A

	Element	Wavelength (Angstrom)	Intensity Arc	Spark	Plasma
	PT	2281.270	4	20	
	RE	2281.310	10		
	TA	2281.310	2	13	
	CO	2281.340	5		
	BI	2281.345	10		
	MN	2281.370		16	
	PD II	2281.370			4
	MO	2281.440		3	
	ZR	2281.470		2	
	NB	2281.506		25	
----	TA	2281.520	2	16	
	TH	2281.560		8	
	RU	2281.609	4	16	
	V	2281.609	2	32	
	IN II	2281.642			20
	RE	2281.660	32	6	
	W	2281.670		4	
	RU	2281.710	13	16	
---	HF	2281.810	5	6	
	NB	2281.825	2	10	
	U	2281.840		3	
	CO	2281.880	3	25	
	IR I	2281.907	25	3	
	FE	2281.990	6		
	SR II	2281.999	8	5	
--	AR	2282.050			40
	PD II	2282.100		32	
	IR I	2282.171	3		
	TA	2282.190	6	16	
	W II	2282.200	4	13	
	MO	2282.210		5	
	RE	2282.210		4	
	SN I	2282.257	10	16	
	*OS II	2282.260	100	130	48 VAF
	AG	2282.270		3	
	CO	2282.366		5	
	PD II	2282.450		16	
	IR	2282.496	8		
--	W	2282.500	6	4	
	CD II	2282.520			3

Table IIa. OS II 2282.260 A
(con't)

Element	Wavelength (Angstrom)	Intensity Arc	Spark	Plasma
AG	2282.620		4	
AR II	2282.640			20
--- ND	2282.730		8	
U	2282.780		25	
OS II	2282.830	25	5	48 VAF
V	2282.857		4	
FE I	2282.863	8		
RH	2282.880		6	
MO	2282.890		20	
AU	2282.900		6	
PD II	2282.977		16	
YB	2282.990	5	50	
-- NB	2283.005	10	25	
OS	2283.040	10		
KR	2283.070			32
FE I	2283.082	8		
PD	2283.090			3
XE	2283.150			4
OS	2283.170		32	
ZR	2283.170		2	
TA	2283.190	2	16	

Table IIb.(2) OS II 2282.260 A

Element	Wavelength (Angstrom)	Intensity Arc	Spark	Plasma
RU	2278.190	79		
OS	2279.110	100	25	
RU	2279.569	100	10	
FE I	2279.924	16	40	
AG II	2279.982	10	130	
IR	2281.021	2	50	
*OS II	2282.260	100	130	48 VAF
OS II	2282.830	25	5	48 VAF
YB	2282.990	5	50	
OS	2283.670	50	16	
RH	2284.080		200	

Table IIb.(2) OS II 2282.260 A
(con't)

Element	Wavelength (Angstrom)	Intensity Arc	Spark	Plasma
ZR	2285.228	100		
RU	2285.380	79		
V	2285.455		100	
CO II	2286.156	40	320	43 VAF
SN	2286.680	63	40	
NI II	2287.084	100	500	

Table IIa.   OS I   2909.061 A

	Element	Wavelength (Angstrom)	Intensity Arc	Spark	Plasma
	YB	2908.090		4	
	CE	2908.097	3		
	TI	2908.140		25	
	ND	2908.140		5	
	MO	2908.164	5	8	
	NB	2908.243	20	200	
	W	2908.261	6	3	
	U II	2908.275	13	32	6 VAF
---	YB	2908.330		5	
	RE	2908.343	20		
	TH	2908.359	13	13	
	TE	2908.360			5
	W	2908.395	6	2	
	U	2908.410	8	8	
	CE	2908.421	32		
	V	2908.438	2	20	
	TB	2908.490	3	20	
	W	2908.493		10	
	ER	2908.535	6		
--	ND	2908.598		5	
	KR II	2908.620			5
	TM	2908.690	5	10	
	CD I	2908.740	5		
--	V II	2908.817	63	400	34 VAF
	HF II	2908.858	3	5	
	FE	2908.859	79	40	
	U	2908.878	2	2	
	MN	2908.879	10		
	NB	2908.881	2	20	
	RU I	2908.883	160	8	
	TA	2908.910	160	10	
	NB	2908.979		5	
	ND	2908.980		5	
	EU	2909.010	40		
	CR I	2909.052	63	13	
	*OS I	2909.061	500	400	
	MO II	2909.116	25	40	
	W	2909.123	8	8	
	YB	2909.190	2	6	
	RU	2909.224	13	2	

Table IIa. OS I 2909.061 A
(con't)

	Element	Wavelength (Angstrom)	Intensity Arc	Spark	Plasma
	U	2909.250	6	16	
--	FE	2909.310	4	2	
	DY	2909.325	2		
	HG II	2909.360			25
	HO	2909.420	40	10	
	TI II	2909.460		4	
	YB	2909.480	2	8	
	FE	2909.503	63	32	
---	IR I	2909.558	20	5	
	ER	2909.577	6		
	FE	2909.590	2		
	CE	2909.614	5		
	W	2909.628	4	3	
	LA	2909.631	2		
	OS	2909.669	13	5	
	RU	2909.740		160	
	U	2909.742	4	4	
----	IR	2909.789	2		
	RE	2909.818	40		
	FE	2909.860	4		
	HF II	2909.912	200	320	
	TI II	2909.925	6	16	
	IR I	2909.943	6		
	V II	2910.019	32	160	

Table IIb.(2) OS I 2909.061 A

Element	Wavelength (Angstrom)	Intensity Arc	Spark	Plasma
TA	2904.074	320	40	
HF I	2904.412	400	63	
HF I	2904.760	320	63	
TA	2905.239	79	100	
V II	2906.457	40	160	
EU	2906.676	320	320	
TI	2906.680		100	
V II	2907.471	40	160	

Table IIb.(2) OS I 2909.061 A
(con't)

Element	Wavelength (Angstrom)	Intensity Arc	Spark	Plasma
NB	2908.243	20	200	
U II	2908.275	13	32	6 VAF
V II	2908.817	63	400	34 VAF
*OS I	2909.061	500	400	
RU	2909.740		160	
HF II	2909.912	200	320	
V II	2910.019	32	160	
ER II	2910.357	20	6	11 VAF
V II	2910.389	32	160	
NB	2910.587	10	100	
V II	2911.064	32	200	
LU II	2911.390	100	320	24 VAF
NB	2911.745	8	100	
FE I	2912.158	160	160	
PT I	2912.256	320	25	
PT I	2913.542	320	25	
SN I	2913.542	100	100	

Table IIa.   P I   2535.650 A

	Element	Wavelength (Angstrom)	Intensity Arc	Spark	Plasma
	W	2534.676	13		
	AR	2534.740			40
	HG I	2534.775	32	32	
	PB	2534.780		2	
	RE	2534.801	100		
	W II	2534.825	2	16	
	V	2534.826	20		
	GA I	2534.830			20
	CE	2534.866	5		
----	PR	2534.920		2	
	RU	2534.944	3	40	
	U	2534.949	13	4	
	TA	2534.967	32		
	ER	2534.979	4		
	LA II	2534.980		6	
	MN	2535.031		3	
	AG II	2535.070			10
	W	2535.102	13		
	FE I	2535.124	2		
---	ZR II	2535.150		5	
	RU	2535.222		32	
	AR II	2535.280			2
	AG II	2535.307	10	25	
	W	2535.332	8		
	FE	2535.364		3	
--	FE II	2535.477		20	
	I	2535.570			32
	W	2535.576		10	
	U	2535.580		8	
	RU	2535.592		100	
	TA	2535.598	50		
	FE I	2535.604	1000		
	MN	2535.645		79	
	*P I	2535.650	100		11 VAF
	TH	2535.870	5	2	
	TI II	2535.871	20	63	
--	U	2535.900	5	2	
	PD	2535.940			3
	TA	2535.960	4		
	CO I	2535.964	10	40	

Table IIa. P I 2535.650 A
(con't)

Element	Wavelength (Angstrom)	Intensity Arc	Spark	Plasma
NI	2535.967	25		
PT	2535.967	25	5	
W	2535.988	4	13	
YB	2536.010		6	
AR II	2536.040			20
MN	2536.080		13	
--- IR	2536.127	2		
RU	2536.216	13	6	
FE	2536.224	3		
TA	2536.227	100		
---- U	2536.238	4	2	
PT I	2536.487	100	10	
CO I	2536.493		2	
HG I	2536.519	2000	1000	49 VAF
TH	2536.558	5	10	
BI	2536.560	5	2	
U	2536.600	3	2	
W	2536.605		13	

Table IIb.(2) P I 2535.650 A

Element	Wavelength (Angstrom)	Intensity Arc	Spark	Plasma
HF II	2531.196	200	1000	
TA	2532.125	79	100	
IR I	2533.131	100	20	
RU I	2533.236	50		
P I	2534.010	50		3 VAF
IR I	2534.457	100	10	
RE	2534.801	100		
TA	2535.598	50		
FE I	2535.604	1000		
*P I	2535.650	100		11 VAF
TA	2536.227	100		
PT I	2536.487	100	10	
HG I	2536.519	2000	1000	49 VAF
HF II	2537.337	50	400	

Table IIb.(2) P I 2535.650 A
(con't)

Element	Wavelength (Angstrom)	Intensity Arc	Spark	Plasma
PT I	2539.205	400	20	
ZR I	2539.650	50		

Table IIa. P I 2136.199 A

	Element	Wavelength (Angstrom)	Intensity Arc	Spark	Plasma
	TA	2135.220		6	
	YB	2135.220		20	
	NI I	2135.335	10		
----	TA	2135.460		6	
	P I	2135.466	13		9 VAF
	CO I	2135.590	3		
	PD II	2135.646		25	
---	TI II	2135.720		3	
	RE	2135.750		6	
	CO	2135.798	4		
	SI	2135.900			5
--	FE	2135.957	10	5	
	CU II	2135.976	25	500	25 VAF
	MO	2136.060		8	
	ZR I	2136.170	5		
	FE	2136.190	2		
	*P I	2136.199	16		39 VAF
	RH	2136.310	2	20	
	YB	2136.330		2	
--	TA	2136.370	13	10	
	ZN	2136.460			10
	CO II	2136.490		2	
	PR	2136.660		32	
---	OS	2136.690	20		
	TM	2136.700		5	
	RE	2136.730		8	
---	OS	2136.810	6		
	IR	2137.020	8		
	NB	2137.054	5	32	
	OS	2137.110	100	32	
	SB	2137.110	13	8	
	W	2137.150	4	13	
	TA	2137.190	2	10	

Table IIb.(3) P I 2136.199 A

Element	Wavelength (Angstrom)	Intensity Arc	Spark	Plasma
BI I	2133.626	100	40	10 VAF
AS I	2133.810	20		
V II	2134.120	32	130	
BI I	2134.308	100	5	
CU II	2134.355	32	40	
HF II	2134.530	16	20	
P I	2135.466	13		9 VAF
CU II	2135.976	25	500	25 VAF
*P I	2136.199	16		39 VAF
OS	2136.690	20		
OS	2137.110	100	32	
BH C	2137.500	20		
CO I	2137.780	16	2	
CU I	2138.507	25		
ZN I	2138.560	790	500	17 VAF
IR	2138.570	16		
CO I	2138.971	16		
HF II	2139.240	32	40	
SB I	2139.760	32	32	16 VAF
V II	2140.087	10	79	20 VAF

Table IIa. P I 2149.108 A

	Element	Wavelength (Angstrom)	Intensity Arc	Spark	Plasma
	AR	2148.150			5
	ZN II	2148.150			50
	IR	2148.220	25	50	
	OS	2148.270	6		
	PD II	2148.280		25	
----	FE I	2148.394	3		
	V II	2148.420	5	32	
	SN I	2148.460	16		
	YB	2148.490		4	
	FE	2148.500	5		
---	OS	2148.620	3	16	
	NB	2148.646	8	16	
	CO I	2148.708	6		
	NB	2148.720	5		
	SN I	2148.740	50	20	
	W	2148.850		8	
--	YB	2148.920		10	
	CD	2148.960			2
	IR	2148.970	10		
	CU II	2148.974	16	25	
	NB	2149.030		2	
	*P I	2149.108	16		39 VAF
	PD	2149.130		32	
	W	2149.147	3	10	
	ZR	2149.150	4		
	FE	2149.170	10		
	OS	2149.200		10	
	AG	2149.210		8	
--	SN II	2149.310			3
	V II	2149.392		6	
	K	2149.420			10
	YB	2149.430		3	
	NB	2149.535	10	20	
---	RE II	2149.590		4	
	FE	2149.620	2		
	PT II	2149.703		20	
	MO	2149.770		10	
---	OS	2149.810	10	2	
	HF	2149.860	6	6	
	OS	2149.970	16	10	

Table IIa. P I 2149.108 A
(con't)

Element	Wavelength (Angstrom)	Intensity Arc	Spark	Plasma
HF II	2150.070	5	5	

Table IIb.(3) P I 2149.108 A

Element	Wavelength (Angstrom)	Intensity Arc	Spark	Plasma
PT II	2144.231	32	100	10 VAF
PT II	2144.231	32	100	
CD II	2144.382	50	1600	12 VAF
FE	2144.450	25		
SB I	2144.860			12 VAF
SB	2145.040	20	20	
TE I	2147.190			14 VAF
NI I	2147.803	16	8	
IR	2148.220	25	50	
SN I	2148.460	16		
SN I	2148.740	50	20	
CU II	2148.974	16	25	
*P I	2149.108	16		39 VAF
OS	2149.970	16	10	
IR	2150.540	20	4	
BH C	2150.600	32		
PT	2150.648	16	8	
FE	2151.700	20		
PT I	2152.080	20	10	
SN	2152.500	32		
IR II	2152.680	50	200	44 VAF
SR II	2152.840	16	16	29 VAF
BI	2152.914	50		
P I	2152.950	13		3 VAF
RE	2153.320	16	2	
BI	2153.532	40		
PT I	2153.550	25	8	
P I	2154.081	16		7 VAF

Table IIa. PB II 2203.505 A

	Element	Wavelength (Angstrom)	Intensity Arc	Spark	Plasma
	PT II	2202.582		13	
	MG II	2202.680	5		
	RH	2202.710		13	
	V I	2202.724	10		
---	AR II	2202.770			2
	BI	2202.830	13	8	
	CO II	2202.957	3	32	
	HF II	2202.990	3	3	
---	RE I	2203.010	40	20	
	YB	2203.070		2	
	BI	2203.120	20		
	ND	2203.130		10	
	NB	2203.168	4	5	
	CE	2203.180		20	
	AR	2203.220			2
--	CO II	2203.430		3	
	FE	2203.463	4		
	PD II	2203.478		25	
	*PB II	2203.505	50	5000	70 VAF
	NI	2203.520		8	
	RH	2203.550	16	50	
	ND	2203.560		32	
	NB	2203.630	16	40	
	SB	2203.660	4	2	
	V I	2203.660	2		
	AG II	2203.662		20	
	IR	2203.680		25	
	N II	2203.720			3
--	OS	2203.750	5		
	W II	2203.790	3	10	
	HF II	2203.850	3	3	
	PT II	2203.882		20	
	CR	2203.910		5	
--	OS	2203.910	25	5	
	W	2204.020	3	10	
	FE	2204.080	2		
	TA	2204.100	3	16	
	I	2204.120			20
--	CD II	2204.180			2
	AG II	2204.400		10	

Table IIa. PB II 2203.505 A
(con't)

Element	Wavelength (Angstrom)	Intensity Arc	Spark	Plasma
W II	2204.480	13	32	49 VAF
V	2204.495		40	

Table IIb.(2) PB II 2203.505 A

Element	Wavelength (Angstrom)	Intensity Arc	Spark	Plasma
GE I	2198.701	10	2	47 VAF
GE I	2198.710			47 VAF
IR	2198.854	50	16	
RE I	2198.914	200	40	
SN I	2199.340	32	63	
CU I	2199.583	50	20	31 VAF
TA	2199.670	25	25	
FE I	2200.390	50		
CO	2200.693	79		
FE I	2200.722	32	8	
SB	2201.400	40	25	
OS	2201.930	25	3	
PT I	2202.220	25	8	
OS	2202.490	25	5	
RE I	2203.010	40	20	
*PB II	2203.505	50	5000	70 VAF
OS	2203.910	25	5	
W II	2204.480	13	32	49 VAF
IR	2205.680	25	5	
OS	2206.270	25	50	
TL I	2207.000	32	3	
TE	2207.170	50		
SI I	2207.980	20	16	23 VAF
IR	2208.088	40	3	

Table IIa. PB I 2833.069 A

	Element	Wavelength (Angstrom)	Intensity Arc	Spark	Plasma
	MO	2832.073		20	
	TI II	2832.160	16	100	
	ND	2832.200	5		
	YB	2832.200		2	
	TI I	2832.260	2		
	CE	2832.307	20		
----	TH II	2832.319	20	25	42 VAF
	SC	2832.340	3		
	KR II	2832.390			2
	FE I	2832.436	320	200	
	CR	2832.460	2	130	
	XE	2832.460			2
	W	2832.480	10	10	
	LA II	2832.550	4	5	
	CE	2832.568	2		
---	RU	2832.625	20		
	U	2832.645	2	2	
	MO	2832.657		5	
	CE	2832.753	2		
	RH I	2832.769	5		
	IR	2832.774	5		
	NB	2832.790		5	
--	I	2832.917			20
	NE I	2832.921			8
	CE	2832.928	2		
	ZN	2832.950			25
	W	2832.952	10	3	
	KR II	2833.000			100
	CL	2833.030			4
	CE	2833.044	3		
	EU	2833.056	3		
	IN	2833.060		3	
	ER	2833.061	25		
	ZR	2833.061	2		
	*PB I	2833.069	500	79	21 VAF
	FE II	2833.100		5	
	K	2833.140	2		2
	IR	2833.236	6	20	
	U	2833.244	8	4	
	EU	2833.250	10	5	

Table IIa. PB I 2833.069 A
(con't)

	Element	Wavelength (Angstrom)	Intensity Arc	Spark	Plasma
	HF	2833.276	25	4	
	NB	2833.304		10	
	CE	2833.309	50		
--	TL II	2833.310			25
	TH	2833.339	8	8	
	TB	2833.370		20	
	CR	2833.393		3	
---	FE	2833.400	10	8	
	CE	2833.580	2		
	W	2833.628	16	13	
	TA	2833.636	320	40	
	IR	2833.671	3		
	GD	2833.753	20	20	
	RU	2833.784		79	
----	MO	2833.792		8	
	TM	2833.820	6	10	
	U	2833.821	16	25	
	ZR II	2833.908	2	5	
	ER	2833.910	5	2	
	CO I	2833.923	40		
	RU	2834.001	32	5	
	RE	2834.060	100		

Table IIb.(3) PB I 2833.069 A

Element	Wavelength (Angstrom)	Intensity Arc	Spark	Plasma
TI II	2828.150	2	200	
TA	2828.579	79	100	
FE II	2828.634		79	
EU	2828.691	200	160	
FE	2828.813	100	63	
TI II	2828.900		160	
NB	2829.752	3	50	
PT I	2830.295	1000	630	
V	2830.402	10	63	
CR	2830.468	16	79	

Table IIb.(3) PB I 2833.069 A
(con't)

Element	Wavelength (Angstrom)	Intensity Arc	Spark	Plasma
YB	2830.980	2	40	
TM	2831.560	10	40	
FE II	2831.562		500	
HO	2831.600		63	
RU	2831.843	10	50	
U	2832.063	32	50	
TI II	2832.160	16	100	
TH II	2832.319	20	25	42 VAF
FE I	2832.436	320	200	
CR	2832.460	2	130	
*PB I	2833.069	500	79	21 VAF
TA	2833.636	320	40	
RU	2833.784		79	
CR	2834.262		130	
LU	2834.350	5	40	
MO	2834.394	20	40	
V	2834.530		40	
CO II	2834.939	2	79	
NB	2835.116	5	100	
MO	2835.331	20	40	
FE I	2835.459	100	100	
CR II	2835.633	100	400	42 VAF
V	2836.520	20	79	
TI II	2836.640		100	
C II	2836.710		200	
CD I	2836.907	200	79	
IN I	2836.919	79	79	
TH II	2837.299	16	10	46 VAF
CU II	2837.547		250	
C II	2837.602		40	
AU	2838.030		79	

Table IIa. PB I 4057.820 A

	Element	Wavelength (Angstrom)	Intensity Arc	Spark	Plasma
	ND	4056.842	8	8	
	CE	4056.900	16	2	
	IN II	4056.936			500
	NB	4056.941	3	5	
	CO I	4056.979	20	2	
	KR II	4057.010			320
	TB	4057.060	4		
----	IN II	4057.070			100
	V I	4057.074	20	10	
	CL	4057.080			10
	CA	4057.100	2	3	
	IN II	4057.188			10
	CO I	4057.199	100		
	CE	4057.303	2		
---	TH	4057.337	10	8	
	FE I	4057.346	20	3	
	NI I	4057.347	2		
	P	4057.390			50
	DY	4057.400	4	4	
	HF	4057.430	3	2	
	MO	4057.438	4	4	
	W	4057.452	6	6	
	XE II	4057.460			100
	HO	4057.550	2	2	
	CE	4057.556	2		
--	MO	4057.584	10	4	
	TI I	4057.624	40	6	
	MG I	4057.632	10		
	SM	4057.653	10	3	
	AR II	4057.669			
	TB	4057.680	2		
	ZN II	4057.710	79		
	ER	4057.819	32		
	*PB I	4057.820	2000	320	11 VAF
	TH	4057.823	2	3	
	V I	4057.825	10	2	
	IN	4057.866	79	10	
	MN	4057.950	79	20	
	U	4057.955		6	
	BH SR	4058.000	3		

Table IIa. PB I 4057.820 A
(con't)

	Element	Wavelength (Angstrom)	Intensity Arc	Spark	Plasma
--	TB	4058.020	2		
	LA II	4058.085	4		
	TA	4058.136	2		
	TI I	4058.144	50	6	
	U	4058.160	10	4	
	PR	4058.190	6	8	
	CO I	4058.190	100		
	SE II	4058.200			20
	FE	4058.229	79	25	
	GD I	4058.231	100	63	
	CE II	4058.244	20		
--	DY	4058.250	2		
	TB	4058.440	5		
	EU	4058.449	2		
	TA	4058.464	10	5	
--	SI	4058.490		3	
	CO I	4058.600	100		
	MO	4058.610		10	
	ZR I	4058.624	10		
	FE I	4058.760	40	10	
	CR	4058.772	79	50	
	PR	4058.778	25	16	
	TB	4058.810	3		

Table IIb.(1) PB I 4057.820 A

Element	Wavelength (Angstrom)	Intensity Arc	Spark	Plasma
GD	4053.302	100	79	
V	4053.587		63	
GD I	4053.648	100	40	
HO	4053.920	400	200	
RU I	4054.051	40	100	
U	4054.313	13	16	
AR I	4054.525			2 VAF
GD I	4054.734	79	20	
PR	4054.845	50	40	

Table IIb.(1) PB I 4057.820 A
(con't)

Element	Wavelength (Angstrom)	Intensity Arc	Spark	Plasma
TI I	4055.019	79	32	
AG I	4055.264	790	500	
MN	4055.543	79	79	
MO	4056.012	25	32	
MO	4056.318	16	16	
PR	4056.543	100	63	
*PB I	4057.820	2000	320	11 VAF
MN	4057.950	79	20	
FE	4058.229	79	25	
GD I	4058.231	100	63	
CR	4058.772	79	50	
PR	4058.778	25	16	
SM II	4058.867	32	20	
MN	4058.930	79	63	
NB I	4058.938	1000	400	
MN	4059.392	20	16	
GD	4059.881	50	20	
TI I	4060.265	63	25	
ND II	4061.085	40	32	31 VAF
TA	4061.399	50	32	
MN	4061.742	79	32	
MO	4062.076	79	79	
PB	4062.144	20	20	
FE I	4062.444	130	100	
U	4062.549	13	20	
CU I	4062.698	500	20	
PR II	4062.817	160	50	6 VAF

Table IIa. PD I 3634.695 A

Element	Wavelength (Angstrom)	Intensity Arc	Spark	Plasma
TH	3633.700		2	
NB	3633.714	8	3	
DY	3633.770	4		
TA	3633.788	32	10	
FE	3633.833	6	3	
IR I	3633.862	3		
CE	3633.865	2		
V I	3633.910	32	8	
RU	3633.921	2	5	
-- CR	3634.014		6	
ZR I	3634.151	25	2	
-- TH	3634.214	5	6	
HE I	3634.235			16
S	3634.250			32
SM II	3634.271	100	25	45 VAF
ND	3634.282	25	32	
FE I	3634.334	16	5	
HE I	3634.367			2
KR II	3634.420			3
CE	3634.432	2		
NB	3634.443	5	16	
-- AR I	3634.461			320
PR	3634.467	20	4	
YB	3634.551	2		
U	3634.560	8		
TH	3634.584	8	4	
NB	3634.600	2		
ER	3634.679	8		
HO	3634.680	6	4	
FE	3634.687	20	2	
*PD I	3634.695	2000	1000	55 VAF
CO	3634.712	63	10	
CS	3634.750			6
TB	3634.750	8		
GD	3634.760	20	20	
RN	3634.800			250
AR II	3634.830			5
ND	3634.871	32	16	
SM II	3634.908	20	2	
RU	3634.929	50	100	

Table IIa. PD I 3634.695 A
(con't)

	Element	Wavelength (Angstrom)	Intensity Arc	Spark	Plasma
--	NI I	3634.941	50	10	
	CR	3634.994	25	13	
	U	3635.010		2	
	ND	3635.106	6	6	
	AU II	3635.133	3	10	
	MO	3635.144	100	10	
---	TB	3635.170	8		
	FE I	3635.196	13	2	
	TI I	3635.202	10	5	
	TH	3635.246	4	5	
	DY	3635.260	10	6	
	CR I	3635.281	25	8	
	PR	3635.284	25	6	
	U	3635.299	10		
	SC	3635.310	3	2	
	NB	3635.324	5	8	
	Y II	3635.334	2	13	
	HO	3635.350		4	
	TH	3635.372	16	10	
	U	3635.403	6	5	
	TB	3635.420	32	8	
	MO	3635.429	25	5	
----	W	3635.434	2		
	NB	3635.463	5	5	
	TI I	3635.463	200	100	
	V I	3635.463	40	5	
	IR I	3635.488	32	4	
	RU I	3635.516	2	8	
	U	3635.579	4	8	
	MO	3635.612	4	3	
	EU	3635.650	6		
	AR	3635.670			3
	MN	3635.682	10	5	

Table IIb.(1) PD I 3634.695 A

Element	Wavelength (Angstrom)	Intensity Arc	Spark	Plasma
EU	3629.797	40	50	
SC II	3630.740	50	63	14 VAF
FE I	3631.464	500	320	
CR II	3631.687	10	63	
FE	3632.040	50	50	
V	3632.124		63	
Y II	3633.123	50	100	36 VAF
SM II	3634.271	100	25	45 VAF
*PD I	3634.695	2000	1000	55 VAF
RU	3634.929	50	100	
TI I	3635.463	200	100	
ZR II	3636.448	32	200	
SB	3637.829	2	63	
FE I	3638.298	100	79	
TB	3638.460	79	50	
PT I	3638.793	250	10	
V I	3639.024	63	63	
CO I	3639.444	200	20	
RH I	3639.512	130	63	
PB I	3639.580	320	50	5 VAF

Table IIa.   PD I   3609.548 A

	Element	Wavelength (Angstrom)	Intensity Arc	Spark	Plasma
	U	3608.686		4	
	EU	3608.720	10		
	RU I	3608.727	2	8	
	GD	3608.758	100	130	
	TM	3608.770	100	20	
	TA	3608.781	16		
-----	CE	3608.847	2		
	FE I	3608.861	500	400	
	K II	3608.871			10
	F II	3608.890			6
	U	3608.960	20	10	
---	TB	3609.060	8		
	N II	3609.090			5
	RU I	3609.107	3	4	
	OS	3609.147	20	13	
	TA	3609.177	8		
	NE I	3609.179			50
	CE	3609.212	2		
	TH	3609.228	5	4	
	LA II	3609.230	2	3	
	DY	3609.250	3		
	V I	3609.289	32	32	
--	CA	3609.300	3	2	
	CU I	3609.307	25	5	
	NI I	3609.314	200	16	
	TA	3609.357	8		
	NB	3609.360		5	
	ER	3609.440	6		
	TH II	3609.447	13	10	27 VAF
	CR	3609.479	20	13	
	SM II	3609.484	63	100	52 VAF
	MO	3609.493	5	10	
	ND	3609.495	25	32	
	TM	3609.540	16	25	
	*PD I	3609.548	1000	630	35 VAF
	TB	3609.550	16		
	TI I	3609.591	13	2	
	ZR	3609.640	3		
	U	3609.682	16	13	
	CE	3609.687	40	10	

Table IIa. PD I 3609.548 A
(con't)

Element	Wavelength (Angstrom)	Intensity Arc	Spark	Plasma
CL II	3609.740			2
CO	3609.758	5	3	
IR	3609.772	32	25	
-- ND	3609.788	16	10	
TB	3609.880	8		
CE	3609.894	2		
TA	3609.931		20	
EU	3609.940	3		
NB	3610.003	3	5	
CL	3610.020			4
--- TH	3610.039	3		
CR	3610.052	20	8	
TI I	3610.156	100	63	
FE I	3610.162	100	100	
LA	3610.245	6	5	
--- CE	3610.257	5		
MN	3610.299	63	40	
XE I	3610.320			16
TH	3610.399	8	4	
CE	3610.446	2		
NI I	3610.462	1000		
U	3610.487	4	8	
RE	3610.493	40		
SE II	3610.500			32
IN	3610.508		20	
CD I	3610.510	1000	500	1 VAF

Table IIb.(1) PD I 3609.548 A

Element	Wavelength (Angstrom)	Intensity Arc	Spark	Plasma
CR I	3605.333	500	400	
FE I	3605.458	320	160	
RH I	3605.863	25	32	
DY	3606.126	200	100	
YB	3606.470	16	63	
FE I	3606.682	200	160	

Table IIb.(1) PD I 3609.548 A
(con't)

Element	Wavelength (Angstrom)	Intensity Arc	Spark	Plasma
V I	3606.689	79	63	
NI I	3606.852	100		
MO	3606.908		32	
TA	3607.406	63	32	
MN	3607.537	79	40	
MN	3608.494	63	40	
GD	3608.758	100	130	
TM	3608.770	100	20	
FE I	3608.861	500	400	
V I	3609.289	32	32	
NI I	3609.314	200	16	
TH II	3609.447	13	10	27 VAF
SM II	3609.484	63	100	52 VAF
ND	3609.495	25	32	
*PD I	3609.548	1000	630	35 VAF
TI I	3610.156	100	63	
FE I	3610.162	100	100	
MN	3610.299	63	40	
NI I	3610.462	1000		
CD I	3610.510	1000	500	1 VAF
Y II	3611.047	40	63	40 VAF
YB	3611.307	13	50	
CS I	3611.520	200		
ZR II	3611.893	16	40	
FE I	3612.074	79	50	
AL	3612.467		79	
RH I	3612.470	200	50	
NI I	3612.741	400	50	
CD I	3612.875	790	500	
ZR II	3613.100	40	40	
W II	3613.790	10	32	
BI	3613.817		32	
SC II	3613.836	40	63	20 VAF
MO	3614.253	50	32	
CD I	3614.450	63	100	

Table IIa. PD II 2488.921 A

	Element	Wavelength (Angstrom)	Intensity Arc	Spark	Plasma
	CD II	2487.937		16	
	W	2487.940	3		
	LU	2487.960		4	
	RU	2488.096	3		
	W II	2488.118	2	13	
	FE I	2488.148	630	100	
----	TA	2488.207	5		
	NB	2488.236	2	10	
	IR	2488.284		5	
	BR	2488.290			2
	BR	2488.370			40
	TA	2488.397	63		
	PD II	2488.412		10	
---	MN	2488.460		13	
	OS I	2488.548	50	16	
	IN II	2488.556			20
	RU	2488.577		16	
	V	2488.620		6	
	IN II	2488.666			40
--	CD II	2488.690			2
	TA II	2488.696	200	160	54 VAF
	PT	2488.736	10	25	
	CL	2488.740			5
	NB	2488.747		40	
	PR	2488.750		32	
	W	2488.771	10	20	
	N II	2488.820			3
	PT	2488.880		25	
	W	2488.910	10	3	
	*PD II	2488.921	10	32	29 VAF
	FE	2488.950	10		
	IN II	2488.951			63
	YB	2488.960		3	
--	XE II	2489.110			25
	TE	2489.200			10
	IR I	2489.204	2		
	W II	2489.233	10	20	41 VAF
	HO	2489.240		10	
	OS	2489.280	10	10	
	CR	2489.293		10	

Table IIa. PD II 2488.921 A
(con't)

	Element	Wavelength (Angstrom)	Intensity Arc	Spark	Plasma
	RU	2489.336		40	
	KR II	2489.390			8
---	BI	2489.400	8	2	
	TM	2489.460	2	100	
	NB	2489.463		3	
	BR	2489.470			3
	FE II	2489.486		40	
	CS	2489.500			2
	NI I	2489.507	40	6	
	W II	2489.510	2	8	
	PD II	2489.611		79	
	TH	2489.640	5	3	
----	CU II	2489.652	3	50	
	W	2489.718	8	2	
	FE I	2489.751	200	2	
	U	2489.782	13	16	
	FE II	2489.822		50	
	RU	2489.912	63		
	W	2489.915	2	6	

Table IIb.(3) PD II 2488.921 A

Element	Wavelength (Angstrom)	Intensity Arc	Spark	Plasma
RU	2483.972	32	20	
W II	2484.003	2	16	
U	2484.008	16	4	
NI I	2484.034	20		
FE I	2484.187	100	5	
U	2484.217	10	16	
FE II	2484.236		20	
NI II	2484.327		50	
W	2484.398	2	16	
BH C	2484.500	13		
W	2484.735	20	6	
MO	2484.753	2	20	
NB	2484.932	2	25	

Table IIb.(3) PD II 2488.921 A
(con't)

Element	Wavelength (Angstrom)	Intensity Arc	Spark	Plasma
TA	2484.954	32		
MO	2485.313	32		
CO II	2485.356	25	79	
CU II	2485.787	5	50	
RE	2485.809	50		
RH	2485.817	2	50	
FE	2485.985	10		
NB	2486.028		20	
MN	2486.163		20	
OS II	2486.244	8	16	13 VAF
W	2486.300	16	2	
FE II	2486.345		79	
FE	2486.371	40		
CO II	2486.440		40	
PD II	2486.528	5	32	15 VAF
FE	2486.693	32	3	
TA	2486.705	20		
RE	2486.776	50		
FE I	2487.064	25	10	
PT I	2487.168	630	20	20 VAF
RE	2487.230	50		
ZR II	2487.288	20	20	
CO II	2487.420	10	20	
RH	2487.470	100	8	
W	2487.492	16	5	
TM	2487.530	16	10	
LA II	2487.588		40	
W	2487.766	10		
CD II	2487.937		16	
FE I	2488.148	630	100	
TA	2488.397	63		
OS I	2488.548	50	16	
RU	2488.577		16	
TA II	2488.696	200	160	54 VAF
PT	2488.736	10	25	
NB	2488.747		40	
PR	2488.750		32	
W	2488.771	10	20	
PT	2488.880		25	
W	2488.910	10	3	

Table IIb.(3) PD II 2488.921 A
(con't)

Element	Wavelength (Angstrom)	Intensity Arc	Spark	Plasma
*PD II	2488.921	10	32	29 VAF
FE	2488.950	10		
W II	2489.233	10	20	41 VAF
OS	2489.280	10	10	
RU	2489.336		40	
TM	2489.460	2	100	
FE II	2489.486		40	
NI I	2489.507	40	6	
PD II	2489.611		79	
CU II	2489.652	3	50	
FE I	2489.751	200	2	
U	2489.782	13	16	
FE II	2489.822		50	
RU	2489.912	63		
PT I	2490.125	320	20	
NB	2490.213		16	
CO	2490.394	5	79	
Y I	2490.420	20	3	
YB	2490.450		20	
FE I	2490.644	200	10	
NI	2490.700	40		
RH II	2490.770	100	1000	52 VAF
W	2490.843	10		
NB	2490.848		20	
FE II	2490.861		16	
RU	2491.099	10	40	
FE I	2491.155	160	10	
NI	2491.187	20		
AU I	2491.240	10	5	
CR I	2491.348	32		
FE II	2491.390		32	
ZN I	2491.480	100	50	
RU	2491.558		16	
TA	2491.591	10		
TM	2491.600	13	5	
OS	2491.689	10	3	
SN	2491.720	13	5	
RU	2491.776	32		
RH	2491.855	2	400	
FE	2491.984	10		

Table IIb.(3) PD II 2488.921 A
(con't)

Element	Wavelength (Angstrom)	Intensity Arc	Spark	Plasma
CU I	2492.146	200	50	
TA	2492.172	16	2	
RH I	2492.299	100	10	
FE II	2492.337		32	
OS	2492.367	50	8	
W	2492.367	13	8	
RE	2492.552	16		
CR I	2492.568	50		
RE	2492.847	20		
AS I	2492.910	25	5	
W II	2492.934	6	25	
NB	2493.022		20	
IR I	2493.080	20	5	
FE II	2493.180	10	100	
FE II	2493.261	10	100	
BH C	2493.300	13		
ZN I	2493.320	25		
W	2493.393	10	2	
V	2493.582		25	
RU	2493.693	20	79	
TL I	2493.910	10		

Table IIa. PR II 4225.327 A

	Element	Wavelength (Angstrom)	Intensity Arc	Spark	Plasma
	U	4224.427	8	10	
	P II	4224.430			16
	CR	4224.514	63	13	
	V II	4224.515	2	3	
	FE I	4224.517	63	16	
	CE	4224.552	8		
	NE II	4224.570			5
----	TH	4224.613	16		
	TH	4224.616	5	3	
	IR	4224.649	3		
	ZR I	4224.650	3		
	DY	4224.680	6		
	N	4224.740			16
	W	4224.765	10	2	
	MO	4224.766	5	3	
	TI I	4224.793	40	8	
---	ND	4224.847	16	5	
	TB	4224.850	5		
	EU	4224.880	4		
	CL II	4224.920			16
	MO	4224.929	5	5	
	I	4224.980			8
	W	4224.988		13	
	GD I	4225.028	16		
--	RU	4225.092	25		
	HF II	4225.100		5	
	CO	4225.109	5		
	F	4225.120			20
	HO	4225.130	3	2	
	GD	4225.148	20	10	
	DY	4225.153	40	8	
	V II	4225.225	3	10	
	MO	4225.248	5	3	
	ZR I	4225.263	4		
	GD	4225.264	3		
	SM II	4225.318	40	32	
	*PR II	4225.327	50	25	7 VAF
	U	4225.369	8	8	
	ZR I	4225.463	6		
	FE I	4225.465	79	20	

Table IIa. PR II 4225.327 A
(con't)

Element	Wavelength (Angstrom)	Intensity Arc	Spark	Plasma
IR	4225.496	16	2	
I II	4225.540			16
ND	4225.556	3	2	
SC I	4225.580	4		
K II	4225.605			40
TE	4225.700			50
CE	4225.746	6	8	
U	4225.750		3	
GD	4225.853	160	50	
AS II	4225.870		10	
FE I	4225.957	79	32	
RN I	4226.060			50
U	4226.065	6	8	
RB	4226.100			8
B	4226.150	4	2	
BR	4226.150			4
TB	4226.150	3		
SM	4226.175	5	2	
CE	4226.196	3		
IR	4226.205	2		
NB	4226.209	3	5	
NB	4226.247	4	3	
MO	4226.291	20	20	

Table IIb.(3) PR II 4225.327 A

Element	Wavelength (Angstrom)	Intensity Arc	Spark	Plasma
FE	4220.348	79	40	
MN	4220.610	63	20	
SM II	4220.653	100	100	
RU I	4220.675	63		
RE I	4221.084	100		
DY	4221.104	63	8	
CR	4221.572	79	32	
P	4222.150	320		160
FE I	4222.221	200	200	

Table IIb.(3) PR II 4225.327 A
(con't)

Element	Wavelength (Angstrom)	Intensity Arc	Spark	Plasma
CE	4222.599	79	20	
CR I	4222.732	100	16	
PR II	4222.980	130	40	6 VAF
SM II	4223.711	50	20	
FE I	4224.176	200	79	
CR	4224.514	63	13	
FE I	4224.517	63	16	
SM II	4225.318	40	32	
*PR II	4225.327	50	25	7 VAF
FE I	4225.465	79	20	
GD	4225.853	160	50	
FE I	4225.957	79	32	
MO	4226.291	20	20	
FE I	4226.430	79	25	
TB	4226.440	50		
GE I	4226.570	200	50	
CA I	4226.728	500	50	2 VAF
CE	4226.734	50	32	
CR	4226.758	130	32	
MO	4227.085	3	25	
GD	4227.145	50	20	
FE I	4227.432	320	250	
RE I	4227.460	200		
ZR I	4227.758	160	8	
NB	4229.149	50	100	
SM II	4229.702	40	32	
GD	4229.808	16	20	
RU I	4230.312	63		

Table IIa. PR II 3908.431 A

Element	Wavelength	Intensity		
	(Angstrom)	Arc	Spark	Plasma
O II	3907.440			20
CE	3907.445	6	2	
FE	3907.470	16	6	
SC I	3907.476	130	25	
V	3907.519		2	
U	3907.558	6	8	
AG	3907.590	3	2	
CE	3907.639	2		
TB	3907.650	3		
AU	3907.650	5	3	
FE	3907.675	2		
--- SB	3907.736			8
IR I	3907.753	4		
CR	3907.778	32	10	
TB	3907.790	3	3	
ND	3907.843	20	13	
IR I	3907.896	10	2	
XE II	3907.910			50
TB	3907.912	5		
--- FE	3907.937	100	63	
CE	3907.964	2		
PR	3908.033	100	50	
TB	3908.076	20		
CE II	3908.094	8		
GD	3908.146	5	3	
MN	3908.162	3	3	
-- RE	3908.202	25		
MO	3908.249	5	3	
SM	3908.258	6		
V	3908.317	50	2	
U	3908.331	10		
CE	3908.408	32	6	
ER	3908.418	10		
*PR II	3908.431	100	63	8 VAF
U	3908.473	8	10	
TH	3908.488	3	2	
CE I	3908.543	20	3	
CE II	3908.543	20	3	
NB	3908.594	3	5	
MO	3908.612		32	

Table IIa. PR II 3908.431 A
(con't)

	Element	Wavelength (Angstrom)	Intensity Arc	Spark	Plasma
--	TB	3908.660	4	3	
	U	3908.682	2	3	
	CR I	3908.755	200	160	
	CE	3908.765	13	2	
	RU	3908.765	13	13	
	IR	3908.834	5		
---	NI I	3908.931	2		
	NB	3908.973	10	8	
	CE	3909.045	6		
	U	3909.065	5	6	
	RU	3909.075	32	50	
	TL II	3909.150			3
	TB	3909.150	16	3	
	PB II	3909.170		40	
----	HF	3909.180	3	3	
	GD	3909.247	3		
	SM	3909.273	5	5	
	ND	3909.276	16	10	
	PR	3909.312	4	2	
	CE II	3909.313	32	3	
	TA	3909.332	16		
	BR	3909.370			2
	AU I	3909.376	10	16	

Table IIb.(2) PR II 3908.431 A

Element	Wavelength (Angstrom)	Intensity Arc	Spark	Plasma
FE I	3903.900	100	79	
TH	3904.089	20	20	
NB	3904.182	10	20	
U	3904.299	8	16	
TI I	3904.785	63	32	
YB	3904.828	13	160	
U	3904.853	8	16	
TH	3905.190	32	32	
SI I	3905.528	20	16	

Table IIb.(2) PR II 3908.431 A
(con't)

Element	Wavelength (Angstrom)	Intensity Arc	Spark	Plasma
GD	3905.651	50	50	
ND	3905.886	40	32	
CO I	3906.294	160		
ER II	3906.316	25	13	14 VAF
HG I	3906.410	25	16	
FE I	3906.482	320	200	
V I	3906.748	50	20	
EU II	3907.110	1000	500	39 VAF
SM	3907.124	10	16	
GD	3907.125	100	100	
SC I	3907.476	130	25	
FE	3907.937	100	63	
PR	3908.033	100	50	
V	3908.317	50	2	
*PR II	3908.431	100	63	8 VAF
MO	3908.612		32	
CR I	3908.755	200	160	
RU	3909.075	32	50	
PB II	3909.170		40	
AU I	3909.376	10	16	
V I	3909.894	50	32	
BA I	3909.917	50	20	
CO I	3909.934	200		
V I	3910.790	32	20	
MO	3911.092	20	20	
ND	3911.169	25	25	
MN	3911.419	16	16	
U	3911.673	20	20	
SC I	3911.810	160	32	
V I	3912.207	50	20	
ND	3912.228	20	20	
TH	3912.285	16	16	
CE I	3912.436	50	5	
CE II	3912.436	50	5	
IN	3912.810		16	
V I	3912.886	40	16	
PR	3912.898	160	79	
TH	3913.012	16	16	

Table IIa. PR II 4143.136 A

	Element	Wavelength (Angstrom)	Intensity Arc	Spark	Plasma
	MO	4142.162	8	4	
	NI I	4142.184	2		
	HO	4142.190	2		
	CR	4142.193	32	8	
	NB	4142.245	4	3	
	W	4142.256	16	10	
	S II	4142.291			160
	NI	4142.320	16		
	CE	4142.343	32	6	
	U	4142.350	3	2	
----	CE	4142.398	32	32	
	PR	4142.434	3		
	TB	4142.460	3		
	TH	4142.482	10	10	
	TI I	4142.492	13	2	
	CR	4142.500	10		
---	TH	4142.703	10	10	
	CE	4142.717	2		
	RE	4142.771	8		
	SM	4142.813	10	2	
	CE	4142.826	6	2	
	Y I	4142.839	100	25	
	HO	4142.860		3	
	V I	4142.882	6	3	
--	SM	4142.958	16		
	ER	4142.960	32	16	
	EU	4142.970	2		
	CU II	4143.039		2	
	TI I	4143.046	32	8	
	DY	4143.100	40	8	
	*PR II	4143.136	200	50	8 VAF
	IR	4143.140	8		
	NB	4143.206	10	10	
	TB	4143.240	4		
--	TI I	4143.280	6		
	FE I	4143.420	200	100	
	TB	4143.530	25		
	MO	4143.549	100	100	
---	U	4143.587	2	3	
	N I	4143.650			32

Table IIa. PR II 4143.136 A
(con't)

Element	Wavelength (Angstrom)	Intensity Arc	Spark	Plasma
TB	4143.700	16		
LA II	4143.742		10	
HE I	4143.759			16
O II	4143.760			16
P	4143.840			50
FE I	4143.871	400	250	
--- U	4143.877	5		
NB	4143.931	5	5	
LA I	4143.936	25	16	
BR I	4143.980			20

Table IIb.(2) PR II 4143.136 A

Element	Wavelength (Angstrom)	Intensity Arc	Spark	Plasma
MO	4138.185	20	16	
MO	4138.543	20	13	
NB	4139.707	50	50	
FE I	4139.923	40	32	
TH	4140.239	16	16	
GD	4140.454	20	20	
SB II	4140.607		16	
PD I	4140.827	100		
MN	4141.063	50	32	
U	4141.228	20	32	
PR	4141.257	160	50	
MO	4141.489		32	
LA II	4141.740	200	200	
CE	4142.398	32	32	
Y I	4142.839	100	25	
ER	4142.960	32	16	
*PR II	4143.136	200	50	8 VAF
FE I	4143.420	200	100	
MO	4143.549	100	100	
FE I	4143.871	400	250	
LA I	4143.936	25	16	
RU I	4144.164	160	200	

Table IIb.(2) PR II 4143.136 A
(con't)

Element	Wavelength (Angstrom)	Intensity Arc	Spark	Plasma
RE I	4144.359	130		
RU I	4145.738	130	160	
CR	4145.804		20	
PR	4146.537	20	13	
RU I	4146.771	100	63	
MO	4146.846		32	
MN	4147.532	40	20	
FE I	4147.673	200	100	
TA	4147.891	40	32	

Table IIa.   PT II 2144.231 A

	Element	Wavelength (Angstrom)	Intensity Arc	Intensity Spark	Intensity Plasma
	MO	2143.250	2	16	
	BI II	2143.350		8	
	BI II	2143.400		13	
	YB	2143.420		5	
	BI II	2143.460		13	
	RE	2143.480	6	2	
---	TE	2143.540			25
	TI	2143.610	5		
	TA	2143.650	2	5	
	CO I	2143.679	3		
	V II	2143.713	3	4	
---	LU	2143.840		3	
	YB	2143.860		4	
	FE	2143.900	8		
	RH	2143.970		25	
--	ZR II	2144.010		3	
	IR	2144.070		5	
	MO	2144.070		20	
	RE II	2144.085	2	32	
	W	2144.090	10	10	
	AS I	2144.100	50		
	HF II	2144.170	6	6	
	NB	2144.200		10	
	PT I	2144.231	32	100	
	*PT II	2144.231	32	100	100 VAF
	PD	2144.250	2	25	
	IR	2144.280	10	3	
	CD II	2144.382	50	1600	120 VAF
	BI	2144.410		5	
--	FE	2144.450	25		
	NB	2144.492	4	2	
	K	2144.500			20
	W	2144.500	8	10	
	IR	2144.580		16	
	HG	2144.670			5
	CU II	2144.730		3	
--	HF II	2144.730	3	3	
	YB	2144.740		32	
	NB	2144.800		4	
	MN	2144.800		10	

Table IIa. PT II 2144.231 A
(con't)

	Element	Wavelength (Angstrom)	Intensity Arc	Spark	Plasma
----	SB I	2144.860			12 VAF
	PT	2145.026	13	4	
	CD	2145.040			20
	SB	2145.040	20	20	
	KR II	2145.080			10
	NI I	2145.180	8	10	
	FE I	2145.188	13	2	

Table IIb.(2) PT II 2144.231 A

Element	Wavelength (Angstrom)	Intensity Arc	Spark	Plasma
HF II	2139.240	32	40	
RH	2139.430		100	
SB I	2139.760	32	32	16 VAF
V II	2139.821	5	25	
YB	2139.960		25	
V II	2140.087	10	79	20 VAF
TA	2140.130	13	32	
PD II	2140.260		32	
TA	2141.500	3	32	
SB	2141.760	32	32	
HF II	2141.840	20	20	
V II	2141.980	10	79	
PD I	2142.130	16		
IR	2142.250	20	10	
TE I	2142.810	630		73 VAF
PR	2142.890		32	
V II	2143.047	10	50	
TA	2143.160		32	
RH	2143.970		25	
RE II	2144.085	2	32	
AS I	2144.100	50		
PT I	2144.231	32	100	
*PT II	2144.231	32	100	100 VAF
PD	2144.250	2	25	
CD II	2144.382	50	1600	120 VAF

Table IIb.(2) PT II 2144.231 A
(con't)

Element	Wavelength (Angstrom)	Intensity Arc	Spark	Plasma
FE	2144.450	25		
YB	2144.740		32	
SB I	2144.860			12 VAF
SB	2145.040	20	20	
AG II	2145.610	8	160	
RE II	2145.900		32	
V II	2145.991	5	40	
TA	2146.870	10	40	
TE I	2147.190			14 VAF
V II	2147.460	2	100	
NI I	2147.803	16	8	
IR	2148.220	25	50	
PD II	2148.280		25	
V II	2148.420	5	32	
SN I	2148.460	16		
SN I	2148.740	50	20	
CU II	2148.974	16	25	
P I	2149.108	16		39 VAF
PD	2149.130		32	

Table IIa.   PT I   2659.454 A

	Element	Wavelength (Angstrom)	Intensity Arc	Spark	Plasma
	FE	2658.470	20	2	
	TM	2658.490	20	6	
	V	2658.506		16	
	HG	2658.510	5	10	
	W	2658.564	4		
	CR II	2658.592	20	32	
	OS	2658.600	50	10	
	SN	2658.609		4	
	BR	2658.650			6
	TH	2658.667	16	10	
	CE	2658.668	4		
	ZR I	2658.687	10		
	RE	2658.688	16		
	PT I	2658.701	40	5	
----	CS	2658.710			2
	W	2658.718		4	
	PB II	2658.722	20	320	
	CL II	2658.740			100
	RU	2658.770	10		
	TA	2658.862	25	50	
	NB	2658.876	3	5	
	MO	2658.902	10		
	W	2658.906	10	2	
	TB	2658.910		500	
	CR	2658.920		5	
	FE	2658.930	10	2	
---	V	2658.975	10	40	
	RH I	2659.011	2	2	
	RE	2659.023	25		
	U	2659.025	8	2	
	NB	2659.052	3	32	
	MN	2659.084		25	
	RH	2659.110		100	
	IR	2659.144	4		
	W	2659.189		10	
--	MN	2659.210		10	
	FE	2659.240	8	2	
	I	2659.270			20
	YB	2659.280	2	5	
	CD II	2659.290			5

Table IIa. PT I 2659.454 A
(con't)

Element	Wavelength (Angstrom)	Intensity Arc	Spark	Plasma
TA	2659.410	20	10	
EU	2659.420	4		
AU	2659.430		5	
*PT I	2659.454	2000	500	37 VAF
RH I	2659.472	3	3	
BH C	2659.600	20		
KR II	2659.600			2
V	2659.606	10	40	
RU I	2659.615	79	13	
TA	2659.655	16		
HF	2659.690		3	
-- W	2659.697		13	
CE	2659.716	2		
CR	2659.746		2	
RE	2659.792	16		
OS	2659.833	32	8	
TH	2659.860	3		
GA I	2659.866	5	13	4 VAF
--- IR I	2659.946	2		
YB	2660.010		3	
NB	2660.035	2	32	
IR I	2660.076	3		
TM	2660.090	32	10	
U	2660.144	16	10	
--- PB II	2660.190			5
MN	2660.340		5	
AL I	2660.393	160	63	
HG	2660.400	5		2
CD I	2660.400	50		5
FE I	2660.402	40	16	

Table IIb.(2) PT I 2659.454 A

Element	Wavelength (Angstrom)	Intensity Arc	Spark	Plasma
NI II	2655.470		400	
NI II	2655.907		500	
NB	2656.078	8	200	
RU	2656.248	20	160	
PB II	2657.563		200	
LU	2657.800	50	160	
HF II	2657.847	100	200	
NB	2658.027		200	
PB II	2658.722	20	320	
TB	2658.910		500	
*PT I	2659.454	2000	500	37 VAF
GA I	2659.866	5	13	4 VAF
AG II	2660.456	32	160	
MO	2660.579	25	130	
RU II	2661.613	79	160	31 VAF
HF II	2661.883	200	500	
PB I	2663.166	320	40	8 VAF

Table IIa. PT II 2036.463 A

	Element	Wavelength (Angstrom)	Intensity Arc	Spark	Plasma
	RE	2035.520	4	6	
	MO	2035.570	3		
	IR	2035.580	10	4	
----	PT	2035.780	16	16	
	CU II	2035.843	13	32	
	W II	2035.890	8	20	
	OS	2035.930	16	4	
	RE I	2035.960	10		
---	IR	2036.000	2	8	
	XE	2036.110			2
	IR	2036.130	2	10	
	RE II	2036.200	3	16	
	MO	2036.210	3		
	SB	2036.210			2
--	HF II	2036.240	3		
	RH	2036.270		25	
	*PT II	2036.463	16	40	54 VAF
	OS	2036.490	10	3	
	SB	2036.510	8	4	
	CO II	2036.590	8	32	
--	RH	2036.700		79	
	CD II	2036.790			32
	IR	2036.840	2	2	
---	K	2036.900			40
	CU II	2036.970			4
	RH	2037.050		32	
	CU II	2037.117	13	32	
---	RE	2037.170	8	6	
	HF II	2037.320	2	2	
	IR	2037.330	3	8	

Table IIb.(3) PT II 2036.463 A

Element	Wavelength (Angstrom)	Intensity Arc	Spark	Plasma
IR	2031.500	4	25	
PD	2032.020		32	
NI II	2032.300	2	25	
PT	2032.410	20	25	
NB	2032.460		25	
CO	2032.728	2	20	
P I	2033.489	16		7 VAF
OS	2033.530	10	25	
IR	2033.570	20	6	
RH	2033.590	4	20	
RE I	2033.780	20	16	
AG II	2033.931		25	
IR	2034.010		32	
NI I	2034.397	16	2	
OS	2034.440	32	5	
RH	2034.630		20	
W II	2035.030	6	13	38 VAF
PT	2035.780	16	16	
CU II	2035.843	13	32	
W II	2035.890	8	20	
OS	2035.930	16	4	
RH	2036.270		25	
*PT II	2036.463	16	40	54 VAF
CO II	2036.590	8	32	
RH	2036.700		79	
RH	2037.050		32	
CU II	2037.117	13	32	
W	2037.584	3	20	
RH	2037.600		50	
MO	2037.680	2	20	
NB	2037.940	3	25	
MO	2038.010	2	20	
IR	2038.110		32	
MO II	2038.440	20		24 VAF
IR	2038.600		20	
OS	2038.720	16	5	
PD	2038.930		20	
IR	2038.990		25	
RE	2039.010	20		
RE	2039.230	20		

Table IIb.(3) PT II 2036.463 A
(con't)

Element	Wavelength (Angstrom)	Intensity Arc	Spark	Plasma
CR	2039.310	25		
RH	2039.350		32	
IR	2039.430	25	13	
SN	2039.500		200	
SB II	2039.600	16	16	7 VAF
RH	2039.670		25	
PT	2039.700	13	20	
IR	2039.790	8	40	
TE	2039.790	320		
SE I	2039.851			40 VAF
RH	2040.190	2	100	
PT	2040.330	16	16	
IR	2040.760	6	20	
RH	2041.040		40	
MO	2041.450	2	32	

Table IIa. RE II 2214.261 A

	Element	Wavelength (Angstrom)	Intensity Arc	Spark	Plasma
	OS	2213.340		16	
	MO	2213.440		3	
----	OS	2213.560	10	5	
	FE II	2213.654	2	32	
	RE	2213.680		20	
	V I	2213.692	6		
	CD	2213.700		2	
	CR	2213.700		32	
---	IR	2213.770		25	
	CO	2213.819	6		
	MN	2213.820	63	10	
	NI	2213.849	5		
	CO	2213.860	4	4	
	PT	2213.930	6		
	KR II	2213.960			5
	MO	2214.010		4	
--	V	2214.016		100	
	NB	2214.033	8		
	PD II	2214.038		40	
	BI I	2214.121	6		
	OS	2214.160	25	10	
	ZR	2214.160	20		
	*RE II	2214.261	500	1000	47 VAF
	MO	2214.270	2	25	
	PT	2214.280		20	
	IR I	2214.305	10	2	
	W	2214.310	4		
	TA	2214.340	3	13	
	FE II	2214.407		4	
--	XE	2214.410			3
	CU I	2214.581	50	5	13 VAF
	ZR II	2214.590		2	
	ZR I	2214.630	3		
	BA II	2214.639	2	6	
	YB	2214.640		8	
	HE II	2214.690			5
---	OS	2214.760	5	32	
	CO II	2214.790	10	13	
	W	2214.800	10	20	
----	SN	2214.920		3	

Table IIa. RE II 2214.261 A
(con't)

Element	Wavelength (Angstrom)	Arc	Intensity Spark	Plasma
CU II	2215.100		16	
AR	2215.200			2
MN	2215.210		6	

Table IIb.(1) RE II 2214.261 A

Element	Wavelength (Angstrom)	Arc	Intensity Spark	Plasma
SN I	2209.660	25	63	16 VAF
AL I	2210.046	13	16	5 VAF
AL I	2210.060			5 VAF
ZN II	2210.870		50	
SI I	2210.912	32	32	47 VAF
PD II	2212.140	5	50	
IR	2212.320		50	
MN	2213.820	63	10	
V	2214.016		100	
*RE II	2214.261	500	1000	47 VAF
CU I	2214.581	50	5	13 VAF
V	2216.034	2	100	
RE II	2216.161		50	
NI II	2216.473	20	40	29 VAF
SI I	2216.685	40	40	72 VAF
V	2217.413		160	
PB	2218.080	50	40	
CU II	2218.100	25	40	17 VAF
V	2218.428		130	

Table IIa. RE II 2275.250 A

	Element	Wavelength (Angstrom)	Intensity Arc	Spark	Plasma
	OS	2274.280		25	
	MO	2274.350		5	
	PT I	2274.380	32	20	
	TA	2274.380		5	
	W	2274.400	6		
	PD II	2274.454		20	
	U	2274.470		2	
	CO	2274.490	10	2	
----	CO I	2274.633	8		
	HF II	2274.640	16	20	
	RE	2274.650	10	6	
	NI I	2274.660	13		
	U	2274.680		2	
	NI II	2274.730		16	
---	NB	2274.763	3		
	ZN	2274.800			3
	PD II	2274.811		10	
	PT	2274.840	25	10	
	CU II	2274.870		3	
	XE	2274.970			2
	RU	2274.989	6	5	
--	MO	2275.000	3	32	
	U	2275.120		2	
	FE	2275.187	10		
	NB	2275.219		20	
	V	2275.229		16	
	*RE II	2275.250	320	320	44 VAF
	AG II	2275.260		50	
	ER	2275.300	6	2	
	ZR II	2275.390		2	
	CO	2275.410		6	
	MO	2275.470		16	
	CA I	2275.471	40	5	
	CR	2275.480		8	
	SR I	2275.480	2		3
--	W	2275.500		5	
	FE I	2275.595	8		
	TA	2275.600	2	25	
	W	2275.630	6	3	
	MO	2275.640		16	

Table IIa. RE II 2275.250 A
(con't)

	Element	Wavelength (Angstrom)	Intensity Arc	Spark	Plasma
	IR	2275.660		25	
---	NI II	2275.689	2	16	
	V	2275.871		2	
	CO I	2275.876	10		
----	W	2275.940		3	
	PD	2276.020		32	
	FE I	2276.024	20	10	
	U	2276.050		20	
	YB	2276.070	2		
	RU	2276.132	4		
	NB	2276.170	2	3	
	MO	2276.210		20	
	RH	2276.210		50	
	NB	2276.214	2		

Table IIb.(2) RE II 2275.250 A

Element	Wavelength (Angstrom)	Intensity Arc	Spark	Plasma
MO	2273.260		100	
NB	2274.131	13	320	
*RE II	2275.250	320	320	44 VAF
BI I	2276.578	100	40	12 VAF
RH	2276.960	20	160	
HF II	2277.166	40	160	
AG II	2279.982	10	130	

Table IIa. RH I 3434.893 A

	Element	Wavelength (Angstrom)	Intensity Arc	Spark	Plasma
	ZR II	3433.906	8	6	
	NB	3433.950		10	
	CU I	3433.972	5		
	TH	3434.000	13	13	
	FE I	3434.025	2		
	V	3434.026		20	
	MO	3434.045	4	4	
	CR I	3434.112	32	25	
	CE	3434.134	5		
	KR I	3434.142			8
----	U	3434.148	6	10	
	HG	3434.160		2	
	RB II	3434.263			63
	KR	3434.280			2
	SR I	3434.280	2		
	U	3434.283	6		
	CE	3434.290	3		
	ER	3434.367	25	8	
	DY	3434.373	79	32	
---	PR	3434.486	3		
	MO	3434.496	5	4	
	TA	3434.500	32	20	
	TB	3434.540	16	3	
	YB	3434.610	5	20	
	U	3434.614	13		
	ER	3434.634	13	2	
--	AG I	3434.646	2		
	RB	3434.720			40
	HG	3434.730		16	
	EU	3434.740	3		
	HO	3434.750		4	
	IR I	3434.757	10		
	PR	3434.757	20	2	
	TH	3434.762	8	8	
	MO	3434.790	50	13	
	U	3434.805	2		
	OS	3434.889	20	3	
	HF	3434.890	5		
	*RH I	3434.893	1000	200	50 VAF
	TB	3434.920	32		

Table IIa. RH I 3434.893 A
(con't)

	Element	Wavelength (Angstrom)	Intensity Arc	Spark	Plasma
	EU	3435.064	13	3	
--	I II	3435.075			6
	RU I	3435.186	63	20	
	U	3435.200	10		
	EU	3435.205	32	2	
	CE	3435.207	20		
	W	3435.243	6	5	
	OS	3435.256	20	10	
	SM II	3435.256	5	3	
	DY	3435.270	4	2	
---	V	3435.373		20	
	RN	3435.400			20
	MO	3435.400		63	
	TI	3435.432	10		
	ND	3435.434	20	4	
	MO	3435.451	6		
	CR	3435.488	8		
	NI I	3435.489	3		
	TB	3435.530	8	3	
	U	3435.530	16		
	SC I	3435.555	13	4	
	NB	3435.587		10	
---	HO	3435.610		4	
	EU	3435.650	4		
	CR I	3435.679	20	13	
	CE	3435.682	3		
	W	3435.715	10	13	
	EU	3435.734	13	10	
	CO I	3435.753	8		
	CR	3435.819	32	13	

Table IIb.(2) RH I 3434.893 A

Element	Wavelength (Angstrom)	Intensity Arc	Spark	Plasma
TM	3429.970	100	100	
ZR II	3430.532	50	50	
RU I	3430.772	63	50	
TA	3430.938	50	63	
TM	3431.200	50	50	
CO I	3431.575	500	40	
NB	3432.701	10	100	
CO I	3433.040	1000	160	
CR II	3433.311	32	160	
PD I	3433.449	1000	500	
NI I	3433.558	790	50	
*RH I	3434.893	1000	200	50 VAF
MO	3435.400		63	
CR I	3436.187	50	50	
RU I	3436.737	320	160	
NB	3436.962	20	50	
NI I	3437.280	630	40	
TA	3437.373	6	320	
ZR II	3438.230	250	200	42 VAF
NB	3438.419		50	
YB	3438.720	4	79	
YB	3438.840	20	100	
TA	3439.001		63	
GD	3439.785	50	50	

Table IIa. RH II 2334.770 A

Element	Wavelength (Angstrom)	Intensity Arc	Spark	Plasma
HF II	2333.770		4	
BI I	2333.795	13		
IR I	2333.839	40	10	
TA	2333.850		10	
RU	2333.907	13	20	
CO	2333.962	3		
MO	2334.050		3	
AU II	2334.060		5	
W II	2334.060	5	2	
PD II	2334.066		4	
CO I	2334.128	5	13	
CA	2334.130		2	
I	2334.130			32
TA	2334.130	2	20	
V	2334.213		250	
OS	2334.250		2	
W	2334.300	4		
TI	2334.330		5	
IR I	2334.330	10	2	
RE	2334.360	20	8	
MO	2334.380		3	
CR	2334.390	5	10	
V I	2334.446	25	40	
SI	2334.480			2
RE	2334.490	10		
IR I	2334.505	25	5	
FE	2334.528	3		
TI	2334.530	4	10	
IN II	2334.570			50
NI II	2334.580	8	20	
ZR	2334.670	6		
*RH II	2334.770	25	500	67 VAF
MO	2334.800		10	
SN I	2334.808	100	100	
NB	2334.810	3	25	
CO	2334.880		4	
TA	2334.880	3	20	
TE	2334.910			5
MO	2334.920		20	
MN	2334.940		25	

Table IIa. RH II 2334.770 A
(con't)

	Element	Wavelength (Angstrom)	Intensity Arc	Spark	Plasma
	IR	2334.958		25	
--	TM	2335.000		8	
	RE	2335.030	10		
	CO I	2335.107	16		
	PT II	2335.190	8	25	
	W II	2335.200	6	8	
	HF II	2335.250		2	
---	BA II	2335.269	63	100	75 VAF
	RH	2335.280		16	
	IR I	2335.285	5		
	NB	2335.314	3	10	
	V	2335.337	2	3	
	MO	2335.380		5	
	YB	2335.419	3	16	
	RB	2335.420			5
	XE II	2335.420			2
	PD	2335.430			3
	I II	2335.450			63
	TH	2335.490		3	
	V	2335.492	3	50	
----	I II	2335.520			16
	W	2335.560		3	
	PD	2335.570			5
	NB	2335.617	2	16	
	ZR	2335.630		2	
	RU	2335.740	13		
	RE	2335.750	40	10	
	TA	2335.750	2	20	

Table IIb.(3) RH II 2334.770 A

Element	Wavelength (Angstrom)	Intensity Arc	Spark	Plasma
V	2330.460	8	320	
PT I	2330.968	25	10	
RE	2331.170	32	10	
AG II	2331.370	20	160	1 VAF
RE	2331.740	25		
V	2331.771		320	
TA II	2331.980	10	20	96 VAF
OS	2332.140	32	4	
PB	2332.426	63	32	
FE II	2332.796	16	40	15 VAF
HF II	2332.968	40	50	
IR I	2333.298	25	8	
IR I	2333.839	40	10	
V	2334.213		250	
V I	2334.446	25	40	
IR I	2334.505	25	5	
*RH II	2334.770	25	500	67 VAF
SN I	2334.808	100	100	
BA II	2335.269	63	100	75 VAF
RE	2335.750	40	10	
RE	2336.130	25	10	
OS II	2336.800	50	79	24 VAF
HF II	2337.330	32	32	
GA I	2338.283	3	4	4 VAF
RE	2338.470	32	6	
OS	2338.630	32	6	
V I	2339.684	25		

Table IIa. RU II 2402.717 A

	Element	Wavelength (Angstrom)	Intensity Arc	Spark	Plasma
	MN	2401.728		5	
	C II	2401.770		6	
	IR I	2401.774	5		
	XE II	2401.790			2
	TE	2401.810			10
	CL	2401.840			3
	NI I	2401.841	40	10	
	W II	2401.859	2	16	
	NB	2401.873	2		
	PT I	2401.874	320	32	
	RU	2401.902	2	5	
	V I	2401.902	16	10	
	MO	2401.927	10	40	
	PB	2401.946	50	40	
----	K	2402.000			10
	HF II	2402.030		2	
	CO I	2402.065	10	3	
	TA	2402.133	10	10	
	CO I	2402.170	32		
---	OS	2402.233	10		
	SE	2402.280			25
	IR I	2402.282	2		
	NB	2402.344	5		
	C II	2402.400		13	
	W	2402.441	16	20	
--	CO I	2402.559	16		
	AG II	2402.570		32	
	FE II	2402.600	10	20	
	NB	2402.657		4	
	AU	2402.710		5	
	*RU II	2402.717	100	160	100 VAF
	W	2402.736	2	3	
	ZN I	2402.820	2		
--	ZR	2402.933	5		
	TE	2403.000			100
	V I	2403.024	2		
	IR	2403.029	2		
	W II	2403.072	2	10	
	PT I	2403.089	400	50	
---	RU	2403.176	8		

Table IIa. RU II 2402.717 A
(con't)

Element	Wavelength (Angstrom)	Intensity Arc	Spark	Plasma
W	2403.223	3	13	
V	2403.250		32	
LA II	2403.280		6	
CU II	2403.335	100	320	
CO I	2403.343	16		
ER	2403.350		6	
V I	2403.362	4		
YB	2403.400		2	
MO	2403.418	6	25	
U	2403.423		32	
-- ZR I	2403.429	10		
W	2403.470		10	
RH	2403.471		10	
RU	2403.536	6		
TL	2403.570			5
HF II	2403.600	2	5	
MO	2403.610	25	25	
CO	2403.643	16	13	
TE	2403.660			32
TA	2403.677	4	8	
LU	2403.690		4	

Table IIb.(2) RU II 2402.717 A

Element	Wavelength (Angstrom)	Intensity Arc	Spark	Plasma
BH B	2398.500	200		
CA I	2398.559	100	20	
IR	2398.746	10	160	
CR I	2399.060	50	2	
LU	2399.140	10	50	
V	2399.683		160	
CU II	2400.112	5	100	
TA II	2400.630	4	2	110 VAF
BI I	2400.884	200	100	
V	2400.903		50	
PT I	2401.874	320	32	

Table IIb.(2) RU II 2402.717 A
(con't)

Element	Wavelength (Angstrom)	Intensity Arc	Spark	Plasma
MO	2401.927	10	40	
PB	2401.946	50	40	
*RU II	2402.717	100	160	100 VAF
PT I	2403.089	400	50	
CU II	2403.335	100	320	
CO II	2404.172	32	50	
V	2404.178		100	
FE II	2404.430	25	40	
FE II	2404.882	50	100	27 VAF
NB	2404.886	5	50	
RE	2405.054	100		
NI II	2405.169		79	
RH	2405.221		50	
NB	2405.340	5	50	
HF II	2405.421	250	400	
RE	2405.602	79		
PT II	2405.728		100	
NB	2405.850	5	50	
TA	2406.549	63		
TM	2406.630	2	63	
CU I	2406.665	160	50	
PD II	2406.743		160	
MO	2407.135	3	40	
CO I	2407.254	100	2	

Table IIa. RU II 2456.568 A

Element	Wavelength (Angstrom)	Intensity Arc	Spark	Plasma
LU	2455.600		4	
IR I	2455.609	32	5	
U	2455.680	10	2	
IN	2455.682		3	
RH	2455.705	16	200	
CS	2455.800			8
-- RE	2455.851	8		
W II	2455.856		13	
IR I	2455.868	10	2	
LA II	2455.880		10	
RU	2455.889	5		
FE II	2455.901		25	
NA I	2455.915	2		
YB	2456.000		3	
-- KR	2456.070			6
W II	2456.079		6	
RH	2456.180	5	160	
W	2456.229	8		
CO I	2456.237	20	6	
PD	2456.268		2	
TH	2456.292	6	4	
-- RU II	2456.438	63	50	25 VAF
TL I	2456.450	5		
OS	2456.462	13	3	
AS I	2456.530	100	8	
W	2456.531	16	5	
ZR I	2456.540	3		
*RU II	2456.568	63	50	100 VAF
FE II	2456.641		16	
NB	2456.675		5	
- IR	2456.769	2		
FE II	2456.819		20	
TH	2456.860	4	2	
W	2456.877		6	
TE	2456.900			5
RU	2456.949	10		
NB	2456.996		200	
PD	2457.020			3
- IR I	2457.030	13	2	
U	2457.166	3	4	

Table IIa. RU II 2456.568 A
(con't)

	Element	Wavelength (Angstrom)	Intensity Arc	Spark	Plasma
	W	2457.180		13	
	RU	2457.181	10	13	
	AL II	2457.200			3
	IR I	2457.227	10	3	
	NB	2457.245		16	
	PD II	2457.260	2	10	
----	I	2457.270			20
	MO	2457.404	16		
	ZR II	2457.436	20	20	
	V	2457.450		32	
	LU	2457.560		5	

Table IIb.(3) RU II 2456.568 A

Element	Wavelength (Angstrom)	Intensity Arc	Spark	Plasma
NB	2451.874	3	100	
TL II	2451.940		32	
MN	2452.526	3	100	
TL	2452.700		100	
MN	2453.175		40	
AG II	2453.328		130	
HF II	2453.338	13	25	
V	2453.351	2	40	
RU	2453.939	10	25	
TA	2454.482	63		
FE II	2454.576		79	
RU	2454.923	63	5	
NI II	2455.526		40	
RU	2455.530	79	100	
RH	2455.705	16	200	
FE II	2455.901		25	
RH	2456.180	5	160	
RU II	2456.438	63	50	25 VAF
AS I	2456.530	100	8	
*RU II	2456.568	63	50	100 VAF
NB	2456.996		200	

Table IIb.(3) RU II 2456.568 A
(con't)

Element	Wavelength (Angstrom)	Intensity Arc	Spark	Plasma
V	2457.450		32	
FE	2457.595	63	32	
MO	2457.766	10	63	
NB	2458.087	2	50	
V	2458.288		63	
RU	2458.621	63	2	
MN	2458.687		79	
FE II	2458.782	6	63	
RH	2458.904	50	320	
CO	2459.490		32	
MO	2459.761		32	
AG II	2460.310		79	
NB	2460.402		200	
HF II	2460.493	20	50	
Y II	2460.620	10	32	
RH II	2461.036	79	200	28 VAF
RE	2461.179	130		
FE II	2461.279	5	50	
V	2461.495		40	

Table IIa. RU I 3798.052 A

	Element	Wavelength (Angstrom)	Intensity Arc	Spark	Plasma
	ER	3797.065	2		
	CR I	3797.126	50	32	
	U	3797.142		3	
	RB II	3797.170			2
	PR	3797.231	16	8	
	RB II	3797.276			2
	SM II	3797.278	16		
	MO	3797.300	5	5	
----	CE	3797.322	2		
	TA	3797.403	6	3	
	SN	3797.420		3	
	TH	3797.516	8	10	
	FE	3797.517	320	200	
	U	3797.520		3	
---	RE I	3797.591	40		
	CR I	3797.716	100	20	
	SM	3797.727	25	5	
	DY	3797.760	4		
	U	3797.773	10		
--	CU II	3797.832		2	
	ND	3797.890	20	10	
	CS	3797.908			4
	TH	3797.910	2	5	
	H I	3797.910			0.1 VAF
	HF II	3797.923	25	25	
	IR	3797.924	2		
	TB	3797.930	16		
	FE I	3797.949	4	2	
	*RU I	3798.052	32	40	
	IR	3798.059	6		
	CE	3798.080	2	2	
	TH	3798.103	5		
	NB	3798.121	50	79	
	YB	3798.170	4		
	LA II	3798.190		2	
	CE	3798.238	2		
	ER	3798.250	5		
	HO	3798.250		4	
	MO I	3798.252	1000	1000	
	U	3798.259	2	8	

Table IIa. RU I 3798.052 A
(con't)

	Element	Wavelength (Angstrom)	Intensity Arc	Spark	Plasma
	IR	3798.272	8		
--	BR	3798.280			6
	TI I	3798.314	10	6	
	YB	3798.440	4		
	F II	3798.460			6
	CE	3798.510	3	3	
	FE I	3798.513	400	320	
---	TM	3798.550	16		
	TB	3798.590	16		
	CE	3798.624	3	3	
	DY	3798.650	2		
	ER	3798.650	4		
	V I	3798.661	6	5	
	HF	3798.662	5	3	
	IR I	3798.666	5		
	EU	3798.710	2		
	TM	3798.760	20	10	
--	CL II	3798.800			50
	U	3798.837	16		
	RU I	3798.901	63	100	
	W	3798.922	6	10	
	TB	3798.950	8	8	
	ZN I	3799.002	5		
	EU II	3799.010	100		
	CE II	3799.038	4		

Table IIb.(3) RU I 3798.052 A

Element	Wavelength (Angstrom)	Intensity Arc	Spark	Plasma
U	3793.104	13	20	
RH I	3793.217	200	63	
CR I	3793.289	50	32	
HF II	3793.372	10	20	
NI I	3793.608	50		
V I	3793.614	32	16	
PR	3793.789	40	32	

Table IIb.(3) RU I 3798.052 A
(con't)

Element	Wavelength (Angstrom)	Intensity Arc	Spark	Plasma
IR	3793.792	32	10	
CR I	3793.879	50	32	
OS I	3793.909	130	320	
SM II	3793.969	10	32	
IR	3794.056	32	6	
TB	3794.230	32		
FE	3794.340	79	50	
SM	3794.347	32	2	
V	3794.358		25	
PR	3794.379	50	25	
CR I	3794.608	50	32	
OS	3794.662	40	10	
LI I	3794.722	63		
LA II	3794.773	400	200	30 VAF
RU	3794.924	20	32	
V I	3794.964	50	50	
FE I	3795.004	500	400	
TM	3795.170	40	6	
TA	3795.185		32	
PR	3795.340	25	25	
NB	3795.542	10	20	
MO	3795.590		20	
OS	3795.667	40	13	
PR	3795.765	32	20	
TM II	3795.765	250	160	26 VAF
RE	3795.800	40		
TA	3796.208		50	
GD	3796.393	160	160	
V I	3796.470	32	13	
MO	3796.570		20	
RE I	3796.600	63		
HO II	3796.730	20	40	12 VAF
PR	3796.925	40	20	
CR I	3797.126	50	32	
FE	3797.517	320	200	
RE I	3797.591	40		
CR I	3797.716	100	20	
H I	3797.910			0.1 VAF
HF II	3797.923	25	25	
*RU I	3798.052	32	40	

Table IIb.(3) RU I 3798.052 A
(con't)

Element	Wavelength (Angstrom)	Intensity Arc	Spark	Plasma
NB	3798.121	50	79	
MO I	3798.252	1000	1000	
FE I	3798.513	400	320	
RU I	3798.901	63	100	
EU II	3799.010	100		
PD I	3799.190	200	160	
MN	3799.259	50	50	
RH I	3799.311	100	25	
RU I	3799.347	63	100	
SM II	3799.546	32	10	
FE I	3799.549	400	320	
PR	3799.679	50	25	
V I	3799.912	63	50	
ND	3800.026	32	20	
IR	3800.123	160	100	
RU I	3800.261	13	40	
PR	3800.303	100	50	
OS	3800.437	50	16	
MN	3800.552	63	63	
SM II	3800.890	20	25	
SN	3800.996	200	160	
NB	3801.154	3	20	
CR	3801.200	32	3	
PR	3801.348	32	10	
ND	3801.373	63	40	
CE II	3801.529	25	3	4 VAF
FE	3801.682	50	25	
MO	3801.837	20	25	
MN	3801.907	20	20	
TB	3802.170	32	3	
PR	3802.763	32	16	
NB	3802.923	50	50	

Table IIa. SB I 2175.810 A

	Element	Wavelength (Angstrom)	Intensity Arc	Spark	Plasma
	W	2174.822		8	
	FE II	2174.862		8	
	CO II	2174.930		2	
	BE I	2174.942	10		3 VAF
	CU II	2174.968	3	40	
	IR	2175.014		50	
----	BE I	2175.069	25	3	3 VAF
	K	2175.120			10
	NI II	2175.156	16	25	
	IR I	2175.245	32	4	
---	RE I	2175.346	20	6	
	HF II	2175.360	25	32	
	W	2175.360		8	
	YB	2175.380		3	
	HG	2175.400			2
	MO	2175.400		5	
	FE I	2175.447	8	25	
	W II	2175.510	6	3	
	NB	2175.551	5		
--	MN	2175.560		40	
	PB	2175.579	40		
	RE I	2175.806	20	10	
	*SB I	2175.810	320	40	68 VAF
	ZR II	2175.820		3	
	V	2175.835	2	5	
	NB	2175.843	5	25	
--	FE	2176.030	3		
	RE I	2176.206	200	50	
	XE	2176.220			2
---	AS I	2176.260	8		
	AU I	2176.330	16		
	FE	2176.400	2		
	W	2176.419	2	13	
	PT I	2176.460	10	2	
	ND	2176.470		16	
	CO	2176.494	4		
----	FE	2176.516		8	
	BI I	2176.618	20	4	
	OS	2176.670	16	3	
	NB	2176.762	8	32	

Table IIb.(3) SB I 2175.810 A

Element	Wavelength (Angstrom)	Intensity Arc	Spark	Plasma
AG II	2170.862		32	
IR	2172.210	32	20	
OS	2172.310	5	20	
PD II	2172.383		20	
PT	2172.388	2	25	
OS	2172.840	16	20	
MO	2172.917	5	20	
CO II	2173.340	2	20	
HF II	2173.440	16	20	
CO II	2174.540	16	25	
NI II	2174.669	13	32	13 VAF
PT I	2174.670	32	32	36 VAF
BE I	2174.942	10		3 VAF
CU II	2174.968	3	40	
IR	2175.014		50	
BE I	2175.069	25	3	3 VAF
NI II	2175.156	16	25	
HF II	2175.360	25	32	
FE I	2175.447	8	25	
MN	2175.560		40	
*SB I	2175.810	320	40	68 VAF
NB	2175.843	5	25	
RE I	2176.206	200	50	
NB	2176.762	8	32	
PT II	2176.872	3	25	
PD II	2176.900		25	
NI II	2177.088	3	20	
RH	2177.130	3	25	
NI II	2177.360	3	20	
RE II	2177.579		20	
RE II	2177.851	2	100	
TA	2178.030	16	32	
FE I	2178.090	100	20	
HF II	2178.900	63	79	
CU I	2178.944	32	13	17 VAF
RH	2178.950	13	32	
IR	2178.980	10	50	
SB I	2179.257	32	40	19 VAF
MO	2179.360		20	
CU II	2179.399	13	32	

Table IIb.(3) SB I 2175.810 A
(con't)

Element	Wavelength (Angstrom)	Intensity Arc	Spark	Plasma
TA	2179.540	6	20	
NI II	2180.470	4	40	
CE	2180.670		100	

Table IIa.   SB I   2068.380 A

	Element	Wavelength (Angstrom)	Intensity Arc	Spark	Plasma
	TA	2067.400	6	13	
	SB	2067.500			25
	PT I	2067.500	20	25	
	HF	2067.600		2	
	RE	2067.600	5	5	
----	GE	2067.630	2		
	OS	2067.710	10	4	
	IR	2067.870	5	4	
	W	2067.870	2	10	
---	RE I	2067.890	20	8	
	S	2067.930			8
	TE	2067.940			16
	ZR II	2068.090		2	
--	RH	2068.140		16	
	PT II	2068.160	2	25	
	IR	2068.250	10	2	
	TL	2068.300		2	
	RH	2068.320	8		
	NI	2068.350	2		
	OS	2068.380	10	10	
	*SB I	2068.380	320	3	91 VAF
	V	2068.530		8	
	HF	2068.580	5		
	SN I	2068.580			
	NI I	2068.606	10		
--	PT II	2068.629	5	16	
	ER	2068.650	5		
	GE I	2068.650	5	200	50 VAF
	PD	2068.670		8	
	IR	2068.730	3	2	
	HF II	2068.760	3		
	BH C	2068.800	32		
	PD I	2068.800	13		
	V	2068.800		25	
---	HF	2068.840	20	4	
	BI II	2068.990	2	10	
	CO I	2068.990	10	3	
	NI	2069.023	16	4	
	IR	2069.070	10	8	
---	TA	2069.120	3	16	

Table IIa. SB I 2068.380 A
(con't)

Element	Wavelength (Angstrom)	Intensity Arc	Spark	Plasma
I	2069.240			20

Table IIb.(3) SB I 2068.380 A

Element	Wavelength (Angstrom)	Intensity Arc	Spark	Plasma
NI I	2063.390	20	2	
SB	2063.440	32	13	
OS	2063.550	25	2	
IR	2063.590	5	16	
ZN II	2063.610	2	20	
W	2063.670		3	
CO II	2063.760	13	32	
IR	2063.890		25	
ZR II	2063.890		3	
PR	2064.070		13	
RH	2064.170	3	100	
NB	2064.210	2	32	
GA	2064.300		2	
NI I	2064.366	16	2	
ZR	2064.370		2	
IR	2064.550	5	5	
V	2064.750		2	
RE I	2064.762	32	5	
IR	2064.770		16	
TA	2064.770	3	13	
HF II	2064.780	16	4	
BI	2064.790	50	4	
IR	2065.090		25	
W	2065.090	10	3	
GE	2065.200	4	50	
ZR II	2065.350		2	
AS I	2065.410	20	3	
CR II	2065.420	50	160	31 VAF
LU	2065.420		32	
CO II	2065.530	10	32	
HF II	2065.550	2	4	

Table IIb.(3) SB I 2068.380 A
(con't)

Element	Wavelength (Angstrom)	Intensity Arc	Spark	Plasma
W II	2065.570	10	20	
IR	2065.590	3	13	
RH	2065.700		25	
NB	2065.710	5	20	
V II	2065.750		25	
IR	2065.790		79	
OS	2065.800	5	2	
AG II	2065.910	4	79	
IR	2066.210		40	
CU II	2066.320		20	
TL	2066.400		2	
NI II	2066.410	2	25	
V	2066.470		3	
RH	2066.780		13	
TL	2066.800		32	
V	2066.830		4	
RE	2066.900	6	5	
PT	2066.920	16	6	
ZR II	2067.080		3	
V	2067.160		3	
IR	2067.170		32	
OS II	2067.210	10	40	22 VAF
TA	2067.400	6	13	
PT I	2067.500	20	25	
HF	2067.600		2	
RE	2067.600	5	5	
OS	2067.710	10	4	
IR	2067.870	5	4	
W	2067.870	2	10	
RE I	2067.890	20	8	
ZR II	2068.090		2	
RH	2068.140		16	
PT II	2068.160	2	25	
IR	2068.250	10	2	
TL	2068.300		2	
OS	2068.380	10	10	
*SB I	2068.380	320	3	91 VAF
V	2068.530		8	
PT II	2068.629	5	16	
GE I	2068.650	5	200	50 VAF

Table IIb.(3) SB I 2068.380 A
(con't)

Element	Wavelength (Angstrom)	Intensity Arc	Spark	Plasma
PD	2068.670		8	
IR	2068.730	3	2	
V	2068.800		25	
HF	2068.840	20	4	
BI II	2068.990	2	10	
CO I	2068.990	10	3	
NI	2069.023	16	4	
IR	2069.070	10	8	
TA	2069.120	3	16	
RE	2069.440	5	10	
IR	2069.480		10	
NI I	2069.499	20	2	
OS	2069.610	25	4	
OS	2069.750	5	2	
AS I	2069.830	20	3	
AG I	2069.830	10	3	
SN	2069.930		2	
TA	2069.960	2	10	
W	2070.020	4	4	
IR	2070.210	2	5	
OS	2070.420	13	2	
IR	2070.510	6	63	
AIR	2070.600		5	
LU	2070.660		3	
OS II	2070.670	16	32	14 VAF
V	2070.780		5	
PT I	2070.930	20	20	
MO	2071.080	6	20	
HF	2071.100		6	
W II	2071.210	10	25	
OS	2071.280	4	10	
PT II	2071.553		20	
RH	2071.760		20	
W	2071.780	10	5	
IR	2071.810	6	4	
PR	2071.880		2	
IR	2071.940	16	8	
RH	2071.940	3	25	
LU	2071.950		2	
PD II	2072.030		25	

Table IIb.(3) SB I 2068.380 A
(con't)

Element	Wavelength (Angstrom)	Intensity Arc	Spark	Plasma
IR	2072.290	4	2	
RH	2072.360	2	2	
V	2072.430		10	
RE	2072.620	6	10	
RH	2072.850		20	
MN	2072.890	10	6	
SN	2072.890	10	6	
IR	2072.960	20	10	
TA	2073.020	3	13	
CO	2073.270	10	2	
RE I	2073.273	20	10	

Table IIa. SC II 3613.836 A

	Element	Wavelength (Angstrom)	Intensity Arc	Intensity Spark	Intensity Plasma
	CE	3612.838	3		
	CL	3612.860			10
	IN	3612.860		16	
	TH	3612.869	5		
	ER	3612.870	10	2	
	CD I	3612.875	790	500	
	TE	3612.900			5
	FE I	3612.945	20	4	
	NB	3613.013	2	2	
	S II	3613.030			13
	DY	3613.060	6	4	
	TB	3613.060	16	3	
	XE I	3613.060			8
	LA I	3613.077	13	3	
----	ER	3613.096	6		
	ZR II	3613.100	40	40	
	FE I	3613.148	6	2	
	U	3613.152	3	2	
	CR	3613.183	5	8	
	NB	3613.239	2	5	
	ND	3613.244	10	4	
	OS	3613.329	32	13	
	HO	3613.330	6	8	
---	MO	3613.374	5	5	
	GD	3613.400	8	6	
	U	3613.403	4	3	
	TI I	3613.435	8		
	ZR II	3613.451	6		
	NB	3613.452	5	3	
	FE	3613.452	10	2	
	SM	3613.521	2	2	
	W	3613.549		6	
--	SM	3613.595	3	2	
	HG	3613.607			40
	FE	3613.609	3		
	HE I	3613.641			32
	MO	3613.642	3	4	
	CR	3613.669	10	8	
	TB	3613.680	16	3	
	PR	3613.700	6	2	

Table IIa. SC II 3613.836 A
(con't)

Element	Wavelength (Angstrom)	Intensity Arc	Spark	Plasma
CE II	3613.701	20	5	
CE I	3613.701	20	5	
ZR I	3613.702	10		
U	3613.748	4		
TI	3613.756	13		
EU	3613.760	2		
CU I	3613.761	63	6	
TH	3613.779	5	3	
W II	3613.790	10	32	
BH LA	3613.800	3		
MG II	3613.800	4		
I	3613.810			10
BI	3613.817		32	
*SC II	3613.836	40	63	200 VAF
SM	3613.902	2		
RH I	3613.939	2		
U	3613.997		2	
AU	3614.000		20	
TH	3614.011	6	8	
CE	3614.029	5		
DY	3614.083	32	10	
EU	3614.084	2	2	
FE	3614.116	10	3	
BH LA	3614.200	3		
TI	3614.207	32	4	
CU I	3614.218	50	6	
CE	3614.233	3		
MO	3614.253	50	32	
W	3614.255	5	2	
CE	3614.362	3		
BH LA	3614.400	4		
GD	3614.410	3		
CD I	3614.450	63	100	
IR	3614.454	2	2	
RH	3614.511	4	2	
FE	3614.561	16	6	
TB	3614.630	32	8	
ER	3614.637	13		
MO	3614.686		20	
EU	3614.700	3		

Table IIa. SC II 3613.836 A
(con't)

Element	Wavelength (Angstrom)	Intensity Arc	Spark	Plasma
DY	3614.707	16	5	
FE	3614.715	6		
SM	3614.743	4		
ZR II	3614.774	40	79	
RH I	3614.775	16	10	
BH LA	3614.800	5		
W	3614.801	6	2	

Table IIb.(3) SC II 3613.836 A

Element	Wavelength (Angstrom)	Intensity Arc	Spark	Plasma
FE I	3608.861	500	400	
V I	3609.289	32	32	
NI I	3609.314	200	16	
TH II	3609.447	13	10	27 VAF
SM II	3609.484	63	100	52 VAF
ND	3609.495	25	32	
PD I	3609.548	1000	630	35 VAF
CE	3609.687	40	10	
TI I	3610.156	100	63	
FE I	3610.162	100	100	
MN	3610.299	63	40	
NI I	3610.462	1000		
RE	3610.493	40		
CD I	3610.510	1000	500	1 VAF
Y II	3611.047	40	63	40 VAF
YB	3611.307	13	50	
TB	3611.330	50	8	
CS I	3611.520	200		
EU	3611.580	40	5	
ZR II	3611.893	16	40	
FE I	3612.074	79	50	
AL	3612.467		79	
RH I	3612.470	200	50	
NI I	3612.741	400	50	
CD I	3612.875	790	500	

Table IIb.(3) SC II 3613.836 A
(con't)

Element	Wavelength (Angstrom)	Intensity Arc	Spark	Plasma
ZR II	3613.100	40	40	
CU I	3613.761	63	6	
W II	3613.790	10	32	
BI	3613.817		32	
*SC II	3613.836	40	63	200 VAF
CU I	3614.218	50	6	
MO	3614.253	50	32	
CD I	3614.450	63	100	
ZR II	3614.774	40	79	
PR	3615.159	40	4	
SM II	3615.244	40	2	
NB	3615.498	32	32	
TB	3615.660	50	16	
EU	3616.154	40	10	
OS I	3616.574	160	20	
V	3616.723	32	32	
HF I	3616.892	2000	200	
RE I	3617.080	50		
IR I	3617.212	50	16	
CS I	3617.410	63		
FE	3617.788	130	79	
DY	3618.522	79	10	
FE I	3618.769	400	400	

Table IIa.   SC II 3572.523 A

	Element	Wavelength (Angstrom)	Intensity Arc	Spark	Plasma
	U	3571.560	3	2	
	TH	3571.575	8	8	
	V I	3571.653	40	32	
	EU	3571.660	2		
	DY	3571.670	3	2	
	F II	3571.680			6
	FE	3571.680	2		
	U	3571.691	2	6	
	RE	3571.706	5		
	I	3571.720			5
	ER	3571.730	16		
	SM	3571.769	5		
----	PR	3571.787	10	3	
	TA	3571.854	20	3	
	NI I	3571.869	1000	40	
	GD	3571.941	25	20	
	FE I	3571.996	100	79	
	PT II	3571.999		16	
	RU	3572.015	3	5	
	W II	3572.020	4	10	
---	TB	3572.070	32	8	
	EU	3572.110	3		
	ER	3572.117	5		
	PR	3572.240	5	2	
--	AR I	3572.290			2 VAF
	TH	3572.399	8	8	
	ND	3572.405	25	6	
	CE	3572.428	13	5	
	ZR II	3572.473	63	79	30 VAF
	W II	3572.476	10	32	
	*SC II	3572.523	32	50	150 VAF
	MO	3572.592		20	
	EU	3572.600	3		
	V I	3572.628	5	2	
	ZN	3572.650	5		
	KR II	3572.680			16
	PB	3572.734	200	20	
	CR I	3572.748	25	5	
	IN	3572.750		6	
	ND	3572.751	10	10	

Table IIa. SC II 3572.523 A
(con't)

	Element	Wavelength (Angstrom)	Intensity Arc	Spark	Plasma
--	CE	3572.754	5		
	RE I	3572.805	2		
	SM	3572.825		2	
	W	3572.860		6	
--	U	3572.931	6		
	DY	3573.030	3		
	ER	3573.077	10		
	ZR II	3573.084	10	10	
	NB	3573.094	2	5	
	ND	3573.183	10	4	
	HO	3573.220	8	10	
	TH	3573.222	5	6	
--	CS	3573.240			4
	RN	3573.300			2
	SM	3573.396	5	2	
	FE	3573.400	50	20	
	W	3573.411	8	6	
	TA	3573.438	16	63	
	EU	3573.450	3		
	TH	3573.511	3	3	
	V I	3573.516	20	20	

Table IIb.(3) SC II 3572.523 A

Element	Wavelength (Angstrom)	Intensity Arc	Spark	Plasma
AR I	3567.657			9 VAF
SC II	3567.701	16	40	37 VAF
LU	3567.840	100	6	
NB	3567.995	2	50	
RE	3568.231	40		
SM II	3568.258	40	50	39 VAF
NB	3568.506	10	50	
TB II	3568.510	50	50	46 VAF
FE I	3568.979	50	32	
DY	3568.994	100		
HF II	3569.041	20	50	

Table IIb.(3) SC II 3572.523 A
(con't)

Element	Wavelength (Angstrom)	Intensity Arc	Spark	Plasma
CO I	3569.379	400	100	
DY	3569.669	40	10	
OS	3569.775	100	32	
FE I	3570.097	320	320	
RH I	3570.182	400	160	
RE	3570.256	32		
FE	3570.258	50	16	
RU	3570.594	13	63	
V I	3571.037	32	20	
PD I	3571.155	40	40	
FE I	3571.227	40	6	
V I	3571.653	40	32	
NI I	3571.869	1000	40	
FE I	3571.996	100	79	
TB	3572.070	32	8	
AR I	3572.290			2 VAF
ZR II	3572.473	63	79	30 VAF
W II	3572.476	10	32	
*SC II	3572.523	32	50	150 VAF
PB	3572.734	200	20	
FE	3573.400	50	20	
TA	3573.438	16	63	
CR I	3573.643	63	16	
IR I	3573.724	8	100	
TI II	3573.737	20	40	
DY	3573.838	79	79	
MO	3573.882	20	25	
FE	3573.888	40	32	
CR I	3574.039	50	16	
TM	3574.060	20	32	
OS	3574.079	32	5	
DY	3574.160	200	100	
V	3574.346		40	
LA I	3574.426	32	8	
GD	3574.738	25	25	
CO I	3574.964	200	25	
ND	3574.977	32	8	
CO I	3575.361	200	25	
FE	3575.375	40	32	
ZR I	3575.792	100	5	

Table IIb.(3) SC II 3572.523 A
(con't)

Element	Wavelength (Angstrom)	Intensity Arc	Spark	Plasma
NB	3575.848	50	79	
FE I	3575.980	79	25	
DY II	3576.250	320		6 VAF
SC II	3576.340	20	50	82 VAF
FE	3576.760	79	40	
NI II	3576.762	2	40	
ZR II	3576.854	16	25	10 VAF
DY	3576.873	200	50	
CE	3577.458	320	13	
PR	3577.465	32	8	

Table IIa. SE I 2039.851 A

	Element	Wavelength (Angstrom)	Intensity Arc	Spark	Plasma
	V	2038.872	10	2	
	PD	2038.930		20	
	IR	2038.990		25	
	GA	2039.000		5	
	RE	2039.010	20		
	V	2039.073		2	
	HF	2039.080	6	2	
----	W	2039.110	4	6	
	IR	2039.140	5	5	
	RE	2039.230	20		
	XE	2039.240			3
	CR	2039.310	25		
	V	2039.310		16	
	RH	2039.350		32	
---	W	2039.380	4	3	
	IR	2039.430	25	13	
	SN	2039.500		200	
	PT	2039.570	5		
	SB II	2039.600	16	16	7 VAF
--	RH	2039.670		25	
	PT	2039.700	13	20	
	IR	2039.790	8	40	
	TE	2039.790	320		
	W	2039.800		10	
	PD I	2039.830	4		
	*SE I	2039.851			40 VAF
	RE	2039.910	3	10	
	AL II	2039.930			16
--	ZN II	2040.000			250
	ZR	2040.110	2		
	PD	2040.180	8		
	RH	2040.190	2	100	
	GA	2040.200		5	
	PT	2040.330	16	16	
---	XE	2040.350			2
	CD II	2040.440			3
----	IR	2040.480	2	5	
	TA	2040.620	2	8	
	OS	2040.680	10	4	
	IR	2040.760	6	20	

Table IIa. SE I 2039.851 A
(con't)

Element	Wavelength (Angstrom)	Intensity Arc	Spark	Plasma
RE II	2040.810	3	10	

Table IIb.(1) SE I 2039.851 A

Element	Wavelength (Angstrom)	Intensity Arc	Spark	Plasma
W II	2035.030	6	13	38 VAF
PT II	2036.463	16	40	54 VAF
MO II	2038.440	20		24 VAF
SB II	2039.600	16	16	7 VAF
TE	2039.790	320		
*SE I	2039.851			40 VAF
GE I	2041.690	5	160	34 VAF
BI	2041.960	100	16	
GE I	2043.760	4	63	35 VAF
IR	2044.190	10	100	30 VAF

Table IIa. SI I 2516.123 A

	Element	Wavelength (Angstrom)	Intensity Arc	Spark	Plasma
	U	2515.147		10	
	V I	2515.149	20	10	
	TA	2515.154	6		
	AU II	2515.160		5	
	TB	2515.230		10	
	RU	2515.285	63	2	
	RH	2515.325	2	25	
	PD II	2515.330			10
	W	2515.330	2	13	
	CE	2515.360	2		
	IR I	2515.364	20	3	
----	NA II	2515.460	3		5
	RE	2515.471	10		
	HF II	2515.486	50	400	
	W	2515.498		5	
	U	2515.524	3	2	
	PT I	2515.576	500	20	
	AR II	2515.600			10
	YB	2515.600		3	
---	V	2515.649	6		
	MO	2515.660	10	10	
	BI I	2515.686	100	25	
	V	2515.723		10	
	U	2515.729	6	2	
	RH I	2515.746	63	10	
	RU	2515.759	6	8	
	LA II	2515.790		4	
	W II	2515.798		10	
	ZN I	2515.807	160	20	
--	RU	2516.006	20	5	
	BR	2516.050			3
	AIR	2516.100		6	
	MO	2516.109	25	25	
	DY	2516.110	5		
	TA	2516.111	6	2	
	RE	2516.116	130		
	V	2516.118	25	100	
	XE	2516.120			6
	*SI I	2516.123	500	500	250 VAF
	FE I	2516.250	2		

Table IIa. SI I 2516.123 A
(con't)

	Element	Wavelength (Angstrom)	Intensity Arc	Spark	Plasma
	CD II	2516.340		25	
--	YB	2516.362	2	5	
	IN	2516.400		3	
	CL	2516.420			2
	TE	2516.490			16
	SE	2516.570			16
	FE I	2516.570	10		
---	W	2516.577	13	3	
	CR	2516.650		6	
	RU	2516.702	6	8	
	TA	2516.717	5		
	MN	2516.730		13	
	AR II	2516.810			20
---	YB	2516.830	2	20	
	HF II	2516.881	32	100	
	CR I	2516.916	32	2	
	U	2516.969	4	2	
	W	2517.011		4	
	CA	2517.030		6	
	FE II	2517.120	10	63	
	U	2517.122	8	2	

Table IIb.(2) SI I 2516.123 A

Element	Wavelength (Angstrom)	Intensity Arc	Spark	Plasma
C II	2512.030		400	
TA	2512.036	5	790	
HF II	2512.702	200	790	150 VAF
HF II	2512.963	100	320	
HF II	2513.033	200	1000	150 VAF
SI I	2514.331	320	200	79 VAF
PD II	2514.478		200	
HF II	2515.486	50	400	
PT I	2515.576	500	20	
*SI I	2516.123	500	500	250 VAF
RH	2517.523		160	

Table IIb.(2) SI I 2516.123 A
(con't)

Element	Wavelength (Angstrom)	Intensity Arc	Spark	Plasma
SI I	2519.207	320	320	61 VAF
CO II	2519.822	40	200	
RH II	2520.533	10	1000	39 VAF

Table IIa. SI I 2881.595 A

Element	Wavelength (Angstrom)	Intensity Arc	Spark	Plasma
W	2880.627	6	6	
CE	2880.637	20		
LA II	2880.644	4	40	
RH I	2880.648	10	2	
NB	2880.715	4	50	
FE II	2880.756	16	50	
CD I	2880.770	200	130	
V	2880.800	2	20	
FE II	2880.827		25	
ZR I	2880.830	5		
CE	2880.849	2		
CR II	2880.869	20	25	
CA	2880.870		2	
CE	2880.928	2		
HO	2880.990	20	10	
U	2881.017	2	2	
DY	2881.058	2		
W	2881.070	5		
CE	2881.131	13		
NA II	2881.140	8		1 VAF
CR	2881.141	25	2	
TH	2881.147	10	10	
IR	2881.158	16	3	
CS	2881.160			20
CD I	2881.230	50		32
TA	2881.232	32	5	
NI II	2881.247		8	
RH I	2881.254	20		
RU	2881.276	32	3	
TB	2881.310	10		
GD	2881.328	2	3	
IR	2881.357	4		
MO	2881.370		10	
CE	2881.420	4		
AL II	2881.463			32
CO	2881.580	4		
DY	2881.590	2		
*SI I	2881.595	500	400	110 VAF
IN	2881.610		2	
IR I	2881.643	6		

Table IIa. SI I 2881.595 A
(con't)

	Element	Wavelength (Angstrom)	Intensity Arc	Spark	Plasma
	O	2881.700			10
	IR	2881.737	3		
	CE	2881.772	2		
--	CA	2881.800		4	
	U	2881.909	4	2	
	YB	2881.920		3	
	CR	2881.931		32	
	MO	2881.937		3	
	TI I	2881.951	2		
	I	2882.010			20
	TH	2882.014	10	13	
	MO	2882.036		3	
---	ZR II	2882.087	2	3	
	RU	2882.116	32	200	
	YB	2882.150		2	
	ZN II	2882.150			25
	W	2882.192	3		
	AG II	2882.200		10	
	CO I	2882.220	32		
	RE	2882.235	6		
	TA	2882.332	3	79	
----	U	2882.345	6	8	
	RH	2882.366	79	10	
	MO	2882.378		25	
	RN	2882.400			3
	NB	2882.474		5	
	V II	2882.501	32	200	
	TH	2882.516	8	8	
	MO	2882.543	6		
	I	2882.557			8
	RU	2882.584	4		
	U	2882.587	8	6	

Table IIb.(2) SI I 2881.595 A

Element	Wavelength (Angstrom)	Intensity Arc	Spark	Plasma
FE II	2876.802		100	
NB	2876.947	40	500	
FE I	2877.301	200	130	
TI II	2877.436	32	100	
PT II	2877.520	40	200	
V	2877.688	16	100	
SB I	2877.915	250	160	5 VAF
CR II	2877.978	32	100	
MO	2879.047	16	100	
V II	2880.030	25	160	
CD I	2880.770	200	130	
NA II	2881.140	8		1 VAF
*SI I	2881.595	500	400	110 VAF
RU	2882.116	32	200	
V II	2882.501	32	200	
U II	2882.741	20	20	6 VAF
CU I	2882.934	320		
NB II	2883.178	100	790	31 VAF
FE II	2883.702		320	
TI II	2884.107	32	130	
V II	2884.785	40	200	

Table IIa. SI I 2124.120 A

	Element	Wavelength (Angstrom)	Intensity Arc	Spark	Plasma
	RH	2123.130	3	10	
	OS	2123.150	5	8	
	IR	2123.210	10		
	MO	2123.300		25	
	YB	2123.300		16	
	S	2123.320			16
	V II	2123.325	5	32	
	OS	2123.340	10		
----	KR	2123.480			3
	ZN II	2123.500			3
	TI	2123.540	6		
	V II	2123.600	2	8	
---	IR	2123.640	2	2	
	PD	2123.669		25	
	RU	2123.670	4		
	HF II	2123.680	40	40	
	PD	2123.760	3		
	SI	2123.770			2
	GE I	2123.820	2		
	OS	2123.840	40	25	
--	IR	2123.920	4		
	RH	2123.990	5	25	
	ZN II	2124.040			2
	TA	2124.080	2	6	
	S	2124.090			25
	CU I	2124.100	2		
	MO	2124.100	6	20	
	NI	2124.110	32		
	V	2124.110	6	5	
	*SI I	2124.120	200	50	180 VAF
	RH	2124.130	40	10	
	PD II	2124.330		4	
--	NB	2124.340	2	8	
	OS	2124.390	25	5	
	RU	2124.390	4		
	HF II	2124.590	50	79	
---	W	2124.600		8	
	LU	2124.640		2	
	XE	2124.670			3
	RU	2124.730	4		

Table IIa. SI I 2124.120 A
(con't)

Element	Wavelength (Angstrom)	Intensity Arc	Spark	Plasma
GE	2124.750	3		
NI I	2124.810	25	3	
PD	2124.818	2	10	
MO	2124.820	2	16	
OS	2124.820	5		
FE	2125.010	32		
CE	2125.040		10	
CU II	2125.098		20	
NI II	2125.110	4	25	
CO	2125.116	10		

Table IIb.(3) SI I 2124.120 A

Element	Wavelength (Angstrom)	Intensity Arc	Spark	Plasma
V	2119.150		25	
YB	2119.250		32	
OS	2119.790	130	32	
TA	2119.920	8	32	
RH	2120.060	3	25	
AG II	2120.450	5	79	
TA	2120.900	8	25	
PR	2122.350		32	
CU II	2122.966	16	50	
MO	2123.300		25	
V II	2123.325	5	32	
PD	2123.669		25	
HF II	2123.680	40	40	
OS	2123.840	40	25	
RH	2123.990	5	25	
*SI I	2124.120	200	50	180 VAF
HF II	2124.590	50	79	
NI II	2125.110	4	25	
NB	2125.206	16	40	
AU	2125.340		32	
RH	2125.370		32	
AG II	2125.520		25	

Table IIb.(3) SI I 2124.120 A
(con't)

Element	Wavelength (Angstrom)	Intensity Arc	Spark	Plasma
NI II	2125.900	4	25	
MO	2125.910	2	32	
CU II	2126.028	16	32	
NB	2126.543	16	50	
YB II	2126.720	40	200	32 VAF
NI II	2126.800	2	25	
IR II	2126.810	25	200	100 VAF
PT II	2127.417	2	25	
OS	2127.970	5	50	
ER	2128.070		250	
NI II	2128.570	5	32	
PT I	2128.610	32	25	24 VAF
HF II	2129.100	63	100	

Table IIa. SM II 3592.595 A

Element	Wavelength (Angstrom)	Intensity Arc	Spark	Plasma
OS	3591.602	2	5	
TB	3591.660	16	3	
SM	3591.680	2		
MO	3591.690		25	
ZR I	3591.725	8		
SM II	3591.741		2	
CE	3591.745	4		
U	3591.747	10	2	
CO I	3591.749	3		
W	3591.769	6	3	
EU	3591.790	3		
NB	3591.792	5	5	
MN	3591.809	16		
DY	3591.821	79	20	
-- U	3591.884	6	10	
GD	3591.907	8	4	
W	3591.972	6	4	
AU	3591.990	16	5	
MO	3592.015	5	3	
V II	3592.024	50	320	
U	3592.065	4	3	
ND	3592.073	8	8	
-- DY	3592.117	79	32	
TA	3592.140	2		
FE	3592.209	6		
HO	3592.220	6	10	
MO	3592.240		16	
U	3592.301	4	4	
OS	3592.317	20	13	
-- BI II	3592.400			5
W II	3592.423	10	32	
CS	3592.480			4
FE	3592.486	3		
TA	3592.495	2		
V	3592.533	40		
ND	3592.595	20	32	
*SM II	3592.595	40	50	69 VAF
MO	3592.650		20	
FE I	3592.692	13	2	
GD	3592.696	50	63	

Table IIa. SM II 3592.595 A
(con't)

	Element	Wavelength (Angstrom)	Intensity Arc	Spark	Plasma
	TH	3592.776	2	6	
	XE I	3592.800			2
--	U	3592.801		6	
	W	3592.846	6	3	
	FE I	3592.890	3		
	SM	3592.905	4		
	Y I	3592.913	79	25	
	PB	3592.921		3	
	EU	3592.927	6		
	HO	3592.950	6		
	HG I	3592.970		2	
	W	3592.979	5	3	
	RU I	3593.022	63	160	
	PR	3593.036	5	2	
---	TI II	3593.093	5	32	
	BH ZR	3593.100	2		
	TB	3593.100	16	8	
	PB	3593.120		32	
	ZR I	3593.129	6		
	EU	3593.130	2		
	HO	3593.130	6	6	
	CE II	3593.134	8		
	DY	3593.150	6	4	
	U	3593.198	10		
	SR II	3593.214	3	2	
	K II	3593.220			5
	BA I	3593.282	3		
	LA	3593.290	2		
	FE I	3593.329	6	2	
	V II	3593.334	32	320	
----	TE	3593.340			5
	RE	3593.397	16		
	SM II	3593.402	2	3	
	GD	3593.432	16	16	
	HG	3593.480			10
	CR I	3593.488	500	400	
	NE I	3593.526			500
	RH I	3593.530	10	2	
	NB	3593.547	16	3	
	W	3593.561	4	3	

Table IIb.(3) SM II 3592.595 A

Element	Wavelength (Angstrom)	Intensity Arc	Spark	Plasma
FE I	3587.758	50	25	
NI I	3587.931	200	13	
NB	3589.107	50	32	
FE I	3589.107	63	32	
RU I	3589.215	63	100	
NB	3589.356	100	100	
FE I	3589.456	50	32	
MO	3589.520		32	
SC II	3589.635	5	13	26 VAF
V II	3589.760	79	630	
PB I	3589.920		40	
SC II	3590.475	20	13	20 VAF
CE	3590.598	50		
NB	3591.198	2	50	
DY	3591.423	200	100	
RB I	3591.590	79	20	
MO	3591.690		25	
DY	3591.821	79	20	
V II	3592.024	50	320	
DY	3592.117	79	32	
W II	3592.423	10	32	
V	3592.533	40		
ND	3592.595	20	32	
*SM II	3592.595	40	50	69 VAF
GD	3592.696	50	63	
Y I	3592.913	79	25	
RU I	3593.022	63	160	
TI II	3593.093	5	32	
PB	3593.120		32	
V II	3593.334	32	320	
CR I	3593.488	500	400	
NB	3593.966	79	50	
IR	3594.391	16	32	
FE I	3594.636	130	100	
CO I	3594.872	200		
DY	3595.046	200	100	
MN	3595.119	50	25	
TA	3595.638	63	5	
TI II	3596.052	50	130	
DY	3596.067	50	20	

Table IIb.(3) SM II 3592.595 A
(con't)

Element	Wavelength (Angstrom)	Intensity Arc	Spark	Plasma
BI I	3596.110	160	50	
RU I	3596.185	320	100	
RH I	3596.194	200	50	
MO	3596.351		32	
TB	3596.380	50	16	
FE	3597.061	40	10	
RH I	3597.147	200	100	
SB	3597.510	2	200	

Table IIa. SM II 3568.258 A

Element	Wavelength (Angstrom)	Intensity Arc	Spark	Plasma
TH	3567.266	2	2	
DY	3567.300	4		
TA	3567.350	4		
TB	3567.350	16	16	
TM	3567.360	10		
HF	3567.364	20	10	
FE I	3567.383	10	2	
MO	3567.465		3	
U	3567.486	3		
TL II	3567.560			10
ER	3567.576	8		
AR I	3567.657			320
W	3567.678	6	6	
SC II	3567.701	16	40	37 VAF
TH	3567.707	3	3	
FE	3567.720	2		
EU	3567.723	3		
BA II	3567.730	2	3	
MO	3567.735	2	3	
LU	3567.840	100	6	
TB	3567.860	8		
TH	3567.978	5	5	
CL	3567.980			6
NB	3567.995	2	50	
IR	3567.995	16	5	
SB II	3568.000		6	
W	3568.040	10	10	
MO	3568.067	3	3	
CE	3568.126	10		
ZR II	3568.135	4	2	
MO	3568.181	3	4	
RE	3568.231	40		
*SM II	3568.258	40	50	39 VAF
PR	3568.290	5	4	
DY	3568.330	4	2	
CR	3568.418	6		
W	3568.418	4	4	
FE I	3568.423	20	4	
CO I	3568.426	2		
AG	3568.435	2		

Table IIa. SM II 3568.258 A
(con't)

	Element	Wavelength (Angstrom)	Intensity Arc	Spark	Plasma
--	NB	3568.506	10	50	
	TB II	3568.510	50	50	46 VAF
	CE	3568.513	2		
	U	3568.525	2	2	
	NE II	3568.530			25
	OS	3568.566	2	5	
	DY	3568.650	4		
	TM	3568.680	5	5	
	U	3568.690	3		
---	NB	3568.725	5	5	
	U	3568.815	6	3	
	FE	3568.822	16	6	
	ZR I	3568.875	13		
	MO	3568.879	4	2	
	ND	3568.879	6	13	
	SM II	3568.891	4		
	CA I	3568.910	2		
	V I	3568.940	20	16	
	FE I	3568.979	50	32	
	TB	3568.980	16	16	
	W	3568.987	10	10	
----	DY	3568.994	100		
	HF II	3569.041	20	50	
	U	3569.062	13	20	
	V I	3569.083	6		
	CR	3569.147	4	4	
	EU	3569.170	4		
	TA	3569.190	2		
	ND	3569.229	20	4	
	W	3569.235	8	6	
	DY	3569.250	2		
	ER	3569.252	8		

Table IIb.(3) SM II 3568.258 A

Element	Wavelength (Angstrom)	Intensity Arc	Spark	Plasma
ND	3563.403	50	10	
NB	3563.497	32	32	
DY	3563.699	50	32	
CO I	3564.951	160		
FE I	3565.381	400	320	
TA	3565.627		50	
TM	3565.903	40	50	
TI II	3565.987	6	25	
CR	3566.163	79	13	
V II	3566.182	25	100	
NI I	3566.372	2000	100	
TM	3566.474	63	20	
PD	3566.626	63		
SB	3566.640		25	
TA	3566.716	50	5	
FE I	3567.038	50	16	
NB	3567.103	5	32	
SC II	3567.701	16	40	37 VAF
LU	3567.840	100	6	
NB	3567.995	2	50	
RE	3568.231	40		
*SM II	3568.258	40	50	39 VAF
NB	3568.506	10	50	
TB II	3568.510	50	50	46 VAF
FE I	3568.979	50	32	
DY	3568.994	100		
HF II	3569.041	20	50	
CO I	3569.379	400	100	
DY	3569.669	40	10	
OS	3569.775	100	32	
FE I	3570.097	320	320	
RH I	3570.182	400	160	
FE	3570.258	50	16	
RU	3570.594	13	63	
PD I	3571.155	40	40	
FE I	3571.227	40	6	
V I	3571.653	40	32	
NI I	3571.869	1000	40	
FE I	3571.996	100	79	
ZR II	3572.473	63	79	30 VAF

Table IIb.(3) SM II 3568.258 A
(con't)

Element	Wavelength (Angstrom)	Intensity Arc	Spark	Plasma
W II	3572.476	10	32	
SC II	3572.523	32	50	15 VAF
PB	3572.734	200	20	

Table IIa. SN I 3034.121 A

	Element	Wavelength (Angstrom)	Intensity Arc	Spark	Plasma
	CE	3033.124	8		
	AU	3033.180	25	32	
	DY	3033.180	8		
	U	3033.193	13	13	
	RN	3033.200			10
	MO	3033.234	8	10	
	I	3033.275			5
	PR	3033.310		4	
	MO	3033.330		32	
	TE	3033.350			5
---	TA	3033.392	2		
	NB	3033.393	2	2	
	SM	3033.393		2	
	RA II	3033.440			160
	U	3033.444		13	
	FE II	3033.445	4	2	
	V	3033.448	20	40	
	RU I	3033.451	63	10	
	ZR	3033.457		2	
	BI II	3033.500			16
	AR II	3033.520			8
	MN	3033.563		2	
	W	3033.575	13	16	
---	IR I	3033.622	25		
	XE II	3033.710			8
	U	3033.770	4	4	
	V II	3033.822	20	100	
	YB	3033.860		3	
--	RU	3033.894		6	
	IR I	3034.039	4		
	U	3034.049	6	5	
	GD II	3034.059		100	63
	RU I	3034.060	63	5	
	TH	3034.069	16	16	
	AG	3034.112	4		
	FE	3034.120	2		
	*SN I	3034.121	200	160	
	IN	3034.122	8		
	KR II	3034.160			2
	SM	3034.174	5		

Table IIa. SN I 3034.121 A
(con't)

	Element	Wavelength (Angstrom)	Intensity Arc	Spark	Plasma
	CR I	3034.190	200	63	
	W	3034.195	10	10	
--	PR	3034.200		3	
	U	3034.375	3		
	CO I	3034.433	79	2	
	NE II	3034.480			20
	SM	3034.492	5	2	
	CA I	3034.535	2		
	RE I	3034.536	32		
	FE I	3034.538	63	40	
	CR	3034.542		32	
	IR I	3034.556	20		
---	TE	3034.620			10
	YB	3034.632	2	6	
	U	3034.661	3	2	
	K I	3034.820	32		
----	SM	3034.843	20	10	
	BI	3034.873	32	32	
	U	3034.895	3		
	TB	3034.910	8	3	
	MO	3034.925	5	25	
	NB	3034.950		100	
	RH	3034.990	2	4	
	CE	3035.008	6		
	NB	3035.025	2	32	
	HG	3035.100			32
	TH	3035.113	16	16	

Table IIb.(2) SN I 3034.121 A

Element	Wavelength (Angstrom)	Intensity Arc	Spark	Plasma
CR I	3029.164	63	50	
FE I	3029.233	79	63	
YB	3029.554		40	
V II	3029.580		40	
TI II	3029.730	13	160	
NB	3029.745	2	100	
SB	3029.807	100	200	
FE	3030.149	320	320	
CR I	3030.245	200	160	
RE I	3030.450	100		
OS I	3030.695	500	40	
YB	3031.110	100	32	
HF II	3031.160	63	100	
FE	3031.215	160	160	
PT II	3031.216	10	40	
FE I	3031.639	200	200	
NI I	3031.870	200		
PD II	3032.200	2	100	
NB	3032.768	3	320	
AS I	3032.840	130	63	
GD II	3032.850	100	100	11 VAF
CR II	3032.927	10	100	
V	3033.448	20	40	
V II	3033.822	20	100	
GD II	3034.059		100	63
*SN I	3034.121	200	160	
CR I	3034.190	200	63	
FE I	3034.538	63	40	
NB	3034.950		100	
FE	3035.740	100	63	
ZN I	3035.781	200	100	
CU I	3036.104	200	50	
PT I	3036.450	200	10	
RU	3036.466	50	160	
Y II	3036.590	10	40	
CR I	3037.044	200	100	
FE I	3037.388	630	400	
TA	3037.505	8	100	
NI I	3037.935	790	100	
V	3038.521		50	

Table IIb.(2) SN I 3034.121 A
(con't)

Element	Wavelength (Angstrom)	Intensity Arc	Spark	Plasma
TI II	3038.706	2	40	
GE I	3039.064	1000	1000	29 VAF

Table IIa. SR II 4077.714 A

Element	Wavelength (Angstrom)	Intensity Arc	Spark	Plasma
U	4076.720	8	10	
RU I	4076.733	63	25	
FE	4076.803	8		
IR	4076.806	2		
SM II	4076.854	10	5	
EU	4076.952	10		
AR	4076.960			10
--- ZR II	4077.046	3		
W	4077.059	6	5	
AS	4077.080		10	
NB	4077.088	3	5	
CR	4077.089	32	10	
ND	4077.150	10	4	
TI I	4077.153	20	2	
--- YB	4077.270	32	100	
LA II	4077.340	630	400	21 VAF
DY	4077.350	4		
Y I	4077.366	50	40	
CO I	4077.406	100	2	
-- CE II	4077.470	20	4	
RH I	4077.570	5	4	
PB	4077.610		2	
ND	4077.620	8	3	
CR	4077.677	32	10	
MO	4077.682	8	10	
PR	4077.690	4	2	
*SR II	4077.714	400	500	72 VAF
CU	4077.716	5		
SN	4077.720	2	3	
TA	4077.721	4	2	
U	4077.786	16	6	
HG I	4077.811	160	160	
-- ER I	4077.880	8		
DY II	4077.974	160	100	8 VAF
V I	4077.977	2		
PR	4077.980	10	2	
HO	4078.000	3	3	
MO	4078.074	4	4	
RE	4078.124	10		
W	4078.124	6	6	

Table IIa. SR II 4077.714 A
(con't)

Element	Wavelength (Angstrom)	Intensity Arc	Intensity Spark	Intensity Plasma
ZN II	4078.140			5
--- PR	4078.160	5		
EU	4078.231	3		
TB	4078.260	3		
ZR I	4078.309	10		
CE	4078.321	16	4	
NB	4078.352	4	3	
FE I	4078.358	79	40	
MO	4078.381	5	3	
---- SB	4078.385		4	
GD	4078.465	16	10	
TB	4078.470	5		
TI I	4078.474	130	50	
CE	4078.515	5		
SC I	4078.584	10	10	
CE	4078.606	5		
GD I	4078.709	20	10	
V I	4078.709	2		

Table IIb.(2) SR II 4077.714 A

Element	Wavelength (Angstrom)	Intensity Arc	Intensity Spark	Intensity Plasma
CE II	4075.714	16	2	4 VAF
CE II	4075.850	4		4 VAF
LA II	4077.340	630	400	21 VAF
*SR II	4077.714	400	500	72 VAF
HG I	4077.811	160	160	
DY II	4077.974	160	100	8 VAF
NB I	4079.729	500	200	
RU	4080.600	130	320	

Table IIa. SR II 4215.524 A

	Element	Wavelength (Angstrom)	Intensity Arc	Spark	Plasma
	IR	4214.524	5		
	TH	4214.552		3	
	RU	4214.557	6		
	ND	4214.603	10	10	
	NB	4214.672		100	
	XE II	4214.690			3
	CE	4214.698	4		
	N I	4214.730			25
	NB	4214.733	40		
---	NB	4214.817	5	10	
	CO	4214.874	2		
	W	4214.959	5		
	U	4215.013	4	4	
	SE	4215.020			160
---	GD	4215.024	200	160	
	TB	4215.130	32		
	PR	4215.140	8	3	
	OS	4215.155	8		
	DY	4215.169	50	8	
--	ZR I	4215.313	4		
	W	4215.382	13	5	
	NA I	4215.400	3		
	FE I	4215.425	63	16	
	RE	4215.506	20		
	U	4215.510	3	2	
	*SR II	4215.524	320	400	39 VAF
	TM	4215.530	10		
	RB I	4215.556	1000	320	
	XE II	4215.600			100
	CL	4215.640			6
--	ZR II	4215.753		3	
	IN	4215.900		5	
	N I	4215.920			5
	TB	4215.950	3		
	ER	4215.964	10		
	FE I	4215.970	2		
	PR	4215.980	4	3	
	U	4215.990	5	3	
--	BH C	4216.000			
	BA II	4216.040			25

Table IIa. SR II 4215.524 A
(con't)

	Element	Wavelength (Angstrom)	Intensity Arc	Spark	Plasma
	W	4216.060		5	
	SC I	4216.101	3		
	FE I	4216.186	200	100	
----	NB	4216.228		10	
	CR I	4216.365	63	25	

Table IIb.(1) SR II 4215.524 A

Element	Wavelength (Angstrom)	Intensity Arc	Spark	Plasma
AG I	4210.936	200	32	
RH I	4211.141	200	16	
ND	4211.286	32	16	
CR	4211.349	100	32	
ER	4211.718	32	5	
DY	4211.719	200	16	
TI I	4211.728	32	6	
MN	4211.748	32	20	
OS	4211.855	160	50	
PR	4211.858	50	25	
GD	4212.019	160	50	
RU I	4212.063	130	79	
CR I	4212.660	79	8	
AG I	4212.680	160	20	
PD I	4212.950	500	320	
CR	4213.179	63	8	
FE	4213.650	100	63	
OS	4213.859	32	3	
ZR I	4213.865	40	3	
CE II	4214.039	32	4	
CE I	4214.039	32	4	
RU	4214.442	100	40	
NB	4214.672		100	
NB	4214.733	40		
GD	4215.024	200	160	
TB	4215.130	32		
DY	4215.169	50	8	

Table IIb.(1) SR II 4215.524 A
(con't)

Element	Wavelength (Angstrom)	Intensity Arc	Spark	Plasma
FE I	4215.425	63	16	
*SR II	4215.524	320	400	39 VAF
RB I	4215.556	1000	320	
FE I	4216.186	200	100	
CR I	4216.365	63	25	
GD	4217.195	100	100	
RU I	4217.268	100	20	
LA II	4217.554	200	100	
FE I	4217.555	200	100	
CR I	4217.626	160	63	
IR	4217.760	40	3	
NB	4217.945	50	50	
DY	4218.091	50	8	
YB	4218.567	3	50	
SB II	4219.070		32	
FE I	4219.364	250	200	
PR	4219.652	32	8	
FE	4220.348	79	40	

Table IIa. TA II 2965.133 A

	Element	Wavelength (Angstrom)	Intensity Arc	Spark	Plasma
	FE II	2964.133		6	
	XE	2964.190			6
	FE	2964.220	3		
	U	2964.247	6	16	
	ER	2964.254	3		
	CD	2964.300			10
	RU	2964.309	8		
	MO	2964.344		6	
----	YB	2964.400		2	
	HO	2964.480		10	
	W	2964.515	13	10	
	ER II	2964.518	20	4	11 VAF
	CE	2964.526	2		
	ZR II	2964.563	2	2	
	OS	2964.615	8	5	
	DY	2964.620	2		
	IR	2964.629	2		
	FE II	2964.630	8	160	
---	CD	2964.700			5
	YB	2964.750	16	25	
	PR	2964.760	10	10	
	TB	2964.760	10	3	
	CE	2964.801	25		
	TA	2964.811	3	3	
--	HF I	2964.885	400	79	
	TH	2964.924	8	6	
	EU	2964.930	2		
	MO	2964.958	5	10	
	Y I	2964.968	6	5	
	CS	2965.000			8
	U	2965.034	8	13	
	FE II	2965.037	4	50	
	CD	2965.100			5
	KR II	2965.110			2
	RE I	2965.125	100		
	*TA II	2965.133	40	790	50 VAF
	CE	2965.135	8		
	YB	2965.160		4	
	RH	2965.161	8		
	RU I	2965.162	79	20	

Table IIa. TA II 2965.133 A
(con't)

	Element	Wavelength (Angstrom)	Intensity Arc	Spark	Plasma
	MG II	2965.190	6		
	TA	2965.200		10	
	IR I	2965.202	25	5	
	ER	2965.220	4		
	TI I	2965.231	25	6	
	FE I	2965.255	400	160	
	CE	2965.269	16		
	MO	2965.270	8	40	
	TB	2965.320	10		
--	U	2965.377	3	2	
	CS	2965.400			2
	GD	2965.420	3	3	
	NB	2965.480	5	2	
	CL	2965.500			10
	TH	2965.501	8	6	
	TA	2965.545	160	50	
---	RU	2965.546	63	200	
	NB	2965.640		5	
	TI I	2965.680	6	2	
	ND	2965.700			5
	TI I	2965.707	16	5	
	RU	2965.711	20		
	NA II	2965.750			5
	RE I	2965.758	160		
	U	2965.785	4	4	
	FE	2965.811	25	16	
	CE	2965.837	6		
	W	2965.860	6	4	
	TM	2965.870	50	100	
--	NB	2965.871		3	
	TA	2965.920	32	79	
	LA II	2966.050		2	
	CR	2966.051	2	63	
	U	2966.120	16	25	
	CE	2966.123	3		
	IR I	2966.127	10		
	KR	2966.130			3

Table IIb.(3) TA II 2965.133 A

Element	Wavelength (Angstrom)	Intensity Arc	Spark	Plasma
EU	2960.230	160		
FE	2960.299	63	32	
OS	2961.012	40	10	
CU I	2961.165	320	320	
RU	2961.689	63	3	
V I	2962.772	63	63	
TA	2963.056	40	10	
LU	2963.320	50	160	
TA	2963.322	320	100	
RU	2963.402	63	160	
RU I	2963.715	63	5	
TA	2963.910	50	10	
ER II	2964.518	20	4	11 VAF
HF I	2964.885	400	79	
RE I	2965.125	100		
*TA II	2965.133	40	790	50 VAF
RU I	2965.162	79	20	
FE I	2965.255	400	160	
TA	2965.545	160	50	
RU	2965.546	63	200	
RE I	2965.758	160		
TM	2965.870	50	100	
FE I	2966.900	1000	630	
HG I	2967.278	100	100	2 VAF
CR I	2967.642	63	32	
RE I	2968.051	40		
RH I	2968.663	130	32	
RU I	2968.954	63	6	
FE I	2969.362	79	79	
TA	2969.470	160	79	
FE I	2969.477	63	63	
TA	2969.902	50	10	
FE I	2970.105	400	200	

Table IIa. TA II 2262.300 A

	Element	Wavelength (Angstrom)	Intensity Arc	Spark	Plasma
	IN	2261.300		5	
	AU II	2261.340		3	
	TA	2261.418	13	16	
	NI I	2261.424	13	3	
	U	2261.470	2	2	
	ND	2261.510		2	
	NB	2261.530	2	3	
	HF	2261.540	3	3	
---	CO	2261.550		8	
	TI II	2261.610		8	
	TA	2261.621	6	13	
	TE	2261.660			16
	CR	2261.670	8	16	
	K	2261.700			2
	NB	2261.727	2	2	
	OS	2261.730	25	50	
	RH	2261.750	2	100	
---	AR II	2261.800			10
	V	2261.850		5	
	W	2261.940		6	
	MO	2261.976	5	16	
--	K	2262.100			10
	NB	2262.131	3	5	
	PD II	2262.141		25	
	HG II	2262.233			100
	YB	2262.250	2	32	
	FE II	2262.261		4	
	BR	2262.290			2
	NB	2262.290		4	
	CD	2262.290	3		
	PD	2262.300			16
	*TA II	2262.300	13	20	120 VAF
	OS	2262.320	25	6	
	W	2262.320		8	
	U	2262.360		2	
	BH C	2262.400	13		
	V	2262.405		5	
	IR	2262.430		6	
	PD II	2262.484		32	
--	SB	2262.538	40	25	

Table IIa. TA II 2262.300 A
(con't)

	Element	Wavelength (Angstrom)	Intensity Arc	Spark	Plasma
	CO I	2262.598	10		
	PT II	2262.662	4	25	
	FE II	2262.680	6	13	
	AU	2262.730		8	
---	U	2262.790		2	
	NI II	2262.900		4	
	PT II	2262.900		5	
	XE II	2262.950			2
----	SE	2262.980			5
	CU I	2263.079	32	10	
	YB	2263.160		3	
	CU II	2263.212		10	
	NB	2263.220		5	
	FE II	2263.228		32	
	U	2263.260		3	
	W	2263.270	3		

Table IIb.(3) TA II 2262.300 A

Element	Wavelength (Angstrom)	Intensity Arc	Spark	Plasma
IR	2257.500	2	50	
NB	2257.537	3	10	
CR	2257.810		10	
NB	2257.881	16		
HF II	2257.890	10	16	
NI	2257.890		13	
NI I	2258.146	16	2	
RH	2258.250		25	
OS	2258.332	13	32	
IR I	2258.508	16	50	
HF II	2258.680	16	20	
TA	2258.710		25	
PT II	2258.717		20	
V	2258.813	25	32	
IR I	2258.856	32	10	
OS	2258.975	3	25	

Table IIb.(3) TA II 2262.300 A
(con't)

Element	Wavelength (Angstrom)	Intensity Arc	Spark	Plasma
TE I	2259.020			17 VAF
W II	2259.068	3	10	
IR	2259.270		32	
PT	2259.450		10	
FE I	2259.511	25		
TA	2259.550	4	13	
W	2259.555	13	6	
NI I	2259.571	320		
FE II	2259.589		20	
CO	2260.010	2	25	
W	2260.070	16	3	
FE II	2260.079	10	16	
OS	2260.110	25	50	
PD II	2260.130		25	
PT II	2260.497		25	
RE	2260.510	10	25	
CU I	2260.528	25	6	
PD	2260.530		25	
ND	2260.540		10	
TA	2260.560		20	
IR	2260.650	3	40	
FE II	2260.853	13	13	
V	2261.085	5	16	
TI II	2261.188	5	32	
GD	2261.195	25	3	
TA	2261.418	13	16	
NI I	2261.424	13	3	
TA	2261.621	6	13	
CR	2261.670	8	16	
OS	2261.730	25	50	
RH	2261.750	2	100	
MO	2261.976	5	16	
PD II	2262.141		25	
YB	2262.250	2	32	
*TA II	2262.300	13	20	120 VAF
OS	2262.320	25	6	
BH C	2262.400	13		
PD II	2262.484		32	
SB	2262.538	40	25	
PT II	2262.662	4	25	

Table IIb.(3) TA II 2262.300 A
(con't)

Element	Wavelength (Angstrom)	Intensity Arc	Spark	Plasma
FE II	2262.680	6	13	
CU I	2263.079	32	10	
CU II	2263.212		10	
FE II	2263.228		32	
PT II	2263.323		25	
OS	2263.367	6	16	
RH	2263.430	5	200	
AL I	2263.453	63	25	5 VAF
W	2263.535	6	25	
CU II	2263.780		25	
OS	2263.910		16	
RE	2263.970	13	2	
IR	2263.970	4	10	
PT II	2263.987		20	
TI	2264.022	16		
ND	2264.080		16	
RH	2264.140	50	25	
IR I	2264.172	13	3	
W II	2264.182	3	13	
PD II	2264.282		25	
MO	2264.350	4	10	
FE I	2264.389	32	5	
NI II	2264.457	160	400	
FE II	2264.591		20	
IR I	2264.607	32	16	
MO	2264.742		20	
CO	2264.874	16		
CE	2264.890		20	
CR	2264.920		10	
CD II	2265.017	25	320	89 VAF
IR	2265.163	5	50	
W II	2265.350	5	10	
IR I	2265.454	13	3	
CU	2265.458		16	
TE I	2265.550			3 VAF
NB	2265.592		50	
YB	2265.650	3	50	
NB	2265.671	5	20	
CD II	2265.810		32	
FE II	2265.994	4	10	

Table IIb.(3) TA II 2262.300 A
(con't)

Element	Wavelength (Angstrom)	Intensity Arc	Spark	Plasma
RE	2266.090	25	2	
MO	2266.220		20	
IR I	2266.331	25	10	
NI I	2266.352	20	2	
PT	2266.417		20	
HF II	2266.520	32	40	
CO	2266.530		32	
OS	2266.570	20	4	
MN	2266.580		10	
AS I	2266.700	13		
NB	2266.726	3	10	
HF II	2266.830	63	79	
FE I	2266.899	16		
PD II	2266.950	3	20	
CO I	2267.101	13		
SN I	2267.190	16	10	
PT	2267.240	5	20	

Table IIa. TB II 3509.170 A

	Element	Wavelength (Angstrom)	Intensity Arc	Spark	Plasma
	BH ZR	3508.200	20		
	PR	3508.212	13	3	
	FE II	3508.213			
	RN I	3508.220			10
	U	3508.306	2		
	CE	3508.335	4		
	HO	3508.350		4	
	ER	3508.395	16	6	
----	LU	3508.420	32	3	
	XE I	3508.420			2
	CE II	3508.467	10		
	MO	3508.476	2	5	
	FE	3508.485	40	20	
	RH	3508.510	2		
	FE I	3508.535	20	10	
	NB	3508.537	5	5	
	IR I	3508.576	6		
	TB	3508.600	8		
	RH I	3508.602	3		
---	SM	3508.706	3	2	
	CE	3508.711	13	2	
	IR I	3508.719	2		
	W	3508.735	10	10	
	EU	3508.740	8		
	ER	3508.811	8		
	CR	3508.836	13		
	U	3508.846	8	10	
	OS	3508.851	8	10	
	EU	3508.859	10		
	CE	3508.865	2		
	XE II	3508.880			10
--	ER	3508.938	5		
	CL II	3508.940			13
	DY	3509.000	6	4	
	W	3509.015	10	6	
	V	3509.041	2	160	
	CE	3509.061	2		
	TH	3509.113	4		
	SM	3509.117	8	10	
	U	3509.122		2	

Table IIa. TB II 3509.170 A
(con't)

	Element	Wavelength (Angstrom)	Intensity Arc	Spark	Plasma
	FE I	3509.132	2		
	*TB II	3509.170	200	200	130 VAF
	TH	3509.197		2	
	RU	3509.201	10	100	
	MO	3509.230		3	
	ND	3509.235	10		
	IR I	3509.241	8	2	
	CE II	3509.254	10	2	
	CE	3509.313	3	2	
	ZR I	3509.323	40	5	
	U	3509.349	3	6	
	HO	3509.350	6	2	
	W	3509.381		13	
	CL II	3509.390			40
--	DY	3509.410	6	2	
	SM	3509.447	4	4	
	CE	3509.531	3		
	U	3509.668	10	16	
--	W	3509.669	8	6	
	V	3509.684		6	
	RU I	3509.717	50	2	
	CE	3509.729	13		
	TB	3509.760	8		
	AR	3509.780			16
	RN	3509.800			5
	SM	3509.801	3		
	CO I	3509.843	400	40	
	TI II	3509.844	8	20	
	U	3509.858	2		
--	FE I	3509.868	16	4	
	CE	3509.934	8		
	MO	3509.986		3	
	LA II	3509.989	10	10	
	W	3510.032	10	10	
	EU	3510.070	4		
	DY	3510.088	2		
	TB	3510.100	50	8	
	SM	3510.151	2		

Table IIb.(2) TB II 3509.170 A

Element	Wavelength (Angstrom)	Intensity Arc	Spark	Plasma
V II	3504.439	63	200	
DY	3504.522	100	3	
OS I	3504.659	320	20	
TI II	3504.892	20	160	
HF II	3505.227	20	50	
GD	3505.520	63	63	
ZR II	3505.669	40	32	12 VAF
CO I	3506.315	400	16	
AR I	3506.490			2 VAF
RH I	3507.316	500	130	
LU II	3507.390	100	160	12 VAF
V	3507.539		50	
NI I	3507.694	100	13	
YB	3507.835	13	63	
V	3509.041	2	160	
*TB II	3509.170	200	200	130 VAF
RU	3509.201	10	100	
CO I	3509.843	400	40	
NB	3510.257	16	200	
NI I	3510.338	1000	50	
TI II	3510.841	40	130	
BI I	3510.853	200	32	
TA	3511.043	100	32	
CR II	3511.836	20	50	
CO I	3512.641	400	100	
CO I	3513.480	320	25	
IR I	3513.645	100	100	
FE I	3513.820	400	320	
NI I	3513.933	200	40	
NB	3514.044		50	

Table IIa. TB II 3848.730 A

	Element	Wavelength (Angstrom)	Intensity Arc	Spark	Plasma
	CR	3847.730	5		
	CE	3847.812	6	6	
	U	3847.835	8	13	
	EU	3847.840	13		
	ND	3847.850	63	50	
	YB	3847.856	40		
	Y	3847.873	8	4	
	TB	3847.880	8		
	ER	3847.890	20		
	O II	3847.890			10
--	TM II	3848.018	400	250	31 VAF
	TA	3848.046	32	5	
	U	3848.069	6	6	
	MG I	3848.090	2	3	
	CE II	3848.105	8		
	IR	3848.160	10	2	
	Y II	3848.194	2	4	
	EU	3848.200	2		
--	ND II	3848.233	10	10	16 VAF
	MG II	3848.240	10	10	
	FE	3848.298	5	2	
	MO	3848.301	25	20	
	ND	3848.309	40		
	TI I	3848.310	10	2	
	PR	3848.332	3	2	
	DY	3848.360	4		
	EU	3848.400	2		
--	W	3848.482		2	
	ND II	3848.524	10	20	16 VAF
	XE II	3848.580			3
	CE	3848.597	20	2	
	U	3848.618	10	8	
	BH B	3848.700	200		
	U	3848.716	4	3	
	*TB II	3848.730	100	200	54 VAF
	MG I	3848.773	2	13	
	SM	3848.806	160	10	
	MG	3848.920		2	
	RU	3848.945	8	13	
-	RE	3848.960	3		

Table IIa. TB II 3848.730 A
(con't)

	Element	Wavelength (Angstrom)	Intensity Arc	Spark	Plasma
	CR I	3848.983	79	50	
	ER	3849.002	13	2	
	RH I	3849.005	2	2	
	BI	3849.007		2	
	SM	3849.012	10	4	
	LA II	3849.013	200	160	12 VAF
	CE	3849.068	4	4	
---	HF	3849.183	16	16	
	ZR I	3849.254	10	4	
	HG	3849.274			2
	V I	3849.324	63	25	
	U	3849.339	2	3	
	CR I	3849.365	40	32	
	EU	3849.400	4	3	
	DY	3849.400	20		
----	TA	3849.420	4	2	
	HF II	3849.524	2	4	
	CR I	3849.534	20	20	
	CE II	3849.569	3		
	NI II	3849.580			
	TB	3849.590	4		
	SE	3849.600			4
	EU	3849.630	2		
	CE	3849.681	2		
	U	3849.707	2	10	

Table IIb.(2) TB II 3848.730 A

Element	Wavelength (Angstrom)	Intensity Arc	Spark	Plasma
MN	3843.983	79	100	
V I	3844.438	100	50	
PR	3844.558	50	10	
FE I	3845.174	100	63	
CO I	3845.471	500	100	
LA II	3845.997	40	50	
BI	3846.027		100	

Table IIb.(2) TB II 3848.730 A
(con't)

Element	Wavelength (Angstrom)	Intensity Arc	Spark	Plasma
FE	3846.415	50	79	
PR	3846.605	63	32	
FE I	3846.803	130	100	
BH B	3847.000	100		
V I	3847.331	100	63	
ND	3847.850	63	50	
TM II	3848.018	400	250	31 VAF
ND II	3848.233	10	10	16 VAF
ND II	3848.524	10	20	16 VAF
BH B	3848.700	200		
*TB II	3848.730	100	200	54 VAF
SM	3848.806	160	10	
CR I	3848.983	79	50	
LA II	3849.013	200	160	12 VAF
V I	3849.324	63	25	
OS I	3849.944	130	20	
FE I	3849.969	500	400	
RU I	3850.432	50	10	
FE I	3850.820	200	79	
PR	3850.825	50	16	
CO I	3850.953	100		
V I	3851.171	50	16	
PR	3851.617	200	160	
ND II	3851.657	20		19 VAF
ND II	3851.748	8	16	19 VAF
CR I	3852.218	63	13	
ND	3852.380	63	50	
GD	3852.497	100	8	
FE I	3852.575	160	100	
PR	3852.805	100	50	
DY	3853.041	100		
OS	3853.439	100	16	

Table IIa. TE I 2385.780 A

	Element	Wavelength (Angstrom)	Intensity Arc	Spark	Plasma
	IR	2384.810	4	79	
	W	2384.817	13	13	
	ZR	2384.827	4		
	CU II	2384.840		16	
	NI	2384.840	8	3	
	NB	2384.847	2	4	
	ND	2384.850		10	
	CO I	2384.860	10	4	
	V	2384.996		6	
	AR	2385.000			40
	YB	2385.006		5	
	FE II	2385.007	8	16	
	NI	2385.010	10	2	
	PD II	2385.010		50	
----	CU II	2385.070		10	
	U	2385.100		3	
	PD	2385.170		16	
	HF	2385.180		4	
	Y	2385.230	13		
	NB	2385.244		6	
	W	2385.260	4	13	
---	MN	2385.280	2	4	
	U	2385.280	2		
	TA	2385.350	2		
	HE II	2385.420			32
	RH	2385.440	3	25	
	OS	2385.490	16	2	
	W II	2385.490	3	10	
	RE	2385.500	13		
--	U	2385.550	3	2	
	FE	2385.579	2		
	RB	2385.600			2
	RH	2385.600		10	
	V	2385.618		10	
	TA	2385.730	8		
	CR I	2385.740	25		
	HF II	2385.780		3	
	*TE I	2385.780	630		17 VAF
	W	2385.790		5	
	CO	2385.816	10		

Table IIa. TE I 2385.780 A
(con't)

	Element	Wavelength (Angstrom)	Intensity Arc	Spark	Plasma
	V	2385.816		100	
	IR	2385.863	20	3	
--	FE I	2385.950	4		
	OS	2386.040	2	20	
	MO	2386.050		20	
	RH	2386.140	79	8	
	XE	2386.140			3
	IR	2386.153	16		
	W	2386.170	8		
	CR I	2386.190	20		
	RU	2386.216	2		
	MO	2386.240		8	
--	PD II	2386.240		5	
	AG II	2386.330		25	
	W	2386.360	3		
	CO II	2386.363	10	25	14 VAF
	FE II	2386.396	2	5	
	RH	2386.400	6	2	
	NB	2386.404	5		
	V I	2386.413	20		
	W	2386.450		8	
	PT II	2386.506	2	25	
	CO I	2386.509	3		
-	BR	2386.510			3
	ER	2386.580	5	4	
	IR	2386.580	25	20	
	NI I	2386.580	20	5	
	CL	2386.640			4
	RU	2386.678	8		
	CO	2386.727	4	16	
	BR	2386.740			25
	OS	2386.750	10		
	ZN II	2386.750			2
	CR I	2386.780	20		

Table IIb.(1) TE I 2385.780 A

Element	Wavelength (Angstrom)	Intensity Arc	Spark	Plasma
AS I	2381.180	79	4	
IR I	2381.622	32	2	12 VAF
FE II	2382.039	40	100	65 VAF
TE I	2383.250	500	320	11 VAF
CO II	2383.460	16	32	21 VAF
SB	2383.630	79	20	
*TE I	2385.780	630		17 VAF
RH	2386.140	79	8	
CO II	2386.363	10	25	14 VAF
TA II	2387.060	20	50	80 VAF
DY II	2387.360	8		6 VAF
RU	2387.900	63	3	
FE II	2388.627	25	32	20 VAF
CO II	2388.915	10	32	50 VAF

Table IIa. TE I 2142.810 A

Element	Wavelength (Angstrom)	Intensity Arc	Spark	Plasma
HF II	2141.840	20	20	
V II	2141.980	10	79	
NB	2142.020	2	6	
OS	2142.130		5	
PD I	2142.130	16		
YB	2142.180		4	
MO	2142.220		13	
IR	2142.250	20	10	
RH	2142.280	3	3	
OS	2142.380	4		
V II	2142.430		4	
MO	2142.440		10	
PT II	2142.499	2	13	
W	2142.511	3	8	
TA	2142.530	8	20	
PD	2142.570	3		
OS	2142.730	4	3	
RE II	2142.736	10	20	
V II	2142.740		3	
LA	2142.810	2	20	
*TE I	2142.810	630		73 VAF
IR	2142.820	3		
PR	2142.890		32	
NB	2142.910		6	
RE	2143.000	10	4	
V II	2143.047	10	50	
TA	2143.160		32	
IR	2143.200		8	
NB	2143.210	4	6	
OS	2143.230	13	3	
MO	2143.250	2	16	
BI II	2143.350		8	
BI II	2143.400		13	
YB	2143.420		5	
BI II	2143.460		13	
RE	2143.480	6	2	
TE	2143.540			25
TI	2143.610	5		
TA	2143.650	2	5	
CO I	2143.679	3		

Table IIa. TE I 2142.810 A
(con't)

Element	Wavelength (Angstrom)	Intensity Arc	Spark	Plasma
V II	2143.713	3	4	

Table IIb.(1) TE I 2142.810 A

Element	Wavelength (Angstrom)	Intensity Arc	Spark	Plasma
ZN I	2138.560	790	500	170 VAF
SB I	2139.760	32	32	16 VAF
V II	2140.087	10	79	20 VAF
*TE I	2142.810	630		73 VAF
PT II	2144.231	32	100	100 VAF
CD II	2144.382	50	1600	120 VAF
SB I	2144.860			12 VAF
TE I	2147.190			14 VAF

Table IIa. TH II 4019.137 A

Element	Wavelength (Angstrom)	Intensity Arc	Spark	Plasma
CR I	4018.202	32	8	
CE II	4018.227	2		
OS	4018.257	63	4	
FE I	4018.275	50	6	
U	4018.282	8	3	
W	4018.309	2	3	
BR I	4018.330			4
ZR II	4018.383	8	5	
CE	4018.400	2		
EU	4018.400	10	4	
RE	4018.409	25		
TB	4018.440	2		
CE	4018.511	2		
SE II	4018.520			63
SM II	4018.539	4	5	
BI II	4018.600		2	
ND	4018.826	16	10	
CA	4018.880		4	
CE	4018.919	2		
U	4018.990	25	16	
CE	4019.044	16	4	
MO	4019.046	2	4	
NI I	4019.046	5		
V	4019.046		6	
TB	4019.120	40	5	
RE I	4019.130	16		
*TH II	4019.137	8	8	36 VAF
CO I	4019.140	5		
CE	4019.193	2		
U	4019.203	6	8	
W	4019.231	20	16	
CE II	4019.280	4		
CO I	4019.302	79		
YB	4019.350		6	
PR	4019.440	10	3	
P II	4019.450			50
DY	4019.480	5	2	
CE	4019.480	6		
SE II	4019.500			10
RU I	4019.553	13	8	

Table IIa. TH II 4019.137 A
(con't)

Element	Wavelength (Angstrom)	Intensity Arc	Spark	Plasma
PB	4019.639	6	6	
TB	4019.660	2		
EU	4019.712	3		
SE I	4019.720			20
GD I	4019.732	16	10	
MO	4019.789	10		
W	4019.805		5	
ND	4019.809	10	8	
SM II	4019.827	8	8	
---- PR	4019.840	16	8	
CE	4019.897	8	3	
SM II	4019.976	32	16	
NE I	4020.015			2
IR	4020.028	79	100	
CL II	4020.060			16
ND	4020.062	6	3	
MN	4020.092	10	5	
TH	4020.096	3	3	
BH SR	4020.100	2		

Table IIb.(3) TH II 4019.137 A

Element	Wavelength (Angstrom)	Intensity Arc	Spark	Plasma
RU	4014.153	13	5	
U	4014.160	8	8	
HO	4014.180	2	4	
PR	4014.350	10	3	
SC II	4014.489	20	8	
FE	4014.534	200	100	
IR	4014.658	8	2	
PR	4014.660	10	2	
CR	4014.668	40	8	
DY	4014.713	20	4	
U	4014.816	13	2	
CE II	4014.899	63	13	
NB	4014.926	5	8	

Table IIb.(3) TH II 4019.137 A
(con't)

Element	Wavelength (Angstrom)	Intensity Arc	Spark	Plasma
W	4014.939	8	6	
OS	4015.037	20	5	
DY	4015.180	10	2	
W	4015.216	25	32	
U	4015.240	10		
TA	4015.247	8	2	
TI I	4015.384	63	10	
PR	4015.389	50	32	
LA I	4015.393	100	2	
TB	4015.500	10		
ND	4015.557	10	8	
ER	4015.579	20		
SM II	4015.765	8	4	
CE	4015.877	20	4	
TB	4015.930	8		
SM	4016.004	16		
U	4016.046	5	6	
AU II	4016.050	10	16	
MO	4016.062		10	
NB	4016.079	5	32	
W	4016.109	6	4	
TI I	4016.284	32	5	
TH	4016.304	5	5	
U	4016.341	13	8	
TB	4016.360	10		
FE I	4016.429	16	4	
W	4016.527	10	13	
MN	4016.665	8	5	
MO	4016.702	5	5	
EU	4016.703	8		
PR	4016.748	25	20	
V	4016.821		16	
U	4016.846	6	8	
CO I	4016.883	10	5	
TI I	4016.975	13	2	
ZR I	4016.985	16		
FE I	4017.153	79	50	
C II	4017.270		5	
V	4017.296		16	
W	4017.351	4	5	

Table IIb.(3) TH II 4019.137 A
(con't)

Element	Wavelength (Angstrom)	Intensity Arc	Spark	Plasma
MO	4017.382	6	6	
U	4017.458	8	6	
NI I	4017.462	16		
TH	4017.495	8	8	
NB	4017.559	3	10	
EU	4017.582	25	25	
CE II	4017.596	10		
GD	4017.717	10	8	
U	4017.723	25	25	
TI I	4017.769	63	8	
MN	4018.102	79	63	
ZR	4018.121	25		
CR I	4018.202	32	8	
OS	4018.257	63	4	
FE I	4018.275	50	6	
U	4018.282	8	3	
ZR II	4018.383	8	5	
EU	4018.400	10	4	
RE	4018.409	25		
SM II	4018.539	4	5	
ND	4018.826	16	10	
CA	4018.880		4	
U	4018.990	25	16	
CE	4019.044	16	4	
MO	4019.046	2	4	
V	4019.046		6	
TB	4019.120	40	5	
RE I	4019.130	16		
*TH II	4019.137	8	8	36 VAF
U	4019.203	6	8	
W	4019.231	20	16	
CO I	4019.302	79		
YB	4019.350		6	
PR	4019.440	10	3	
RU I	4019.553	13	8	
PB	4019.639	6	6	
GD I	4019.732	16	10	
MO	4019.789	10		
W	4019.805		5	
ND	4019.809	10	8	

Table IIb.(3) TH II 4019.137 A
(con't)

Element	Wavelength (Angstrom)	Intensity Arc	Spark	Plasma
SM II	4019.827	8	8	
PR	4019.840	16	8	
CE	4019.897	8	3	
SM II	4019.976	32	16	
IR	4020.028	79	100	
MN	4020.092	10	5	
PD I	4020.225	16		
NB	4020.237	2	10	
PR	4020.280	10	2	
SC I	4020.399	50	20	
MO	4020.455	10	10	
TB	4020.470	20	3	
ER	4020.519	20		
ND	4020.872	16	16	
DY	4020.897	10	4	
CO I	4020.905	500		
PR	4020.991	40	32	
RU I	4020.995	16	13	
U	4021.005		4	
MO	4021.015	16	25	
TB	4021.120	10		
U	4021.254	6	6	
ND	4021.330	13	13	
TM	4021.390	4	10	
ND	4021.795	13	10	
TI I	4021.827	100	20	
FE I	4021.870	200	100	
V I	4021.925	8	2	
TH	4022.093	20	16	
W	4022.119	13	10	
RU I	4022.161	40	100	
PR	4022.199	8	4	
CR	4022.263	79	40	
CE	4022.272	16	4	
AS II	4022.320		5	
GD	4022.345	16		
NB	4022.388	2	10	
CU I	4022.657	400	25	
SM II	4022.714	6	5	
PR	4022.738	13	6	

Table IIb.(3) TH II 4019.137 A
(con't)

Element	Wavelength (Angstrom)	Intensity Arc	Spark	Plasma
W	4022.833	4	5	
TB	4022.872	16		
RE I	4022.965	25		
ND	4023.002	16	16	
RH I	4023.144	10	5	
GD	4023.154	20	10	
U	4023.170	6	6	
SC I	4023.223	63		
SM II	4023.223	32	25	
GD I	4023.350	20	10	
RE I	4023.353	40		
CE	4023.370	8		
V II	4023.389	10	32	
CO I	4023.403	200		
CR I	4023.430	16		
MO	4023.560		25	
LA II	4023.588	50	16	
U	4023.597	3	6	
SC I	4023.688	100	25	
DY	4023.722	10	2	
CR	4023.739	40	16	
ND	4023.824	13	3	
RU I	4023.833	25	63	
ZR I	4023.981	32	2	
TB	4024.070	40		
MO	4024.091	32	25	
FE I	4024.104	8	2	
SB	4024.137		4	

Table IIa.   TH II 2837.299 A

	Element		Wavelength (Angstrom)	Intensity Arc	Spark	Plasma
	RN		2836.300			25
	MN		2836.313	16		
	FE		2836.322	8	5	
	O	II	2836.350			5
	IR	I	2836.404	25	10	
	TI	I	2836.412	4		
	TH		2836.438	5	2	
	RA	II	2836.460			25
	CR		2836.479	3	20	
	ZR	I	2836.494	5		
	FE	II	2836.509	3	13	
	V		2836.520	20	79	
--	RU		2836.573	32		
	TI	I	2836.612	5	2	
	TA		2836.618	79	2	
	TI	II	2836.640		100	
	HG		2836.670		5	
	RH	I	2836.690	63		
	V	I	2836.698	10		
	MO		2836.705		25	
	C	II	2836.710		200	
	FE	II	2836.716		20	
--	GA		2836.870		2	
	CD	I	2836.907	200	79	
	U		2836.918	5	10	
	IN	I	2836.919	79	79	
	I		2836.922			40
	GD		2836.955	2		
	MO		2836.965	8		
	DY		2836.994	2		
	RE		2837.015	2		
	FE		2837.030		2	
--	ND		2837.100	5		
	ER		2837.106	5		
	PD	II	2837.120			40
	CO	I	2837.151	79		
	U		2837.187	10	8	
	SE		2837.210			32
	PT	I	2837.230	2		
	ZR	I	2837.232	100		

Table IIa. TH II 2837.299 A
(con't)

	Element	Wavelength (Angstrom)	Intensity Arc	Spark	Plasma
	RU	2837.274	20		
	CE	2837.289	50		
	SC	2837.293	4	6	
	FE II	2837.299		25	
	*TH II	2837.299	16	10	46 VAF
	SB	2837.311		3	
	MO	2837.320	5		
	IR I	2837.327	20	10	
	U	2837.328	10	8	
	W	2837.344	13	10	
	OS	2837.422	25	10	
	ND	2837.510	5		
	CU II	2837.547		250	
--	RE	2837.549	40		
	CE	2837.598	8		
	C II	2837.602		40	
	DY	2837.613	2		
	PD II	2837.655		10	
	U	2837.726	6	6	
	AG II	2837.760		5	
---	W	2837.762	10	10	
	CR	2837.877	2	32	
	CE	2837.896	10		
	MO	2837.901	16		
	TA	2837.942	32	32	
	AL	2837.950			8
	CE	2837.994	8		
	ZR I	2838.022	3		
----	AU	2838.030		79	
	V I	2838.058	6	2	
	CS	2838.090			20
	FE I	2838.120	160	160	
	RE	2838.120	8		
	OS	2838.173	32	13	
	U	2838.204	3	2	
	FE II	2838.235		8	
	TA	2838.240	2	160	

Table IIb.(3) TH II 2837.299 A

Element	Wavelength (Angstrom)	Intensity Arc	Spark	Plasma
CE	2832.307	20		
TH II	2832.319	20	25	42 VAF
FE I	2832.436	320	200	
CR	2832.460	2	130	
W	2832.480	10	10	
LA II	2832.550	4	5	
RU	2832.625	20		
MO	2832.657		5	
NB	2832.790		5	
ER	2833.061	25		
PB I	2833.069	500	79	21 VAF
FE II	2833.100		5	
IR	2833.236	6	20	
EU	2833.250	10	5	
HF	2833.276	25	4	
NB	2833.304		10	
CE	2833.309	50		
TH	2833.339	8	8	
TB	2833.370		20	
FE	2833.400	10	8	
W	2833.628	16	13	
TA	2833.636	320	40	
GD	2833.753	20	20	
RU	2833.784		79	
MO	2833.792		8	
TM	2833.820	6	10	
U	2833.821	16	25	
ZR II	2833.908	2	5	
CO I	2833.923	40		
RU	2834.001	32	5	
RE	2834.060	100		
RH I	2834.117	63	32	
HF	2834.134	20	13	
TI II	2834.161		8	
W II	2834.212		32	
CR	2834.262		130	
LU	2834.350	5	40	
MO	2834.394	20	40	
TA	2834.408		8	
CO I	2834.428	50		

Table IIb.(3) TH II 2837.299 A
(con't)

Element	Wavelength (Angstrom)	Intensity Arc	Spark	Plasma
TH	2834.484	6	5	
V	2834.530		40	
NI I	2834.553	40	16	
Y II	2834.570	2	20	
RE I	2834.615	32		
PT I	2834.713	79	5	
FE	2834.755	16	10	
MO	2834.780	16	10	
CO II	2834.939	2	79	
YB	2834.970		8	
HO	2834.990		10	
NB	2835.116	5	100	
LU	2835.250		10	
MO	2835.331	20	40	
V	2835.347		10	
AU	2835.430		8	
FE I	2835.459	100	100	
U	2835.572	5	6	
CR II	2835.633	100	400	42 VAF
W	2835.636	13	10	
TI I	2835.643	8	6	
U	2835.803	8	10	
AU	2835.890		10	
MO	2835.914	16		
FE I	2835.955	16	10	
TH	2836.049	10	8	
TB	2836.110		20	
RU	2836.147	20		
FE II	2836.188	3	10	
NB	2836.241	3	5	
W	2836.252	10	10	
MO	2836.295	2	10	
MN	2836.313	16		
FE	2836.322	8	5	
IR I	2836.404	25	10	
CR	2836.479	3	20	
FE II	2836.509	3	13	
V	2836.520	20	79	
RU	2836.573	32		
TA	2836.618	79	2	

Table IIb.(3) TH II 2837.299 A
(con't)

Element	Wavelength (Angstrom)	Intensity Arc	Spark	Plasma
TI II	2836.640		100	
HG	2836.670		5	
RH I	2836.690	63		
MO	2836.705		25	
C II	2836.710		200	
FE II	2836.716		20	
CD I	2836.907	200	79	
U	2836.918	5	10	
IN I	2836.919	79	79	
CO I	2837.151	79		
U	2837.187	10	8	
ZR I	2837.232	100		
RU	2837.274	20		
CE	2837.289	50		
SC	2837.293	4	6	
FE II	2837.299		25	
*TH II	2837.299	16	10	46 VAF
IR I	2837.327	20	10	
U	2837.328	10	8	
W	2837.344	13	10	
OS	2837.422	25	10	
CU II	2837.547		250	
RE	2837.549	40		
C II	2837.602		40	
PD II	2837.655		10	
U	2837.726	6	6	
AG II	2837.760		5	
W	2837.762	10	10	
CR	2837.877	2	32	
MO	2837.901	16		
TA	2837.942	32	32	
AU	2838.030		79	
FE I	2838.120	160	160	
OS	2838.173	32	13	
FE II	2838.235		8	
TA	2838.240	2	160	
LA II	2838.438	2	5	
FE I	2838.450	8	5	
V	2838.535		10	
RU	2838.621	32		

Table IIb.(3) TH II 2837.299 A
(con't)

Element	Wavelength (Angstrom)	Intensity Arc	Spark	Plasma
U	2838.623	6	13	
OS I	2838.626	100	100	
W	2838.674	10	5	
ND	2838.702		5	
ER	2838.714	16	3	
TB	2838.720	3	20	
RU	2838.846		10	
IN	2838.870		6	
W	2838.890	10	6	
TM	2838.940	16	40	
NI I	2838.958	25		
TM	2839.110	10	6	
IR I	2839.158	25	16	
MO	2839.164		25	
CR	2839.235		20	
TH	2839.243	8	6	
W	2839.337	10	10	
ZR II	2839.339	6	6	
CE	2839.364	20		
RU	2839.374		6	
V I	2839.437	13	6	
FE II	2839.529	4	25	
MO	2839.585	25		
NB	2839.598		10	
ND	2839.690		5	
NB	2839.800	2	5	
TI II	2839.800		100	
W II	2839.810		10	
FE II	2839.819		10	
TM	2839.850	5	10	
U	2839.890	20	20	
PD II	2839.892		100	
W	2839.920		10	
SN I	2839.989	320	320	27 VAF
MN	2840.001	20		
CR	2840.021	25	130	
V	2840.106	2	25	
TH	2840.156	8	6	
IR I	2840.219	16	10	
GD	2840.236	50	63	

Table IIb.(3) TH II 2837.299 A
(con't)

Element	Wavelength (Angstrom)	Intensity Arc	Spark	Plasma
FE II	2840.342		8	
RE I	2840.348	40		
TA	2840.390	2	50	
FE I	2840.423	130	20	
CR	2840.438		6	
LA II	2840.500	3	25	
RU	2840.539	63	8	
V	2840.599	2	13	
FE II	2840.647		63	
PB II	2840.660		20	
FE II	2840.762		32	
Y II	2840.960	6	20	
PD II	2841.029		100	
V	2841.039	16	32	
W	2841.110	2	10	
RU	2841.125		130	
NB	2841.146	10	100	
TH	2841.168	8	8	
W	2841.570	13	13	
RU	2841.677	50	200	
NA II	2841.721	20		79
MO	2841.773		16	
TI II	2841.938	40	130	
HF	2841.950		5	
PR	2841.970		8	
PT	2842.030		10	
HO	2842.040		10	
V	2842.041		6	
FE II	2842.076		5	
U	2842.088	16	13	
RH	2842.126	16		
MO	2842.151	2	40	
U	2842.240	5	8	
V	2842.289		10	

Table IIa. TI II 3349.406 A

	Element	Wavelength (Angstrom)	Intensity Arc	Spark	Plasma
	TI I	3348.535	6		
	TB	3348.540	16		
	HO	3348.580		4	
	ND	3348.586	2		
----	OS	3348.662	32	16	
	SM	3348.682	10	10	
	U	3348.691	4		
	RU I	3348.705	50	2	
	CS I	3348.720	16		
	RB I	3348.720	100		
	ER	3348.740	10		
	TH	3348.766	5	4	
	CE	3348.775	6		
	NB	3348.784		32	
	TI II	3348.844	13	13	
	SM	3348.871	3		
	DY	3348.880	2		
---	MO	3348.940		32	
	TH	3348.961	4	6	
	TI II	3349.035	130	790	40 VAF
	U	3349.036	4		
	NB	3349.059	79	100	
	CR	3349.072	130	40	
--	HF II	3349.170	5	4	
	MO	3349.193	6	5	
	SC I	3349.220	4		
	CO I	3349.222	3		
	CU I	3349.292	63	40	
	CR	3349.322	32	50	
	TH	3349.340	5	10	
	W	3349.340	2	10	
	NB	3349.348	5	100	
	AU	3349.400	16	5	
	CE	3349.405	4	5	
	*TI II	3349.406	100	400	79 VAF
	TB	3349.420	32	50	
	U	3349.430	4	3	
	CS	3349.445			10
	CU II	3349.463		2	
	NB	3349.523	32	5	

Table IIa. TI II 3349.406 A
(con't)

	Element	Wavelength (Angstrom)	Intensity Arc	Spark	Plasma
	CO	3349.526	6		
--	W	3349.538	4	6	
	FE	3349.738			
	I	3349.793			10
	LA I	3349.834	4		
	SM	3349.835	3		
--	TH	3349.866		2	
	RE I	3349.910	20		
	CE II	3349.967	32		
	TM	3349.990	20		
	ER	3350.063	20	6	
	GD	3350.099	3	3	
--	RU	3350.107	13		
	CA I	3350.209	100	10	
	I	3350.240			16
	RU	3350.251	10	2	
	ER	3350.255	10	2	
	PR	3350.275	10	3	
	U	3350.280	4		
	FE I	3350.285	8	4	
	MO	3350.299	5	6	
	TH	3350.330	2	2	
	CA I	3350.360	16	3	
	CO I	3350.376	4		
	MN	3350.401	4		

Table IIb.(2) TI II 3349.406 A

Element	Wavelength (Angstrom)	Intensity Arc	Spark	Plasma
CA I	3344.513	100	6	
RU I	3344.532	63	6	
LA II	3344.560	320	200	
MO	3344.746	50	40	
CE II	3344.761	50	8	
ZN I	3345.020	790	320	2 VAF
RU	3345.317	63	5	

Table IIb.(2) TI II 3349.406 A
(con't)

Element	Wavelength (Angstrom)	Intensity Arc	Spark	Plasma
ZN I	3345.572	500	100	0.4 VAF
V	3345.900		130	
ZN I	3345.934	160	50	
RE I	3346.199	100		
TI II	3346.728	63	63	
CR	3346.742	160	79	
CO I	3346.935	100	2	
RU I	3347.613	63	6	
CR II	3347.837	32	130	
FE I	3347.927	160	100	
RU	3348.012	50	3	
CO I	3348.113	79		
RU I	3348.705	50	2	
RB I	3348.720	100		
TI II	3349.035	130	790	40 VAF
NB	3349.059	79	100	
CR	3349.072	130	40	
CU I	3349.292	63	40	
NB	3349.348	5	100	
*TI II	3349.406	100	400	79 VAF
CA I	3350.209	100	10	
GD II	3350.470			14 VAF
GD II	3350.482	160	200	14 VAF
RB I	3350.890	160		
SR I	3351.246	320	16	
FE I	3351.521	63	63	
OS	3351.735	50	13	
FE	3351.745	79	63	
RU I	3351.932	50	4	
CR I	3351.966	50	50	
SN	3352.435		100	
RU	3353.648	50	4	
SC II	3353.734	50	63	80 VAF
V	3353.775	2	100	
CO I	3354.377	200		

Table IIa.   TI II 3349.035 A

	Element	Wavelength (Angstrom)	Intensity Arc	Spark	Plasma
	TB	3348.070	32	3	
	MO	3348.075	6		
	CO I	3348.113	79		
	ER	3348.140	16	2	
	ND	3348.161	10	2	
	CE	3348.190	8		
	NB	3348.284		50	
--	W	3348.293	2	16	
	U	3348.330	5		
	V	3348.374		20	
--	TI I	3348.535	6		
	TB	3348.540	16		
	HO	3348.580		4	
	ND	3348.586	2		
	OS	3348.662	32	16	
	SM	3348.682	10	10	
	U	3348.691	4		
	RU I	3348.705	50	2	
	CS I	3348.720	16		
	RB I	3348.720	100		
	ER	3348.740	10		
	TH	3348.766	5	4	
	CE	3348.775	6		
	NB	3348.784		32	
--	TI II	3348.844	13	13	
	SM	3348.871	3		
	DY	3348.880	2		
	MO	3348.940		32	
	TH	3348.961	4	6	
	*TI II	3349.035	130	790	40 VAF
	U	3349.036	4		
	NB	3349.059	79	100	
	CR	3349.072	130	40	
	HF II	3349.170	5	4	
	MO	3349.193	6	5	
	SC I	3349.220	4		
-	CO I	3349.222	3		
	CU I	3349.292	63	40	
	CR	3349.322	32	50	
	TH	3349.340	5	10	

Table IIa. TI II 3349.035 A
(con't)

	Element	Wavelength (Angstrom)	Intensity Arc	Spark	Plasma
	W	3349.340	2	10	
	NB	3349.348	5	100	
	AU	3349.400	16	5	
	CE	3349.405	4	5	
	TI II	3349.406	100	400	79 VAF
	TB	3349.420	32	50	
	U	3349.430	4	3	
	CS	3349.445			10
	CU II	3349.463		2	
	NB	3349.523	32	5	
---	CO	3349.526	6		
	W	3349.538	4	6	
----	FE	3349.738			
	I	3349.793			10
	LA I	3349.834	4		
	SM	3349.835	3		
	TH	3349.866		2	
	RE I	3349.910	20		
	CE II	3349.967	32		
	TM	3349.990	20		

Table IIb.(2) TI II 3349.035 A

Element	Wavelength (Angstrom)	Intensity Arc	Spark	Plasma
RH I	3344.198	100	20	
RE I	3344.347	160		
CA I	3344.513	100	6	
LA II	3344.560	320	200	
ZN I	3345.020	790	320	2 VAF
ZN I	3345.572	500	100	0.4 VAF
ZN I	3345.934	160	50	
RE I	3346.199	100		
CR	3346.742	160	79	
CO I	3346.935	100	2	
FE I	3347.927	160	100	
CO I	3348.113	79		

Table IIb.(2) TI II 3349.035 A
(con't)

Element	Wavelength (Angstrom)	Intensity Arc	Spark	Plasma
RB I	3348.720	100		
*TI II	3349.035	130	790	40 VAF
NB	3349.059	79	100	
CR	3349.072	130	40	
TI II	3349.406	100	400	79 VAF
CA I	3350.209	100	10	
GD II	3350.470			14 VAF
GD II	3350.482	160	200	14 VAF
RB I	3350.890	160		
SR I	3351.246	320	16	
FE	3351.745	79	63	
SC II	3353.734	50	63	80 VAF

Table IIa.   TI II 3361.213 A

	Element	Wavelength (Angstrom)	Intensity Arc	Spark	Plasma
	MO	3360.280		25	
	TB	3360.290	8	3	
	CE	3360.291	3		
	CR II	3360.295	50	200	
	FE	3360.313	3		
	IR	3360.313	3		
	W	3360.325	4	13	
	U	3360.337	5		
	CE	3360.366	2		
	TH	3360.381	4	5	
	CE	3360.405	2		
	ZR I	3360.455	20		
----	P	3360.490			32
	CE	3360.541	32	4	
	NE II	3360.630			20
	SM	3360.640	2		
	DY	3360.650	2	2	
---	CE	3360.714	5		
	GD	3360.716	25	25	
	W	3360.741	4	2	
	RU	3360.792	4		
	RH I	3360.803	2		
	IN	3360.850		6	
	HO	3360.860		4	
	PR	3360.900	8	2	
	NB	3360.902	3	50	
	FE I	3360.928	6		
--	TI I	3360.990	10		
	CE	3360.993	3		
	ER	3361.020	13		
	TM	3361.040	10	16	
	C II	3361.090		13	
	CO I	3361.093	5		
	W II	3361.107	10	16	
	RE I	3361.142	25		
	RU	3361.148	32		
	OS	3361.149	79	20	
	BI	3361.209	3	2	
	AU	3361.210	3	4	
	*TI II	3361.213	100	630	57 VAF

Table IIa. TI II 3361.213 A
(con't)

	Element	Wavelength (Angstrom)	Intensity Arc	Spark	Plasma
	*U	3361.213	2		
	EU	3361.214	20	2	
	CE	3361.232	8	5	
	TB	3361.240	3	8	
	NI I	3361.241	10		
	TI I	3361.263	79	50	
	CO I	3361.267	20		
	SC II	3361.270	25	10	17 VAF
	MO	3361.373	32	32	
--	SM	3361.435	10	2	
	RB	3361.490			10
	V	3361.508		200	
	CE II	3361.555	10		
	NI I	3361.556	500	20	
	CO I	3361.558	79	2	
	PB I	3361.580		16	
	EU	3361.615	2	3	
	TA	3361.640	130	50	
---	ER	3361.670	6		
	U	3361.727	16	8	
	AR II	3361.730			5
	TH	3361.739	4	4	
	KR I	3361.740			2
	C II	3361.750		6	
	CE	3361.762	25		
	CR II	3361.770	10	100	
	ND	3361.773	8	2	
	RE	3361.833	4		
	TI I	3361.835	13		
	W	3361.852	6	3	
	CE II	3361.853	8		
	CA I	3361.918	130	10	
	SC II	3361.935	25	10	
---	FE	3361.952	6		
	W	3361.971	4	3	
	Y II	3362.000	13	25	
	RU I	3362.003	63	8	
	LA I	3362.040	10	3	
	U	3362.050	2		
	TI I	3362.100	6	3	

Table IIa. TI II 3361.213 A
(con't)

Element	Wavelength (Angstrom)	Intensity Arc	Spark	Plasma
CA I	3362.131	16	5	
DY	3362.170	3	2	
NB	3362.172		100	
RH I	3362.184	100	20	
TH	3362.196	3	4	
CR I	3362.213	79	8	

Table IIb.(2) TI II 3361.213 A

Element	Wavelength (Angstrom)	Intensity Arc	Spark	Plasma
V I	3356.352	130	63	
CO I	3356.472	160	2	
ZR II	3357.264	50	40	
OS	3357.969	100	16	
MO	3358.124	63	32	
NB	3358.417	100	100	
CR II	3358.501	40	200	
TA	3358.527	63	25	
GD II	3358.628	100	100	14 VAF
RU I	3359.095	63	20	
NI I	3359.106	63		
CO I	3359.286	100	2	
LU	3359.560	160	16	
SC II	3359.679	50	25	
RH I	3359.895	100	50	
CR II	3360.295	50	200	
OS	3361.149	79	20	
*TI II	3361.213	100	630	57 VAF
TI I	3361.263	79	50	
SC II	3361.270	25	10	17 VAF
V	3361.508		200	
NI I	3361.556	500	20	
CO I	3361.558	79	2	
TA	3361.640	130	50	
CA I	3361.918	130	10	
RU I	3362.003	63	8	

Table IIb.(2) TI II 3361.213 A
(con't)

Element	Wavelength (Angstrom)	Intensity Arc	Spark	Plasma
RH I	3362.184	100	20	
CR I	3362.213	79	8	
GD II	3362.244	160	200	15 VAF
RU	3362.335	50	5	
TM II	3362.610	250	200	26 VAF
CO I	3362.802	79		
NI I	3362.806	100		
CO I	3363.760	79		
OS	3364.122	100	13	
CU I	3365.349	63	32	
V I	3365.553	130	79	
NI I	3365.766	400	13	
NI I	3366.168	400	13	

Table IIa.   TL I   3775.720 A

	Element	Wavelength (Angstrom)	Intensity Arc	Spark	Plasma
	SI	3774.740		4	
	FE I	3774.827	100	40	
	DY	3774.836	4		
	U	3774.907	10		
	ND	3774.928	10	6	
	TB	3774.960	8		
----	P II	3775.020			32
	V I	3775.187	25	5	
	IR	3775.206	8		
	PR	3775.211	10	4	
---	ND	3775.240	16	8	
	TB	3775.260	16	8	
	U	3775.262	6	3	
	TH	3775.319	20	5	
	CE	3775.363	2	4	
	EU	3775.410	4		
	U	3775.442	4	4	
	W	3775.446	8	6	
	NB	3775.448	5	10	
	AR I	3775.450			10
	SM	3775.459	10	3	
	ZR I	3775.461	6		
	RE	3775.464	10		
--	ND	3775.501	10	20	
	CE	3775.564	4		
	NI I	3775.572	500	40	
	U	3775.608	13	5	
	MO	3775.647	16	20	
	ER	3775.665	8	2	
	EU	3775.690	4	4	
	HG	3775.697			2
	RH I	3775.716	8	3	
	V I	3775.719	32	2	
	TI I	3775.720	3200	1000	
	*TL I	3775.720			13 VAF
	BI	3775.745	2		
	SM	3775.849	2		
	CA	3775.860	2	2	
	FE I	3775.868	2		
--	TH	3775.942	20	10	

Table IIa. TL I 3775.720 A
(con't)

	Element	Wavelength (Angstrom)	Intensity Arc	Spark	Plasma
	CE	3775.986	13	16	
	U	3775.991	10	13	
	BA	3776.000	3	6	
	DY	3776.010	2		
	TB	3776.020	8		
	TI II	3776.059	8	63	
	SM	3776.060	16		
	PR	3776.089	8	3	
	MO	3776.104	5	5	
	CE	3776.146	5	5	
	V I	3776.157	50	2	
	NB	3776.161	2	5	
---	EU	3776.200	10		
	OS	3776.254	50	16	
	HG II	3776.260			32
	TH	3776.276	8		
	CD	3776.320		3	
	ND	3776.338	16	13	
---	FE	3776.459	130	63	
	TB	3776.490	100	100	
	EU	3776.520	4	4	
	MN	3776.527	25	25	
	MO	3776.550	5	8	
	Y II	3776.561	13	13	
	NB	3776.601	3	3	
	CE II	3776.608	6	3	
	W	3776.690		2	
	CE II	3776.711	2		

Table IIb.(1) TL I 3775.720 A

Element	Wavelength (Angstrom)	Intensity Arc	Spark	Plasma
SM	3771.350	50	2	
TI I	3771.655	63	32	
PR	3772.854	79	20	
PR	3774.064	100	50	
EU	3774.080	50	10	
Y II	3774.332	13	100	57 VAF
OS	3774.400	63	16	
CO I	3774.605	200		
OS	3774.619	63	13	
FE I	3774.827	100	40	
NI I	3775.572	500	40	
TI I	3775.720	3200	1000	
*TL I	3775.720			13 VAF
V I	3776.157	50	2	
OS	3776.254	50	16	
FE	3776.459	130	63	
TB	3776.490	100	100	
OS	3776.992	160	20	
CO I	3777.542	200		
RU I	3777.586	63	50	
HF I	3777.658	2000	200	
FE	3778.513	63	25	
V I	3778.684	63	4	
FE I	3779.446	100	63	
LA II	3780.674	50	50	

Table IIa. TM II 3462.200 A

	Element	Wavelength (Angstrom)	Intensity Arc	Spark	Plasma
	CE	3461.217	4		
	TH	3461.218	10	4	
	PR	3461.244	4		
	XE II	3461.260			50
	DY	3461.310	3		
	CE	3461.344	16		
	HO	3461.360		4	
	W	3461.364	6	6	
	EU	3461.380	25	2	
	ER	3461.400	10		
	SM	3461.405	5	4	
	CE	3461.417	2		
---	TI II	3461.500	79	130	
	RB II	3461.574			200
	V	3461.576		3	
	IR	3461.582	4		
	NB	3461.611	5	2	
	NI I	3461.652	790	50	
	W	3461.655	3	3	
	EU	3461.665	3		
---	YB	3461.710		3	
	CE	3461.793	10		
	W	3461.814	10	5	
	PR	3461.873	6	2	
	RU	3461.924	32		
--	GD	3461.956	5	5	
	HO	3461.960	20	20	
	RH I	3462.040	1000	160	
	RU	3462.040	5		
	MO	3462.060		25	
	PR	3462.100	6		
	SC I	3462.183	4	2	
	OS	3462.191	20	10	
	*TM II	3462.200	250	200	37 VAF
	CE	3462.234	3		
	LA	3462.320	2	2	
	PR	3462.360	3		
	FE I	3462.361	10	3	
--	CE	3462.433	8		
	NA II	3462.494	2		16

Table IIa. TM II 3462.200 A
(con't)

	Element	Wavelength (Angstrom)	Intensity Arc	Spark	Plasma
	TB	3462.510	8	3	
	ER	3462.580	10		
	CA	3462.620	2	4	
	HF II	3462.641	16	13	
	NB	3462.647	5	3	
---	SM II	3462.688	5	2	
	CR II	3462.730	2	8	
	CE	3462.767	10		
	CO I	3462.803	1000	79	
	NI I	3462.804	6	10	
	TH	3462.855	10	13	
	MN II	3462.878			10
----	TA	3462.894	2		
	TB	3462.970	16	3	
	GD	3463.000	6	4	
	ZR II	3463.017	20	40	
	EU	3463.030	8		
	NB	3463.030	5	32	
	MO	3463.032		20	
	V	3463.078		32	
	RN	3463.100			2
	PR	3463.111	8		
	IR	3463.120	4		
	CE	3463.128	3		
	RU I	3463.144	63	4	
	SM	3463.184	2		

Table IIb.(2) TM II 3462.200 A

Element	Wavelength (Angstrom)	Intensity Arc	Spark	Plasma
CR	3457.630	4	130	
RH I	3457.926	130	10	
YB	3458.280	13	100	
OS	3458.385	200	13	
NI I	3458.474	790	50	
FE I	3459.918	79	50	

Table IIb.(2) TM II 3462.200 A
(con't)

Element	Wavelength (Angstrom)	Intensity Arc	Spark	Plasma
MN II	3460.328	63	500	
RE I	3460.470	1000		3 VAF
PD I	3460.774	320	630	16 VAF
TI II	3461.500	79	130	
NI I	3461.652	790	50	
RH I	3462.040	1000	160	
*TM II	3462.200	250	200	37 VAF
CO I	3462.803	1000	79	
NB	3463.811	32	50	
GD	3463.984	100	130	
YB	3464.370	200	50	
SR II	3464.457	200	200	13 VAF
TI II	3465.562	6	63	
CO I	3465.800	2000	25	
FE I	3465.863	500	400	
CD I	3466.201	1000	500	1 VAF
FE I	3466.500	32	63	

Table IIa. TM II 3131.260 A

	Element	Wavelength (Angstrom)	Intensity Arc	Spark	Plasma
	V II	3130.267	50	200	
	FE	3130.278	5	4	
	IR	3130.285	4		
	TA	3130.290	16		
	CE	3130.334	32		
	U	3130.373	3		
	TI I	3130.376	2		
	HO	3130.380		4	
	P II	3130.380			32
	XE II	3130.400			2
	BE II	3130.416	200	200	11 VAF
	W	3130.456	10	8	
	SI	3130.480		5	
----	CE	3130.516	5		
	U	3130.564		16	
	CR	3130.565		13	
	FE II	3130.567	4	4	
	BA I	3130.570	2	3	
	IR	3130.578	3		
	TA	3130.578	100	32	
	CS	3130.700			4
	S	3130.710			16
	U	3130.732	10	6	
	EU	3130.740	100	100	
---	NB II	3130.786	1000	100	60 VAF
	RH I	3130.790	63	2	
	AR I	3130.800			20
	TI II	3130.800	25	100	
	GD	3130.814	3		
	CE	3130.872	32	2	
	CE	3130.919	5		
	HO	3130.990	6	6	
--	ER	3131.070	3		
	TH	3131.070	13	10	
	BE II	3131.072	200	160	41 VAF
	ZR I	3131.110	6		
	OS	3131.115	130	32	
	MO	3131.195		5	
	CR	3131.211	20	6	
	TA	3131.216	5	25	

Table IIa. TM II 3131.260 A
(con't)

Element	Wavelength (Angstrom)	Intensity Arc	Spark	Plasma
*TM II	3131.260	400	500	58 VAF
U	3131.325	8	6	
TB	3131.350	8	3	
FE	3131.455	2		
OS	3131.481	20	10	
-- CE	3131.505	2		
U	3131.534	4	2	
HG I	3131.546	400	320	
U	3131.618		2	
CE	3131.679	4		
FE II	3131.718		32	
EU	3131.750	20	5	
-- SR II	3131.750		10	
HF	3131.814	40	10	
CO I	3131.830	8		
HG I	3131.833	200	100	
-- U	3131.987	8	6	
NB	3132.011		5	
RE I	3132.025	2		
ER	3132.030	6	2	
CE	3132.043	16		
CR II	3132.058	25	130	
ZR I	3132.066	10	2	
DY	3132.122	5	2	
NB	3132.137		2	
LA II	3132.140	2	3	
EU	3132.160	16		
U	3132.177	2		
CO I	3132.216	40		
NE II	3132.220			4

Table IIb.(2) TM II 3131.260 A

Element	Wavelength (Angstrom)	Intensity Arc	Spark	Plasma
CR II	3128.699	32	160	
V II	3130.267	50	200	
BE II	3130.416	200	200	11 VAF
NB II	3130.786	1000	100	60 VAF
BE II	3131.072	200	160	41 VAF
*TM II	3131.260	400	500	58 VAF
HG I	3131.546	400	320	
HG I	3131.833	200	100	
CR II	3132.058	25	130	
MO	3132.594	1000	320	
YB II	3132.628	16	160	
TA	3132.643	250	25	
CD I	3133.167	200	320	
V II	3133.328	50	200	
TM II	3133.890	200	200	22 VAF
NI I	3134.108	1000	160	
FE I	3134.111	200	130	
HF	3134.718	79	130	
V	3134.931	32	160	

Table IIa. U II 3859.580 A

	Element	Wavelength (Angstrom)	Intensity Arc	Spark	Plasma
	TA	3858.608	10	2	
	U	3858.682	3	6	
	V I	3858.685	50	16	
	RU	3858.686	4	4	
	SM	3858.741	20		
	KR II	3858.780			5
	CA	3858.810		2	
--	MO	3858.840		16	
	W	3858.846	5	6	
	CR I	3858.891	32	20	
	NB	3858.950	20	50	
	U	3859.014	8	4	
--	PR	3859.142	4	4	
	MG I	3859.211	2		
	FE I	3859.216	100	100	
	EU	3859.220	2		
	W	3859.294	10	13	
--	AL II	3859.330			10
	V I	3859.341	20	8	
	AU II	3859.370		3	
	PR	3859.416	3	3	
	ND	3859.422	10	13	
	EU	3859.520	4		
	*U II	3859.580	20	32	12 VAF
	ND	3859.670	25	25	
	CA	3859.675	6	5	
	RU	3859.712	6	16	
	CE	3859.792	3		
--	TA	3859.798	10	2	
	TH	3859.832	5	10	
	FE I	3859.913	1000	630	
	CE	3859.942	3	5	
--	W	3859.983	16	32	
	SM	3860.144	5		
	S II	3860.150			8
	CE	3860.178	2	4	
	SM II	3860.280	4	5	
--	LA II	3860.310		2	
	CE	3860.401	6	3	
	CU I	3860.472	32	6	

Table IIb.(3) U II 3859.580 A

Element	Wavelength (Angstrom)	Intensity Arc	Spark	Plasma
U II	3854.655	20	32	6 VAF
OS	3854.705	32	13	
CR	3854.788	20	16	
LA II	3854.910		40	
PR II	3854.963	79	32	
CR I	3855.286	32	32	
V I	3855.370	50	50	
NB	3855.500		50	
CR I	3855.571	32	32	
RU	3855.578	4	16	
SC	3855.620	16	16	
V I	3855.841	200	200	
SM II	3855.894	5	16	
RE	3855.935	20		
IR I	3856.072	25	4	
CR I	3856.281	20	16	
FE I	3856.373	500	320	
RU	3856.459	50	8	
RH I	3856.515	50	20	
MN	3856.535	16	32	
V	3856.669	20		
NB	3856.678	20	5	
CO	3856.800	79	2	
ND	3856.950	20	16	
OS I	3857.089	160	16	
ND	3857.140	32	20	
MO	3857.203		63	
RU	3857.551	50	25	
CR I	3857.631	50	25	
ND	3857.850	20	20	
TI I	3858.144	40	6	
PR	3858.256	20	8	
CO	3858.299	20	4	
NI I	3858.301	790	63	
V I	3858.685	50	16	
SM	3858.741	20		
MO	3858.840		16	
CR I	3858.891	32	20	
NB	3858.950	20	50	
FE I	3859.216	100	100	

Table IIb.(3) U II 3859.580 A
(con't)

Element	Wavelength (Angstrom)	Intensity Arc	Spark	Plasma
V I	3859.341	20	8	
*U II	3859.580	20	32	12 VAF
ND	3859.670	25	25	
RU	3859.712	6	16	
FE I	3859.913	1000	630	
W	3859.983	16	32	
CU I	3860.472	32	6	
SM	3860.619	25	10	
CE	3860.626	8	16	
V	3860.628	20		
U	3860.630		16	
RU I	3860.723	20	4	
CO I	3861.165	320	16	
EU	3861.180	32	32	
FE	3861.342	79	50	
HO	3861.680	40	20	
CU I	3861.746	50	2	
V I	3862.223	79	20	
ND	3862.487	20	13	
CR	3862.548	25	20	
RU	3862.646	2	63	
TI I	3862.826	32	4	
NB	3863.052	3	20	
ND II	3863.327	10	4	23 VAF
TH	3863.391	20	20	
U	3863.403	5	20	
ND II	3863.409	20	20	23 VAF
CO I	3863.607	32		
FE I	3863.742	63	32	
V II	3863.866	25	16	
V I	3863.866	25	16	
ZR I	3863.874	20	6	
MO I	3864.110	1000	500	
BI	3864.200		160	
V I	3864.300	32		
ZR I	3864.335	50	20	
LA II	3864.488	100	160	

Table IIa. U II 3670.072 A

	Element	Wavelength (Angstrom)	Intensity Arc	Spark	Plasma
	CE	3669.084	2		
	EU	3669.130	6		
	FE I	3669.157	50	32	
	TB	3669.210	8	3	
	NI I	3669.241	160	10	
	LA	3669.270		3	
	CE	3669.313	3		
----	W	3669.328	4	6	
	MO	3669.344	8	8	
	NB	3669.346	2	3	
	U	3669.360		3	
	DY	3669.390	2		
	MN	3669.399	3	3	
	V	3669.408	20	320	
	TH	3669.413	4	4	
	RE	3669.433	16		
	RU	3669.494	50	63	
	HO	3669.520	6	4	
	FE I	3669.523	200	160	
	PR	3669.523	10	3	
---	AR	3669.620			10
	TB	3669.620	32		
	RN	3669.700			5
	YB	3669.710	50	79	
	CE	3669.719	2		
	NB	3669.737	5	8	
	FE	3669.757	2		
	RE I	3669.774	25		
	EU	3669.810	8		
	PR	3669.817	10	2	
--	MN	3669.838	32	32	
	SM II	3669.886	8	2	
	XE I	3669.910			10
	W	3669.920		6	
	TH	3669.958	5		
	CE	3669.962	3		
	ND	3670.022	4	2	
	RE I	3670.025	16		
	FE I	3670.028	100		
	NB	3670.053		16	

Table IIa. U II 3670.072 A
(con't)

Element	Wavelength (Angstrom)	Intensity Arc	Spark	Plasma
CO I	3670.058	20		
TH	3670.062	3	4	
FE	3670.071	200	200	
CE	3670.072	2		
*U II	3670.072	16	20	10 VAF
CA	3670.210	2	4	
LA II	3670.230	2	3	
PR	3670.263	8	3	
U	3670.265	3	2	
CL	3670.290			8
HO	3670.290	6	6	
RE	3670.361	16		
LI I	3670.400	5		
NI I	3670.427	160	20	
SI	3670.490		4	
CE	3670.490	10		
SM II	3670.503	10	3	
ND	3670.510	10	4	
MN	3670.517	25	16	
RE I	3670.522	20		
U	3670.530	6		
TB	3670.550	16	3	
TH	3670.635	5		
AR I	3670.640			320
SM	3670.657	10	5	
MO	3670.668		25	
CE II	3670.672	6		
YB	3670.680	2	10	
W	3670.767	6	6	
W	3670.786		5	
PR	3670.794	8	2	
FE I	3670.812	20	6	
SM II	3670.817	100	50	40 VAF
EU	3670.830	4		
OS I	3670.891	200	20	
ND	3670.915	4		
AR II	3671.010			3
ZR	3671.042	3		

Table IIb.(3) U II 3670.072 A

Element	Wavelength (Angstrom)	Intensity Arc	Spark	Plasma
V I	3665.142	100	50	
ND	3665.180	25	13	
HF II	3665.346	20	25	
TB	3665.600	32	3	
CU I	3665.735	20	5	
MO	3665.747	16	16	
TM	3665.810	40	20	
CR	3665.980	20	16	
CE	3666.024	16		
TA	3666.100	20		
RH I	3666.215	63	32	
FE I	3666.249	20	6	
OS	3666.309	40	16	
NB	3666.533	8	10	
SC II	3666.537	16	8	
CR I	3666.642	20	16	
MO	3666.715	8	10	
HF II	3666.775	25	4	
TA	3666.890	16		
NB	3667.001	8	16	
HO	3667.050	6	10	
FE	3667.262	79	25	
CD II	3667.320		10	
TB	3667.350	16		
NB	3667.662	3	10	
W	3667.719	10	10	
V I	3667.741	79	25	
SM II	3667.907	25	10	
U	3667.975	16	8	
CE	3667.981	79	16	
FE I	3667.999	63	10	
CR I	3668.029	16	4	
TB	3668.070	16	3	
TM	3668.080	79	20	
FE I	3668.214	16	4	
GD	3668.310	5	10	
ZR II	3668.447	10	10	
Y II	3668.489	6	20	
ER	3668.491	25	3	
TB	3668.500	16	3	

Table IIb.(3) U II 3670.072 A
(con't)

Element	Wavelength (Angstrom)	Intensity Arc	Spark	Plasma
W	3668.665	10	13	
ND	3668.789	25	4	
PR	3668.830	100	40	
TI I	3668.966	100	40	
NB	3669.007	10	10	
FE I	3669.157	50	32	
NI I	3669.241	160	10	
V	3669.408	20	320	
RE	3669.433	16		
RU	3669.494	50	63	
FE I	3669.523	200	160	
TB	3669.620	32		
YB	3669.710	50	79	
RE I	3669.774	25		
MN	3669.838	32	32	
RE I	3670.025	16		
FE I	3670.028	100		
NB	3670.053		16	
CO I	3670.058	20		
FE	3670.071	200	200	
*U II	3670.072	16	20	10 VAF
RE	3670.361	16		
NI I	3670.427	160	20	
MN	3670.517	25	16	
RE I	3670.522	20		
TB	3670.550	16	3	
MO	3670.668		25	
YB	3670.680	2	10	
FE I	3670.812	20	6	
SM II	3670.817	100	50	40 VAF
OS I	3670.891	200	20	
V I	3671.205	100	63	
GD	3671.216	160	100	
ZR II	3671.269	40	32	
PB	3671.390		63	
PB	3671.503	50	6	
TI I	3671.673	160	63	
PR	3671.915	20	5	
AS II	3671.920		16	
CU I	3671.953	20	3	

Table IIb.(3) U II 3670.072 A
(con't)

Element	Wavelength (Angstrom)	Intensity Arc	Spark	Plasma
PT I	3671.994	79	10	
LA I	3672.016	25	3	
HF	3672.269	25	3	
ER	3672.303	20	2	
DY	3672.312	100	100	
ND	3672.363	32	13	
RE I	3672.400	20		
V I	3672.403	100	40	
U	3672.579	8	16	
W	3672.586		20	
CE	3672.789	16	5	
MO	3672.820	20	20	
TM	3673.140	10	16	
DY	3673.152	25	5	
EU	3673.196	20	2	
NB	3673.228	5	10	
TH	3673.262	10	10	
V I	3673.404	160	79	
ND	3673.542	20	20	
TH	3673.799	10	10	
PT I	3674.045	79	4	
TB	3674.050	16	3	
SM II	3674.053	50	10	
ND	3674.055	32	10	
GD	3674.063	100	32	
ER	3674.087	20	2	
DY	3674.093	100	50	
NI I	3674.150	200	50	
EU	3674.180	16		
W	3674.580	10	13	
ND	3674.645	16	6	
EU	3674.670	25		
NB	3674.678	20	4	
V	3674.685	2	20	
ZR II	3674.718	100	40	
FE	3674.768	40	25	
HO	3674.770	8	16	
NB	3674.778	20	5	
TA	3674.827	20	3	
IR I	3674.984	100	50	

Table IIa. V 3710.300 A

	Element	Wavelength (Angstrom)	Intensity Arc	Spark	Plasma
	TB	3709.300	16		
	V	3709.333		25	
	NB	3709.417	5	10	
	U	3709.460	3		
	I II	3709.514			3
	SM	3709.526	10	6	
	FE	3709.533	6	6	
--	CE	3709.588	5		
	NE II	3709.640			40
	FE	3709.665	4	4	
	PR	3709.665	3		
	TH	3709.673	5	3	
	NB	3709.742	2	2	
--	U	3709.878	2	2	
	AR	3709.900			5
	CE	3709.933	25	10	
	RE I	3709.937	40		
	TI I	3709.961	79	25	
	EU	3710.010	5		
	PR	3710.012	6	2	
--	DY	3710.080	20	5	
	CR	3710.090	4	2	
	MO	3710.142	20	16	
	U	3710.168		5	
	TI I	3710.186	6		
	CE	3710.253	2		
	W	3710.289	4	5	
	EU	3710.290	20		
	Y II	3710.290	79	160	86 VAF
	*V	3710.300			
	SM	3710.300	25	16	
	U	3710.308	8		
	YB	3710.310	4	6	
	RU	3710.316	4		
	F	3710.365			10
	NB	3710.448	16	20	
	P II	3710.450			32
	SB II	3710.519	2	5	
--	U	3710.534	3	4	
	LA II	3710.590	3	2	

Table IIa. V 3710.300 A
(con't)

Element	Wavelength (Angstrom)	Intensity Arc	Spark	Plasma
CR	3710.600	2	2	
DY	3710.730	4	2	
HO	3710.750	8	8	
CA	3710.753		3	
CS	3710.774			4
U	3710.785	4	2	
--- TA	3710.790	6		
SM II	3710.869	100	6	
EU	3710.881	32	32	
U	3710.908	3	8	
LU	3710.950	3		
---- CE	3711.001	2		
NA II	3711.074	8		63
PR	3711.099	8	3	
V	3711.123		79	
TA	3711.150	5		
IR I	3711.153	6		
PR	3711.218	3	2	
FE I	3711.225	79	50	
CR	3711.282		13	
HO	3711.300	6	4	

Table IIb.(2) V 3710.300 A

Element	Wavelength (Angstrom)	Intensity Arc	Spark	Plasma
FE I	3705.567	630	500	
LA II	3705.818	50	79	
TI II	3706.226	32	130	
FE I	3707.049	160	100	
NB	3707.918	3	100	
FE I	3707.925	79	63	
V I	3708.721	100	63	
FE I	3709.249	630	400	
Y II	3710.290	79	160	86 VAF
*V	3710.300			
V	3711.123		79	

Table IIb.(2) V 3710.300 A
(con't)

Element	Wavelength (Angstrom)	Intensity Arc	Spark	Plasma
TB	3711.740	200	32	
GD	3712.714	200	250	
SM II	3712.759	100	100	
CR II	3712.949	13	130	
NB	3713.014	100	79	
RH I	3713.022	100	100	
LA II	3713.544	200	63	
GD	3713.583	100	79	

Table IIa. V II 3093.108 A

	Element	Wavelength (Angstrom)	Intensity Arc	Spark	Plasma
	CE	3092.197	3		
	CL II	3092.220			50
	ZR	3092.240	3		
	HF II	3092.245	20	20	
	W	3092.285	6	3	
	TE	3092.320			16
----	CD II	3092.393	10	16	
	FE	3092.399	4	5	
	IR	3092.401	5		
	XE II	3092.410			10
	SC	3092.420	3		
	TA	3092.444	50	16	
	SC II	3092.519	3	4	
	YB	3092.560		32	
---	MO	3092.700	20	2	
	Y	3092.712	8		
	AL I	3092.713	1000	1000	13 VAF
	U	3092.720	5	3	
	V	3092.720	100	50	
	CE	3092.724	4		
	NA II	3092.729	50		200
	FE	3092.778	50	32	
	CE	3092.818	4		
	AL I	3092.842	50	20	13 VAF
--	NB	3092.886		5	
	NE II	3092.910			4
	ND	3092.915	8	6	
	MO	3092.920		10	
	MG I	3092.991	130	20	
	TA	3092.994	20		
	RN	3093.000			20
	CL	3093.000			4
	U	3093.012	20	20	
	TH	3093.055	13	6	
	DY	3093.108	16	5	
	*V II	3093.108	100	400	60 VAF
	TM	3093.120	32	63	
	CE	3093.243	6		
	SI	3093.280		6	
	AU II	3093.300		5	

Table IIa. V II 3093.108 A
(con't)

	Element	Wavelength (Angstrom)	Intensity Arc	Spark	Plasma
	ZR I	3093.321	3		
	CE II	3093.339	13		
--	FE	3093.357	63	40	
	U	3093.370	4	3	
	AR II	3093.410			50
	YB	3093.440		2	
	RU	3093.447	4		
	CA	3093.460	2	3	
	RH I	3093.480	2	3	
	CR	3093.488		100	
	W	3093.512	13	10	
	BI I	3093.580	10	8	
---	OS	3093.587	130	16	
	CE	3093.614	20		
	RE I	3093.652	63		
	MO	3093.680	10		
	CD II	3093.740	3	2	
	ER	3093.750	6		
	Y II	3093.770	10	16	
	V I	3093.792	32		
	FE I	3093.806	50	40	
	TI I	3093.813	5		
---	DY	3093.820	10		
	U	3093.868	3	4	
	TA	3093.869	50	16	
	YB	3093.879	3	16	
	FE	3093.883	40	32	
	ND	3093.883	6		
	RU	3093.901	32	100	
	CR	3093.947		25	
	CE	3093.950	4		
	CU I	3093.992	160	50	
	IR I	3094.009	20	10	
	W	3094.032	8	10	
	OS	3094.074	32	8	
	NE II	3094.080			13

Table IIb.(3) V II 3093.108 A

Element	Wavelength (Angstrom)	Intensity Arc	Spark	Plasma
TH	3088.473	20	20	
BH B	3088.600	100		
PB I	3089.093		32	
TI II	3089.401	13	100	
CO I	3089.595	100		
TI II	3090.051		100	
OS	3090.085	100	16	
OS	3090.296	100	13	
FE I	3091.578	320	200	
TL II	3091.610		25	
TL II	3091.660		50	
MO	3092.074	32	100	
HF II	3092.245	20	20	
YB	3092.560		32	
AL I	3092.713	1000	1000	13 VAF
V	3092.720	100	50	
FE	3092.778	50	32	
AL I	3092.842	50	20	13 VAF
MG I	3092.991	130	20	
U	3093.012	20	20	
*V II	3093.108	100	400	60 VAF
TM	3093.120	32	63	
FE	3093.357	63	40	
CR	3093.488		100	
OS	3093.587	130	16	
FE I	3093.806	50	40	
FE	3093.883	40	32	
RU	3093.901	32	100	
CR	3093.947		25	
CU I	3093.992	160	50	
NB II	3094.183	100	1000	83 VAF
V	3094.199	20	130	
RU	3094.558	6	50	
MO	3094.664	160	25	
TA	3095.393	63	20	
CR	3095.859	130	3	
W	3095.873	2	20	
CR	3096.130		100	
FE II	3096.296	2	32	
TI II	3096.424	3	32	

Table IIb.(3) V II 3093.108 A
(con't)

Element	Wavelength (Angstrom)	Intensity Arc	Spark	Plasma
RU	3096.568	63	63	
FE	3096.836	32	20	
MG I	3096.899	160	25	
TM	3096.970	16	50	
MN	3097.063	79	40	
NI I	3097.118	200	50	
NB	3097.122	3	100	
TI II	3097.186	20	160	
EU	3097.456	200	5	
RU	3097.605	2	20	
MO	3097.689	10	32	
TH	3097.960		50	
TM	3098.010	5	20	

Table IIa.   V II 3110.706 A

	Element	Wavelength (Angstrom)	Intensity Arc	Spark	Plasma
	NB	3109.736	2	2	
	AR	3109.750			3
	PR	3109.757	10	2	
	RE I	3109.759	20		
	DY	3109.768	40	20	
	YB	3109.780		2	
	CR	3109.790	2		
	HG	3109.840			20
	HO	3109.930	6	2	
----	SM	3109.965		10	
	CO I	3110.021	63	2	
	TH	3110.022	16	16	
	U	3110.038	6	3	
	TI II	3110.095	3	32	
	TA	3110.103	3		
	FE	3110.188	6	6	
	SM	3110.193	16	4	
---	U	3110.226	3	3	
	IR	3110.227	4		
	SC I	3110.238	3	5	
	EU	3110.240	20		
	CE	3110.278	32		
	FE	3110.281	40	32	
--	U	3110.525	3	8	
	RU I	3110.549	63	6	
	PR	3110.580	10		
	OS	3110.618	32	10	
	TI II	3110.620	8	20	
	MO	3110.644	3		
	AR I	3110.660			3
	MN	3110.677	32	32	
	*V II	3110.706	63	320	30 VAF
	FE	3110.710	2		
	U II	3110.710			30 VAF
	DY	3110.757	10	5	
	U	3110.784	3		
	NB	3110.800		5	
	TA	3110.815		63	
	CO I	3110.825	63	2	
	U	3110.832	6	8	

Table IIa. V II 3110.706 A
(con't)

	Element	Wavelength (Angstrom)	Intensity Arc	Spark	Plasma
	FE	3110.841	20	10	
	BE I	3110.842	20		
	MO	3110.844		32	
	CR	3110.860	25	8	
	RE I	3110.862	100		
	ER	3110.870	13	3	
	HF II	3110.874	32	40	
	ZR II	3110.878	10	8	
--	BI II	3110.880			5
	BE I	3110.965	16		
	CE	3110.998	2		
	CR	3111.000	8		
	BI II	3111.020			6
	OS	3111.090	100	20	
	W	3111.123	10	16	
	ZR II	3111.160	2		
	CE	3111.165	20		
---	BI II	3111.200			10
	MO	3111.224		16	
	CE	3111.234	10		
	TI I	3111.283	16	8	
	PR	3111.337	16	2	
	CO I	3111.339	20	3	
	CR	3111.340	6		
	PB	3111.380		2	
	BI II	3111.410			16
	RB	3111.430			32
	EU	3111.432	200		
	KR	3111.450			2
---	NB	3111.450	4	5	
	U	3111.462	3	3	
	PD II	3111.520		3	
	DY	3111.548	4		
	RE I	3111.569	25		
	NB	3111.612		4	
	U	3111.621	16	16	
	MO	3111.640		25	
	BI II	3111.670			25
	FE	3111.684	8	3	

Table IIb.(2) V II 3110.706 A

Element	Wavelength (Angstrom)	Intensity Arc	Spark	Plasma
CO I	3105.924	32		
OS	3105.992	160	20	
TI II	3106.234	25	160	
RU I	3106.840	50	3	
CO I	3107.044	63	3	
OS	3107.378	40	10	
CR	3107.572	2	130	
RU I	3107.715	63	5	
OS	3107.864	32	8	
RU I	3108.426	32		
CA I	3108.580	32	3	
RE I	3108.811	130		
OS	3108.981	130	16	
HF II	3109.117	50	100	
CR	3109.336	32	13	
OS	3109.381	130	20	
CO I	3109.510	63		
DY	3109.768	40	20	
CO I	3110.021	63	2	
CE	3110.278	32		
FE	3110.281	40	32	
RU I	3110.549	63	6	
OS	3110.618	32	10	
MN	3110.677	32	32	
*V II	3110.706	63	320	30 VAF
U II	3110.710			30 VAF
CO I	3110.825	63	2	
RE I	3110.862	100		
HF II	3110.874	32	40	
OS	3111.090	100	20	
EU	3111.432	200		
RU I	3111.912	50	5	
FE	3112.081	32	20	
MO	3112.124	40	10	
RU	3112.304	32		
RU	3112.677	50	3	
RU I	3113.397	50		
CO I	3113.479	100	4	
V	3113.567	6	100	
TA	3113.903	50	32	

Table IIb.(2) V II 3110.706 A
(con't)

Element	Wavelength (Angstrom)	Intensity Arc	Spark	Plasma
PD I	3114.040	400	500	
NI I	3114.124	320	50	
OS	3114.814	50	13	
RH I	3114.909	100	2	
BI	3115.423		500	
MN	3115.465	50	25	

Table IIa. V I 4379.238 A

	Element	Wavelength (Angstrom)	Intensity Arc	Spark	Plasma
	PR	4378.263	16	3	
	CR	4378.322	5		
	ER	4378.342	10		
	CE	4378.407	3		
	O II	4378.410			10
	CU II	4378.430		2	
----	W	4378.493	25	13	
	U	4378.527	6	2	
	GD	4378.570	40		
	CE	4378.576	5		
	PR	4378.635	3		
	TB	4378.700	10		
---	CE	4378.818	4		
	TA	4378.822	40	2	
	PR	4378.836	3		
	BR	4378.970			4
--	CE	4379.081	3		
	ND	4379.111	2		
	HO	4379.150	3		
	HF	4379.167	10	4	
	*V I	4379.238	200	200	
	AG	4379.250	5		
	TB	4379.260	4		
	Y I	4379.324	4		
	PR	4379.335	100	2	
	BI II	4379.400	25	20	
	DY	4379.410	2		
--	XE II	4379.440			5
	NE II	4379.500			100
	GD	4379.520	2		
	NB	4379.525	2	3	
	O	4379.550			16
	PD I	4379.561	6	2	
	BH LA	4379.600	10		
	TB	4379.600	3		
	HO	4379.640		2	
	U	4379.644	2	2	
---	BH LA	4379.700	20		
	AR	4379.740			79
	ZR II	4379.776	10	8	

Table IIa. V I 4379.238 A
(con't)

Element	Wavelength (Angstrom)	Intensity Arc	Spark	Plasma
BR	4379.780			6
CR I	4379.782	16	2	
EU	4379.813	5	2	
CE	4379.841	2		
ND	4379.876	4		
CL I	4379.910			16
-- RH I	4379.917	63	25	
CE	4380.060	32	2	
CO I	4380.071	5	3	
KR	4380.110			2
W	4380.115	8	2	
EU	4380.167	3		
DY	4380.230	5	2	

Table IIb.(2) V I 4379.238 A

Element	Wavelength (Angstrom)	Intensity Arc	Spark	Plasma
SC II	4374.455	100	25	17 VAF
RH I	4374.800	1000	500	
Y II	4374.935	160	160	46 VAF
MN	4374.947	160	20	
SM II	4374.975	200	200	
FE I	4375.932	500	200	
IR	4377.006	100	4	
MO	4377.765	5	200	
CU I	4378.200	200	32	
SM II	4378.230	100	100	
*V I	4379.238	200	200	
PR	4379.335	100	2	
GD	4380.640	100	130	
MO	4381.640	160	160	
EU I	4383.158	100	20	
LA II	4383.450	10	50	
FE I	4383.547	1000	790	

Table IIa. W II 2764.266 A

	Element	Wavelength (Angstrom)	Intensity Arc	Spark	Plasma
	OS	2763.273	25	10	
	MO	2763.298		20	
	RE I	2763.306	40		
	TA	2763.370	25	63	
	NB	2763.384	5	5	
	FE	2763.410	2		
	U	2763.415	8	6	
	RU	2763.419	50	16	
----	CR	2763.590		32	
	NB	2763.597	2	10	
	CE	2763.598	5		
	TH	2763.613	10	10	
	MO	2763.620	25	50	
	CU	2763.690	3	2	
	U	2763.692	2	2	
	CE	2763.722	2		
---	HE I	2763.800			20
	RE	2763.803	50		
	CL II	2763.880			10
	ZN II	2763.880			3
	CD I	2763.890	100	50	
	RU	2763.903	32		
	FE II	2763.907		25	
	TI II	2763.930	2	4	
	MO	2763.934	25		
	OS	2763.935	20	8	
	CR	2763.972		32	
	W	2764.004	10	2	
--	OS	2764.058	2		
	GD	2764.080	25	32	
	CD	2764.110	50	25	
	CE	2764.142	2		
	CO I	2764.188	100		
	U	2764.249	10	10	
	*W II	2764.266	20	63	
	V	2764.293		10	
	FE	2764.330	63	40	
	CR I	2764.350	200	6	
	CE	2764.397	3		
	YB	2764.410		8	

Table IIa. W II 2764.266 A
(con't)

	Element	Wavelength (Angstrom)	Intensity Arc	Spark	Plasma
--	CS	2764.420			20
	NB	2764.562	3	10	
	CE	2764.633	5		
	TH	2764.643	8	8	
	AR II	2764.660			10
	CE	2764.696	5		
	ER	2764.701	3		
	U	2764.707	2	8	
	RU	2764.725	50		
--	TA	2764.753	10	3	
	FE II	2764.784		20	
	TI II	2764.821	16	63	
	RH	2764.830	16	130	
	HF II	2764.885	10	10	
	CL	2764.950			4
	BA	2764.980		8	
--	ND	2764.982	10		
	MO	2765.096	16		
	CE	2765.117	8		
	TH	2765.127	10	8	
	RU	2765.133		79	
	W	2765.205	8		
	U	2765.206		4	
	TI II	2765.210	2	3	

Table IIb.(3) W II 2764.266 A

Element	Wavelength (Angstrom)	Intensity Arc	Spark	Plasma
CR	2759.391	10	32	
MO	2759.582	20		
HG I	2759.712	20	16	
FE I	2759.817	100	63	
V II	2760.124	20	32	
RU	2760.167	20	13	
MO	2760.526		32	
NI II	2760.671		40	

Table IIb.(3) W II 2764.266 A
(con't)

Element	Wavelength (Angstrom)	Intensity Arc	Spark	Plasma
V	2760.698	25	100	
MN	2760.930	79		
OS	2761.082	20	5	
RH	2761.260		50	
TI II	2761.294	10	32	
CO I	2761.372	79	5	
CE	2761.415	20		
OS	2761.418	50	10	
MO	2761.533	40	20	
TA	2761.547	79		
HF	2761.631	25	3	
TA	2761.676	200	160	
CR I	2761.755	320	32	
FE I	2761.785	200		
FE II	2761.813	50	200	
RE	2761.932	25		
FE I	2762.032	100	63	
TA	2762.050		40	
CE	2762.215	20		
RU	2762.306	50	3	
W	2762.345	20	10	
CR II	2762.593	40	100	15 VAF
FE	2762.680	50		
MO	2762.697	25	2	
FE I	2762.778	25	10	
MO	2763.027	25		
CR	2763.063	32	4	
PD II	2763.092	320	32	
FE I	2763.108	100	63	
RU	2763.142	32	5	
OS	2763.273	25	10	
RE I	2763.306	40		
TA	2763.370	25	63	
RU	2763.419	50	16	
CR	2763.590		32	
MO	2763.620	25	50	
RE	2763.803	50		
CD I	2763.890	100	50	
RU	2763.903	32		
MO	2763.934	25		

Table IIb.(3) W II 2764.266 A
(con't)

Element	Wavelength (Angstrom)	Intensity Arc	Spark	Plasma
OS	2763.935	20	8	
CR	2763.972		32	
GD	2764.080	25	32	
CD	2764.110	50	25	
CO I	2764.188	100		
*W II	2764.266	20	63	
FE	2764.330	63	40	
CR I	2764.350	200	6	
RU	2764.725	50		
TI II	2764.821	16	63	
RH	2764.830	16	130	
RU	2765.133		79	
RU	2765.441	50	160	
V	2765.668	50	200	
LU	2765.740	20	3	
CR	2765.863		32	
RU	2765.874		50	
TA	2766.117	40	20	
CO	2766.221	50	50	
RU	2766.228	32		
MO	2766.258	32	3	
CU I	2766.371	500	25	
CO I	2766.387	50	3	
RE	2766.403	50		
V II	2766.455	40	100	
RH	2766.540	5	160	
CR II	2766.540	40	320	22 VAF
RU	2766.549		100	
MO	2766.722	20		
FE I	2766.912	100	40	
CE	2767.007	20		
V	2767.151	2	32	
MO	2767.224	20		
FE II	2767.503	10	400	
AG II	2767.523	32	200	
FE I	2767.523	320		
CR	2767.540	32	8	
RH I	2767.728	100	4	
TL I	2767.870	400	320	25 VAF
FE	2768.112	32	8	

Table IIb.(3) W II 2764.266 A
(con't)

Element	Wavelength (Angstrom)	Intensity Arc	Spark	Plasma
NB	2768.128	10	100	
RH	2768.229	50	4	
FE	2768.439	25	5	
V	2768.556	32	160	
CR	2768.591		63	
CO	2768.689	20		
NI II	2768.785		250	
RE I	2768.856	25		
CU I	2768.878	25		
RU	2768.926	63	200	
FE II	2768.934	40	100	
TA	2768.934	2	32	
V I	2768.937	32	4	
W	2768.985	20	10	
CO	2769.085		32	

Table IIa. W I 4008.753 A

	Element	Wavelength (Angstrom)	Intensity Arc	Intensity Spark	Intensity Plasma
	DY	4007.770	13	2	
	PR	4007.780	8	3	
	S II	4007.780			5
	SE II	4007.900			160
	U	4007.934	8	3	
	CO I	4007.943	3		
	ER	4007.967	32	6	
	EU	4007.980	6		
--	MN	4008.020	16	5	
	IR I	4008.052	13	2	
	MO	4008.054	4	5	
	TI I	4008.062	50	6	
	KR II	4008.080			25
	SM II	4008.091	10	3	
	V II	4008.169	2	10	
	ER	4008.185	8		
	TH	4008.216	10	8	
--	RU I	4008.269	20	20	
	NB	4008.280	5	10	
	SM	4008.330	8	10	
	GD I	4008.331	16	10	
	ND	4008.416	5	3	
	CE	4008.446	6		
	HF II	4008.460	5	8	
	KR II	4008.480			10
	DY	4008.490	5		
--	CE	4008.664	8		
	MO	4008.667		20	
	U	4008.700	2		
	PR II	4008.714	160	50	5 VAF
	*W I	4008.753	50	50	
	ND	4008.754	13	10	
	BR	4008.760			20
	EU	4008.872	8	2	
	FE	4008.873	5		
	U	4008.918	8		
	GD	4008.922	20	3	
	TI I	4008.928	79	32	
--	IR I	4008.967	5		
	CE	4009.063	13		

Table IIa. W I 4008.753 A
(con't)

	Element	Wavelength (Angstrom)	Intensity Arc	Spark	Plasma
	TH	4009.066	10	8	
	ER	4009.165	16		
	U	4009.170	8	16	
	TB	4009.193	5		
	GD	4009.210	50	2	
---	PR	4009.240	5	2	
	TB	4009.270	2		
	HE I	4009.270			10
	MO	4009.366	20	25	
	ND	4009.367	8	5	
	ZR	4009.387	3		
	S II	4009.390			3
----	U	4009.410	4	8	
	TB	4009.540	10		
	TH	4009.546	3		
	CE	4009.550	2		
	AL II	4009.580			4
	TI I	4009.663	63	25	
	NB	4009.714	5	10	
	FE I	4009.717	130	100	
	SR	4009.750		2	

Table IIb.(3) W I 4008.753 A

Element	Wavelength (Angstrom)	Intensity Arc	Spark	Plasma
FE	4003.767	32	79	
TI I	4003.806	63	50	
OS	4004.024	50	6	
PR	4004.714	20	25	
FE I	4005.246	250	200	
TH	4005.549	20	32	
TB	4005.550	100	130	
RU I	4005.640	25	32	
V II	4005.712	10	32	
FE	4006.315	63	32	
AS II	4006.340		50	

Table IIb.(3) W I 4008.753 A
(con't)

Element	Wavelength (Angstrom)	Intensity Arc	Spark	Plasma
FE I	4007.273	79	50	
SM II	4007.486	50	25	
TI I	4008.062	50	6	
PR II	4008.714	160	50	5 VAF
*W I	4008.753	50	50	
TI I	4008.928	79	32	
GD	4009.210	50	2	
MO	4009.366	20	25	
TI I	4009.663	63	25	
FE I	4009.717	130	100	
MO	4011.966	25	25	
NB	4012.168		100	
ND II	4012.250	79	40	58 VAF
CE II	4012.388	63	20	4 VAF
CE I	4012.388	63	20	
TI II	4012.391	32	50	
CR	4012.469	63	63	
EU	4012.816	20	25	
MO	4013.210		40	
TI I	4013.584	63	6	

Table IIa. W II 2079.106 A

	Element	Wavelength (Angstrom)	Intensity Arc	Spark	Plasma
	YB	2078.120		10	
	TA	2078.210	8		
	W	2078.350	6	25	
----	IR	2078.370		25	
	OS	2078.490	10		
	ND	2078.530		10	
---	CU II	2078.652		40	
	RE	2078.680	8		
	NI II	2078.760		13	
--	IR	2078.880		25	
	V	2078.960		4	
	OS	2078.990	5	20	
	*W II	2079.106	13	32	100 VAF
	HF II	2079.110	3		
	OS	2079.210	10		
	IN II	2079.260		16	4 VAF
	CR	2079.310		2	
	TA	2079.310	2	13	
	PD	2079.320	2		
--	IR	2079.330	16	5	
	RH	2079.370	3	25	
	AS I	2079.410	2		
	CU I	2079.460	16		
---	PT II	2079.496		16	
	SB	2079.630	10		
	ND	2079.640		16	
	RH	2079.660	5		
	W	2079.660	2	6	
	IR	2079.710		40	
	HF	2079.720	6		
	PT II	2079.762		16	
	ZN II	2079.820			2
----	TA	2079.850	4	6	
	OS	2079.970	40	10	
	NB	2080.050	3	16	
	CU II	2080.060		8	

Table IIb.(3) W II 2079.106 A

Element	Wavelength (Angstrom)	Intensity Arc	Spark	Plasma
NI II	2074.120		16	
MO	2074.290	4	16	
AS	2074.500		20	
TE	2074.740	32		
SE I	2074.793			10 VAF
V	2074.870		16	
OS	2074.950	16	5	
IR	2075.050	6	16	
PT II	2075.386	5	25	
IR	2075.430		25	
W	2075.590	10	20	
IR	2076.070		20	
TA	2076.320	13	32	
IR	2076.760	2	40	
RH	2076.830		100	
V	2076.850	5	25	
OS	2076.950	25	13	
RE	2077.300	25	8	
OS	2077.330	16	3	
PT II	2077.435	10	20	
V	2077.800		20	
RH	2077.970		16	
OS	2078.090	25	10	
W	2078.350	6	25	
IR	2078.370		25	
CU II	2078.652		40	
IR	2078.880		25	
OS	2078.990	5	20	
*W II	2079.106	13	32	100 VAF
IN II	2079.260		16	4 VAF
IR	2079.330	16	5	
RH	2079.370	3	25	
CU I	2079.460	16		
PT II	2079.496		16	
ND	2079.640		16	
IR	2079.710		40	
PT II	2079.762		16	
OS	2079.970	40	10	
NB	2080.050	3	16	
RH	2080.240		32	

Table IIb.(3) W II 2079.106 A
(con't)

Element	Wavelength (Angstrom)	Intensity Arc	Spark	Plasma
IR	2080.640	2	63	
NI II	2080.849	3	32	
IR	2080.950	8	16	
TE I	2081.030	400		
PD I	2081.120	20		
PD II	2081.420		25	
OS	2081.460	20	10	
RH	2081.520		32	
SI I	2082.010	16		
AU II	2082.090			70 VAF
RE	2082.260	6	20	
PD I	2082.270	20	2	
PT I	2082.510	16	6	
V I	2082.510	25		
OS	2082.540	20	5	
CO	2082.680	3	25	
HF II	2082.800	16	8	
NI I	2082.856	25	2	
NB	2082.890	6	25	
OS	2083.200	16	2	
IR	2083.220	25	10	
PT	2083.340	20	10	
RE II	2083.342	4	20	
CE	2083.350		25	
IR	2083.430		25	
OS	2083.580	16	2	
HF II	2083.800	40	32	
IR	2083.850	16	5	
RE I	2083.925	40	8	

Table IIa.   Y II 3710.290 A

	Element	Wavelength (Angstrom)	Intensity Arc	Spark	Plasma
	TB	3709.300	16		
	V	3709.333		25	
	NB	3709.417	5	10	
	U	3709.460	3		
	I II	3709.514			3
	SM	3709.526	10	6	
	FE	3709.533	6	6	
--	CE	3709.588	5		
	NE II	3709.640			40
	FE	3709.665	4	4	
	PR	3709.665	3		
	TH	3709.673	5	3	
	NB	3709.742	2	2	
--	U	3709.878	2	2	
	AR	3709.900			5
	CE	3709.933	25	10	
	RE I	3709.937	40		
	TI I	3709.961	79	25	
	EU	3710.010	5		
	PR	3710.012	6	2	
--	DY	3710.080	20	5	
	CR	3710.090	4	2	
	MO	3710.142	20	16	
	U	3710.168		5	
	TI I	3710.186	6		
	CE	3710.253	2		
	W	3710.289	4	5	
	EU	3710.290	20		
	*Y II	3710.290	79	160	86 VAF
	V	3710.300			
	SM	3710.300	25	16	
	U	3710.308	8		
	YB	3710.310	4	6	
	RU	3710.316	4		
	F	3710.365			10
	NB	3710.448	16	20	
	P II	3710.450			32
	SB II	3710.519	2	5	
--	U	3710.534	3	4	
	LA II	3710.590	3	2	

Table IIa. Y II 3710.290 A
(con't)

	Element	Wavelength (Angstrom)	Intensity Arc	Spark	Plasma
	CR	3710.600	2	2	
	DY	3710.730	4	2	
	HO	3710.750	8	8	
	CA	3710.753		3	
	CS	3710.774			4
	U	3710.785	4	2	
---	TA	3710.790	6		
	SM II	3710.869	100	6	
	EU	3710.881	32	32	
	U	3710.908	3	8	
	LU	3710.950	3		
----	CE	3711.001	2		
	NA II	3711.074	8		63
	PR	3711.099	8	3	
	V	3711.123		79	
	TA	3711.150	5		
	IR I	3711.153	6		
	PR	3711.218	3	2	
	FE I	3711.225	79	50	
	CR	3711.282		13	

Table IIb.(2) Y II 3710.290 A

Element	Wavelength (Angstrom)	Intensity Arc	Spark	Plasma
FE I	3705.567	630	500	
LA II	3705.818	50	79	
CA II	3706.026	16	40	
V I	3706.035	50	50	
MN	3706.078	79		
TI II	3706.226	32	130	
OS	3706.556	50	16	
FE I	3707.049	160	100	
TI I	3707.531	100	10	
FE I	3707.824	79	50	
NB	3707.918	3	100	
FE I	3707.925	79	63	

Table IIb.(2) Y II 3710.290 A
(con't)

Element	Wavelength (Angstrom)	Intensity Arc	Spark	Plasma
TI I	3708.648	50	3	
SM II	3708.660	50	25	
V I	3708.721	100	63	
CO I	3708.823	100		
FE I	3709.249	630	400	
ZR II	3709.257	50	32	
RE I	3709.937	40		
TI I	3709.961	79	25	
*Y II	3710.290	79	160	86 VAF
SM II	3710.869	100	6	
V	3711.123		79	
FE I	3711.225	79	50	
FE	3711.410	50	25	
TB	3711.740	200	32	
CO I	3712.180	40	8	
GD	3712.714	200	250	
SM II	3712.759	100	100	
OS	3712.838	50	13	
CR II	3712.949	13	130	
NB	3713.014	100	79	
RH I	3713.022	100	100	
EU	3713.459	50	2	
LA II	3713.544	200	63	
GD	3713.583	100	79	
OS	3713.726	100	20	
V I	3713.957	63	10	
MO	3713.960		40	
PR	3714.057	50	20	
LA II	3714.858	63	40	
EU I	3714.911	40	2	

Table IIa.   Y II 4177.552 A

	Element	Wavelength (Angstrom)	Intensity Arc	Intensity Spark	Intensity Plasma
	RE I	4176.560	13		
	FE I	4176.572	100	50	
	MN	4176.602	100	40	
	EU	4176.620	8	2	
	DY	4176.640	3		
	CR	4176.686	16	2	
	CE	4176.703	20	2	
	EU	4176.725	5		
	TL II	4176.750			2
	V I	4176.795	5	2	
----	TB	4176.860	4		
	RE	4176.900	5		
	MO	4176.900	10	10	
	U	4176.910	3	3	
	TA	4176.986	16	40	
	KR	4177.020			3
---	V I	4177.072	16	3	
	MO	4177.257	20	20	
--	ND II	4177.321	16	25	22 VAF
	CE	4177.349	2	3	
	TI I	4177.357	10	2	
	IR	4177.358	5		
	U	4177.390	2	5	
	NA I	4177.400	3		
	TL II	4177.460		8	
	LA I	4177.486	16		
	PR	4177.500	4	2	
	HF II	4177.505	5	8	
	*Y II	4177.552	50	50	26 VAF
	EU	4177.570	8	2	
	FE I	4177.596	100	25	
	RH I	4177.635	25	5	
	DY	4177.754	6		
--	CU I	4177.758	63		
	GD	4177.830	5		
	W	4177.830	10	3	
	U	4177.838	4	3	
	TB	4177.850	2		
	NB	4177.867		5	
	PR	4177.890	13	3	

Table IIa. Y II 4177.552 A
(con't)

	Element	Wavelength (Angstrom)	Intensity Arc	Spark	Plasma
	CR	4177.899	40		
	MO	4177.917		16	
	TA	4177.919	20	16	
	RA I	4177.980			10
	SM II	4178.006	3	10	
	HG	4178.025			50
--	FE	4178.051	3		
	TH	4178.064	10	13	
	DY	4178.072	13		
	CE	4178.084	3		
	TB	4178.140	2		
	CE	4178.153	3		
	MO	4178.272	25	20	
--	CE	4178.283		2	
	IR	4178.342	4		
	P	4178.360			320
	V II	4178.389	3	10	
	AR	4178.390			20
	CE	4178.390		2	
	NB	4178.406		10	
	ND	4178.441		5	
	U	4178.525	8	8	
	ND	4178.532	5		
	MO	4178.535	5	2	

Table IIb.(3) Y II 4177.552 A

Element	Wavelength (Angstrom)	Intensity Arc	Spark	Plasma
IR I	4172.559	160	13	
OS	4172.570	63	3	
FE I	4172.750	63	10	
HO	4173.230	50		
OS	4173.234	100	6	
TI II	4173.549	13	40	
GD	4173.561	32	32	
FE I	4173.926	50	5	

Table IIb.(3) Y II 4177.552 A
(con't)

Element	Wavelength (Angstrom)	Intensity Arc	Spark	Plasma
Y I	4174.133	100	8	
CR I	4174.795	100	40	
FE I	4174.917	100	25	
TA	4175.206	100	40	
GD	4175.539	50		
W II	4175.590	6	25	
OS	4175.627	100	4	
FE I	4175.640	100	79	
FE I	4176.572	100	50	
MN	4176.602	100	40	
TA	4176.986	16	40	
ND II	4177.321	16	25	22 VAF
*Y II	4177.552	50	50	26 VAF
FE I	4177.596	100	25	
CU I	4177.758	63		
TB	4178.970	50	2	
GE	4179.041		25	
CR	4179.257	100	40	
PR II	4179.422	200	40	7 VAF
AC	4179.930		63	
YB	4180.828	10	100	
TI I	4180.868	100	20	
SM II	4181.099	8	25	
TA	4181.152	40	25	
FE I	4181.757	200	160	
FE I	4182.386	79	32	
RU	4182.457	20	32	
IR	4182.467	50	6	

Table IIa.   YB II 3694.203 A

Element	Wavelength (Angstrom)	Intensity Arc	Spark	Plasma
MO	3693.232	3		
U	3693.320	4	6	
PR	3693.363	32	8	
CO I	3693.364	20		
NB	3693.368	8	10	
MO	3693.375	5	3	
CE	3693.422	10		
-- CO I	3693.478	32	10	
XE I	3693.490			40
PR	3693.493	2		
TB	3693.560	32		
RU	3693.591	20	5	
MN	3693.667	50	63	
TH	3693.698	6	4	
-- U	3693.705	4	20	
CE I	3693.708	8		
CE II	3693.708	8		
NB	3693.765		2	
IN II	3693.788			25
EU	3693.820	25		
DY	3693.840	4		
IN II	3693.899			32
NI I	3693.932	50		
TH	3693.940	2	3	
-- SM II	3693.996	100	160	
FE I	3694.010	400	320	
GD	3694.011	25	25	
IN II	3694.029			40
MN	3694.115	5	5	
TB	3694.120	8		
CE	3694.173	4		
ER	3694.193	25	16	
NE II	3694.197			250
*YB II	3694.203	500	1000	100 VAF
HO	3694.240	10	20	
LA II	3694.270	4	10	
SM	3694.314	16	8	
PR	3694.321	3		
U	3694.325	6		
TH	3694.355	5		

Table IIa. YB II 3694.203 A
(con't)

	Element	Wavelength (Angstrom)	Intensity Arc	Spark	Plasma
	DY	3694.360	6	4	
	MO	3694.421		20	
--	TI I	3694.447	79	20	
	IR	3694.454	2		
	W	3694.508	10	20	
	TA	3694.519	6	20	
	AG	3694.620	2		
	V I	3694.622	63	5	
	NB	3694.666	10	10	
---	PR	3694.695	40	4	
	DY	3694.750	20	32	
	TB	3694.750	50	8	
	TM	3694.760	20	20	
	ND	3694.792	32	20	
	NB	3694.795		5	
	HO	3694.810		4	
	ER	3694.811	16	6	
	U	3694.816		3	
	SM	3694.817	16	3	
	TH	3694.894	5		
	CE I	3694.911	16	2	
	CE II	3694.911	16	2	
	MO	3694.945	40	32	
	W	3694.946	2		
----	RH I	3694.953	3	2	
	CR	3694.974		6	
	FE	3695.054	200	160	
	ER	3695.091	2		
	V	3695.158		3	
	LA	3695.200	2	2	

Table IIb.(1) YB II 3694.203 A

Element	Wavelength (Angstrom)	Intensity Arc	Spark	Plasma
FE I	3689.463	200	160	
RE I	3689.522	100		
TI I	3689.911	100	40	
RU	3690.029	5	100	
V I	3690.281	200	130	
PD I	3690.341	320	1000	
YB	3690.577	10	50	
RH I	3690.704	130	50	
CO I	3690.723	63	10	
FE	3690.730	79	63	
AR I	3690.896			2 VAF
TB	3691.150	50	50	
RE I	3691.500	100		
SM II	3692.221	100	40	
V I	3692.225	200	160	
RH I	3692.357	500	160	35 VAF
RH I	3692.357	500		
MO	3692.645	3	160	
ER II	3692.652	20	13	16 VAF
MN	3692.812	50	50	
CO I	3693.112	79	16	
MN	3693.667	50	63	
NI I	3693.932	50		
SM II	3693.996	100	160	
FE I	3694.010	400	320	
*YB II	3694.203	500	1000	100 VAF
TI I	3694.447	79	20	
V I	3694.622	63	5	
TB	3694.750	50	8	
FE	3695.054	200	160	
V I	3695.335	130	63	
BI	3695.550		50	
V I	3695.865	160	100	
NB	3695.898	4	200	
MN	3696.568	100	50	
RU I	3696.587	50	16	
NB	3696.677	2	50	
FE I	3697.432	100	63	
RE	3697.704	160		
GD	3697.739	200	200	

Table IIb.(1) YB II 3694.203 A
(con't)

Element	Wavelength (Angstrom)	Intensity Arc	Spark	Plasma
NB	3697.846	50	50	
ZR II	3698.167	50	79	

Table IIa. YB II 3289.370 A

	Element	Wavelength (Angstrom)	Intensity Arc	Spark	Plasma
	TI II	3288.428	6	8	
	EU	3288.430	2		
	V I	3288.433	8	2	
	HO	3288.460	6	6	
	TA	3288.466	20		
	EU	3288.530	5		
	CS I	3288.560	4		
	CE	3288.561	2		
	TI II	3288.575	13	20	
--	DY	3288.635	5		
	PR	3288.650	13		
	FE	3288.654	16	6	
	CE	3288.770	16		
	ZR II	3288.803	10	6	
	OS	3288.837	32	16	
	I II	3288.858			8
--	TI I	3288.897	2		
	CE	3288.941	2		
	FE	3288.967	32	16	
	V	3288.982	10	16	
	MO	3289.015	40	32	
--	CS I	3289.130	2		
	RH I	3289.138	160	50	
	AG	3289.170		5	
	RU	3289.245	5		
	CE	3289.280	10		
	U	3289.307	4	4	
	DY	3289.340	2	2	
	FE II	3289.346		40	
	*YB II	3289.370	500	1000	170 VAF
	EU	3289.380	2		
	V II	3289.389	10	63	
	AR I	3289.390			3
	FE	3289.436	10	4	
	NB	3289.450	2		
	CE	3289.523	2		
--	NB	3289.548		10	
	RH	3289.636	50	5	
	CL	3289.720			13
	HF II	3289.740	5	3	

Table IIa. YB II 3289.370 A
(con't)

Element	Wavelength (Angstrom)	Intensity Arc	Spark	Plasma
U	3289.749	2		
CE	3289.790	2		
TA	3289.838	25		
MO	3289.844	10	10	
--- Y	3289.850	16	10	
CE	3289.944	4		
AR I	3289.950			3
NB	3290.008	10	10	
FE	3290.041	3	2	
U	3290.042	8		
O II	3290.080			20
---- RE I	3290.098	5		
U	3290.121	8		
TH	3290.127	6	5	
PT I	3290.220	160	10	
V	3290.238	2	63	
IR I	3290.258	2		
OS	3290.263	200	20	
SM	3290.282	10	4	
CE II	3290.341	20		

Table IIb.(1) YB II 3289.370 A

Element	Wavelength (Angstrom)	Intensity Arc	Spark	Plasma
FE I	3284.588	200	130	
FE II	3285.415	63	40	
FE I	3286.755	500	400	
NI I	3286.946	100		
CO I	3287.194	63		
PD I	3287.249	320	25	
TA	3287.265	50		
TI II	3287.655	40	200	
RH I	3289.138	160	50	
*YB II	3289.370	500	1000	170 VAF
V II	3289.389	10	63	
RH	3289.636	50	5	

Table IIb.(1) YB II 3289.370 A
(con't)

Element	Wavelength (Angstrom)	Intensity Arc	Spark	Plasma
PT I	3290.220	160	10	
V	3290.238	2	63	
OS	3290.263	200	20	
MO	3290.823	40	100	
FE I	3290.989	130	79	
TM II	3291.000	130	79	19 VAF
NB	3291.059	10	100	
CR	3291.762	10	200	
NB	3292.020	3	100	
FE	3292.023	160	130	
TI I	3292.075	63	40	
MO	3292.312	10	320	
TA	3292.482	63	3	
FE I	3292.590	320	160	
TB II	3293.070	50	100	29 VAF
V	3293.150		50	
TA	3293.930	63	10	
TB	3294.040	50	3	
RU I	3294.110	63	200	
RH I	3294.278	63	25	
NB	3294.360	2	100	

Table IIa. ZN I 2138.560 A

	Element	Wavelength (Angstrom)	Intensity Arc	Spark	Plasma
	K	2137.560			32
	RE	2137.560	6		
	RH	2137.620		5	
	W II	2137.652	3	3	
	ZR II	2137.684	10	6	
	YB	2137.710		16	
	RE	2137.770		8	
	CO I	2137.780	16	2	
----	C II	2137.930		6	
	OS	2137.940	8		
	FE	2138.010	2	3	
	NI II	2138.050		5	
---	W II	2138.150	10	25	
	V II	2138.160	8	50	
	HG	2138.250			5
--	YB	2138.320		10	
	OS	2138.400	3	4	
	CU I	2138.507	25		
	AS	2138.530	2		
	NB	2138.552	2		
	*ZN I	2138.560	790	500	170 VAF
	IR	2138.570	16		
	NI II	2138.580	10	16	
	FE I	2138.589	8		
	OS	2138.610	8		
	RH	2138.660		25	
--	OS	2138.750	3		
	NB	2138.880	2	8	
	CO I	2138.971	16		
	NI II	2139.060		3	
---	RE	2139.060		20	
	CR	2139.150		20	
	W II	2139.160	6	13	
	OS	2139.170		10	
	LU	2139.190		6	
	AR	2139.200			5
	IR	2139.220	4		
	HF II	2139.240	32	40	
	TI II	2139.250		2	
	YB	2139.250		2	

Table IIa. ZN I 2138.560 A
(con't)

Element	Wavelength (Angstrom)	Intensity Arc	Spark	Plasma
TL	2139.290		3	
-- W	2139.300		5	
RH	2139.430		100	
MO	2139.440	2	13	
RE	2139.480		3	
IR	2139.560	4		

Table IIb.(1) ZN I 2138.560 A

Element	Wavelength (Angstrom)	Intensity Arc	Spark	Plasma
BI I	2133.626	100	40	10 VAF
V II	2134.120	32	130	
BI I	2134.308	100	5	
CU II	2134.355	32	40	
PD II	2135.078		25	
P I	2135.466	13		9 VAF
PD II	2135.646		25	
CU II	2135.976	25	500	25 VAF
P I	2136.199	16		39 VAF
PR	2136.660		32	
NB	2137.054	5	32	
OS	2137.110	100	32	
PD II	2137.240		40	
V II	2137.310	10	79	
NB	2137.545	5	40	
W II	2138.150	10	25	
V II	2138.160	8	50	
*ZN I	2138.560	790	500	170 VAF
RH	2138.660		25	
HF II	2139.240	32	40	
RH	2139.430		100	
SB I	2139.760	32	32	16 VAF
V II	2139.821	5	25	
YB	2139.960		25	
V II	2140.087	10	79	20 VAF
TA	2140.130	13	32	

Table IIb.(1) ZN I 2138.560 A
(con't)

Element	Wavelength (Angstrom)	Intensity Arc	Spark	Plasma
PD II	2140.260		32	
TA	2141.500	3	32	
SB	2141.760	32	32	
V II	2141.980	10	79	
TE I	2142.810	630		73 VAF
PR	2142.890		32	
V II	2143.047	10	50	
TA	2143.160		32	

Table IIa. ZN II 2025.510 A

	Element	Wavelength (Angstrom)	Intensity Arc	Spark	Plasma
	P I	2024.546	13		13
	PT	2024.630	8		
	ZN II	2024.660			25
---	RH	2024.930	3	25	
	OS	2025.120	6	2	
---	SE	2025.130			4
	CU	2025.150	2		
	RH	2025.150	2	40	
	PT II	2025.159	2	10	
--	NB	2025.310	10	32	
	ER	2025.330	5		
	ZN II	2025.330		4	
	NI I	2025.380	10	6	
	IN	2025.440		13	
	CU II	2025.475	8	32	
	IR	2025.490	4	10	
	*ZN II	2025.510	200	200	75 VAF
	U	2025.530	2	3	
	OS	2025.690	6		
--	CO II	2025.750	10	32	
	NI	2025.810	6		
	MG I	2025.820	8		1 VAF
	PT	2025.860	10	3	
	HF	2025.900		4	
---	IR	2025.960		25	
	W II	2026.080	6	25	
	IR	2026.100	4		
	OS	2026.160	5		
	RH	2026.170		3	
	ZN II	2026.190			2
	BH C	2026.200	25		
---	TL	2026.200		2	
	NI	2026.360	8		
	SE	2026.510			4

Table IIb.(2) ZN II 2025.510 A

Element	Wavelength (Angstrom)	Intensity Arc	Spark	Plasma
FE	2020.520	2	100	
IR	2021.510		50	
CO II	2022.340	20	79	
IR	2022.830	4	50	
P I	2023.471	16		4 VAF
RE II	2023.660	10	50	3 VAF
RH	2023.740		63	
*ZN II	2025.510	200	200	75 VAF
MG I	2025.820	8		1 VAF
CO II	2027.020	16	50	
IR	2028.650	5	50	
NB	2029.320	10	50	
W II	2029.983	10	32	40 VAF

Table IIa.   ZR II 3438.230 A

	Element	Wavelength (Angstrom)	Intensity Arc	Spark	Plasma	
	NI I	3437.280	630	40		
	TH	3437.310	6	2		
	CE	3437.324	13			
	PB I	3437.360		10		
	TA	3437.373	6	320		
---	SB	3437.490	2	8		
	IR I	3437.498	32	3		
	PR	3437.603	6			
	SM	3437.614	3	2		
	ER	3437.640	10			
	TM	3437.640	10	16		
	U	3437.671	6			
	CO I	3437.692	160			
	RE I	3437.720	100			
---	XE	3437.730			2	
	V I	3437.773	4	2		
	CE	3437.813	10			
	V I	3437.873	2			
	U	3437.934	6	10		
	FE	3437.952	16	6		
	Y	3437.953	3			
--	SM II	3438.054	10	3		
	CE	3438.066	16			
	IR I	3438.094	6			
	W	3438.210	6	3		
	*ZR II	3438.230	250	200	42	VAF
	CE	3438.235	3			
	HF II	3438.235	25	25		
	FE	3438.306	10	3		
	EU	3438.310	5			
	PR	3438.310	25			
	ER	3438.320	13			
	IN II	3438.339			50	
	RU I	3438.368	63	32		
	U	3438.407	3	3		
	NB	3438.419		50		
	HF	3438.432	13	3		
	ER	3438.473	10	2		
--	ND	3438.474	6			
	IN II	3438.519			10	

Table IIa. ZR II 3438.230 A
(con't)

	Element	Wavelength (Angstrom)	Intensity Arc	Spark	Plasma
	TB	3438.570	16	8	
	OS	3438.611	4	10	
	U	3438.697	2	3	
	CO I	3438.713	79		
	YB	3438.720	4	79	
---	TE	3438.730			10
	W	3438.815	10	6	
	YB	3438.840	20	100	
	MO	3438.871	20	20	
	KR II	3438.880			3
	CE	3438.881	3		
	CO I	3438.909	32		
	U	3438.924	3		
	DY	3438.952	25	10	
	TH	3438.953	10	13	
	W	3438.966	5	4	
	ND	3438.970	6		
	NE II	3438.970			4
	MN II	3438.974	20	20	
----	SM	3438.978	8		
	TA	3439.001		63	
	TB	3439.050	16		
	ND	3439.210	32	13	
	GD	3439.212	63	32	
	MO	3439.230	2	2	

Table IIb.(2) ZR II 3438.230 A

Element	Wavelength (Angstrom)	Intensity Arc	Spark	Plasma	
CR II	3433.311	32	160		
PD I	3433.449	1000	500		
NI I	3433.558	790	50		
RH I	3434.893	1000	200	50	VAF
MO	3435.400		63		
CR I	3436.187	50	50		
RU I	3436.737	320	160		

Table IIb.(2) ZR II 3438.230 A
(con't)

Element	Wavelength (Angstrom)	Intensity Arc	Spark	Plasma
NB	3436.962	20	50	
NI I	3437.280	630	40	
TA	3437.373	6	320	
CO I	3437.692	160		
*ZR II	3438.230	250	200	42 VAF
NB	3438.419		50	
YB	3438.720	4	79	
YB	3438.840	20	100	
TA	3439.001		63	
GD	3439.785	50	50	
NB	3439.924	4	50	
GD II	3439.990	63	50	10 VAF
TA	3440.237	20	50	
NB	3440.590	16	79	
FE I	3440.610	500	320	
FE I	3440.991	320	200	
PD I	3441.396	790	2	
CR I	3441.439	79	100	
TM II	3441.510	160	79	18 VAF
MN II	3441.988	79	79	7 VAF
CO I	3442.926	400	16	

Table IIa. ZR II 3391.975 A

	Element	Wavelength (Angstrom)	Intensity Arc	Spark	Plasma
	HG	3391.040		2	
	U	3391.040	2	2	
	NI I	3391.050	400	40	
	YB	3391.100	10	40	
	W	3391.100	10	10	
	SM II	3391.122	8	2	
	DY	3391.130	3	3	
----	RE	3391.237	2		
	RN I	3391.270			8
	OS	3391.286	2	5	
	NB	3391.330	5	3	
	CR	3391.372	13		
	U	3391.398	6		
	CR II	3391.434	4	160	
---	W	3391.531	10	10	
	MO	3391.536	4		
	LU	3391.550	10	2	
	NB	3391.591		5	
	CE	3391.593	10		
	TH	3391.717		3	
	TB	3391.720	16	3	
--	RH I	3391.730	3		
	SM	3391.841	2	3	
	MO	3391.851		32	
	CE	3391.884	3		
	RU	3391.890	50	6	
	ND	3391.948	6		
	*ZR II	3391.975	320	400	39 VAF
	ER II	3391.989	32	13	9 VAF
	DY	3391.990	4		
	EU	3391.992	40	5	
	TB	3392.010	16	8	
	FE	3392.014	20	6	
	CU I	3392.016	6		
	TH II	3392.040	10	16	30 VAF
	HO	3392.050	6	8	
	MO	3392.173	16	2	
	TI	3392.197	2		
--	RN	3392.200			2
	CE	3392.256	3		

Table IIa. ZR II 3391.975 A
(con't)

	Element	Wavelength (Angstrom)	Intensity Arc	Spark	Plasma
	AR I	3392.310			3
	FE I	3392.313	130	79	
	NB	3392.338	20	32	
	RE	3392.384	3		
---	IR	3392.475	2	4	
	GD	3392.534	40	25	
	RU	3392.537	100	40	
	IR I	3392.589	3		
	K II	3392.630			10
	FE I	3392.657	320	200	
---	TI I	3392.713	20	8	
	V I	3392.735		32	
	NE II	3392.780			20
	CE II	3392.784	10		
	AR I	3392.810			100
	CA	3392.810		2	
	HF	3392.811	20	3	
	CL	3392.870			16
	LA II	3392.940	4	3	
	TH	3392.970	3	3	

Table IIb.(2) ZR II 3391.975 A

Element	Wavelength (Angstrom)	Intensity Arc	Spark	Plasma
TI	3387.837	63	130	
ZR II	3387.872	100	100	
CO II	3388.173	250	13	
CO I	3388.173	250	13	
NI I	3391.050	400	40	
CR II	3391.434	4	160	
*ZR II	3391.975	320	400	39 VAF
ER II	3391.989	32	13	9 VAF
TH II	3392.040	10	16	30 VAF
FE I	3392.657	320	200	
CR II	3392.987	10	100	
NI I	3392.992	630		

Table IIb.(2) ZR II 3391.975 A
(con't)

Element	Wavelength (Angstrom)	Intensity Arc	Spark	Plasma
CR II	3393.839	16	130	
RH I	3393.850	1000	500	24 VAF
CR	3394.295	16	160	
TI II	3394.575	63	200	
FE I	3394.589	160	79	
CO I	3395.375	400	50	
CR	3395.610	2	100	
NB	3396.372	2	160	

Explanation of Table III

This table presents a compilation of lines below 2000 angstroms. In a few isolated cases, the best analysis line for a particular element occurs in this region. The symbol "#" precedes the element symbol whenever a transition has been reported to be analytically useful. Data in this region are as complete as we could make them from currently available tables.[1,2] As in Table II, all wavelengths are reported to seven digits regardless of significance.

The table is arranged in order of wavelength value from 1850 to 2000 angstroms. A linear range plasma intensity is given in column three, with the appropriate reference in column four.

References

1. R.K. Winge, V.J. Peterson, and V.A. Fassel, *Applied Spectroscopy,* **33,** 3, pp. 206–219, (1979).
2. R.L. Kelley and L.J. Palumbo, *Atomic and Ionic Emission Lines below 2000 Angstroms—Hydrogen through Krypton,* NRL Report 7599 (Naval Research Laboratory, Washington, D.C., 1973).

Table III
Lines Below 2000A

Element	Wavelength (Angstrom)	Plasma Intensity	Source
MN II	1850.000	20	NRL
KR II	1850.093	80	NRL
ZN II	1850.140	10	NRL
AS I	1850.240	40	NRL
MN II	1850.430	3	NRL
SE	1850.510	60	NRL
AL	1850.560	40	NRL
MN II	1850.600	1	NRL
P II	1850.610	20	NRL
SI I	1850.672	400	NRL
CA II	1850.691	400	NRL
KR II	1850.773	80	NRL
MN II	1850.880	100	NRL
ZN	1850.894	60	NRL
P I	1851.220	240	NRL
FE I	1851.390		NRL
MN II	1851.440	60	NRL
CO I	1851.490	80	NRL
FE II	1851.517	20	NRL
MN II	1851.610	10	NRL
CO II	1851.639	10	NRL
SI I	1851.782	700	NRL
N II	1851.810		NRL
V II	1852.017	75	NRL
SE	1852.020	60	NRL
P I	1852.090	120	NRL
CR II	1852.130	500	NRL
P II	1852.270	20	NRL
CR II	1852.370	60	NRL
FE II	1852.458	10	NRL
SI I	1852.472	250	NRL
CO I	1852.520	150	NRL
NI II	1852.522	2	NRL
CO II	1852.564		NRL
KR II	1852.603	20	NRL
P II	1852.650	10	NRL
CO I	1852.710	300	NRL
N II	1852.721		NRL
MN II	1852.780	60	NRL
NI II	1852.875	2	NRL

Table III
Lines Below 2000A
(con't)

Element	Wavelength (Angstrom)	Plasma Intensity	Source
MN II	1853.050	2	NRL
GE I	1853.134	80	NRL
SI I	1853.152	350	NRL
C	1853.200	1	NRL
AS I	1853.210	20	NRL
NE II	1853.220	65	NRL
MN II	1853.272	300	NRL
CO II	1853.726	10	NRL
FE II	1853.754	6	NRL
AS I	1853.950	3	NRL
V II	1854.064		NRL
NE II	1854.110	65	NRL
MN II	1854.240	2	NRL
CO I	1854.280	80	NRL
CR II	1854.460	20	NRL
P II	1854.590	150	NRL
CR II	1854.680	60	NRL
MN II	1854.890	40	NRL
AR II	1854.986	50	NRL
CO I	1855.050	400	NRL
CR II	1855.140	400	NRL
MN II	1855.140	8	NRL
SE	1855.200	600	NRL
AS I	1855.390	10	NRL
P I	1855.480	1	NRL
MN II	1855.540	8	NRL
FE I	1855.580	100	NRL
ZN	1855.670	10	NRL
AL II	1855.928	400	NRL
MN II	1855.947	5	NRL
AL	1856.100	20	NRL
CO I	1856.130	150	NRL
FE I	1856.230		NRL
AS I	1856.240	2	NRL
AL	1856.270	40	NRL
MN II	1856.700	12	NRL
GE II	1856.810	10	NRL
FE II	1856.928	1	NRL
CU II	1856.929		NRL
MN II	1857.010	30	NRL

Table III
Lines Below 2000A
(con't)

Element	Wavelength (Angstrom)	Plasma Intensity	Source
P I	1857.040	45	NRL
ZN II	1857.274	5	NRL
N II	1857.870	300	NRL
MN II	1857.923	200	NRL
FE II	1857.935	240	NRL
AL II	1858.031	800	NRL
CO I	1858.260	30	NRL
V II	1858.500	20	NRL
CR II	1858.540	400	NRL
N II	1858.545	200	NRL
CR II	1858.720	300	NRL
SE	1858.840	500	NRL
P I	1858.910	450	NRL
MN II	1859.110	10	NRL
FE I	1859.260	40	NRL
N II	1859.260	500	NRL
P I	1859.430	450	NRL
MN II	1859.440	8	NRL
CO II	1859.622		NRL
N II	1859.636		NRL
SE	1859.690	100	NRL
FE II	1859.741	300	NRL
AL II	1859.990	240	NRL
FE II	1860.055	400	NRL
GE I	1860.086	50	NRL
CR II	1860.120	240	NRL
CO II	1860.214	40	NRL
AS II	1860.342	350	NRL
AS I	1860.400	80	NRL
MN II	1860.420	3	NRL
AS I	1860.460	80	NRL
MN II	1860.510	100	NRL
P I	1860.630	45	NRL
NI II	1860.689	1	NRL
NI II	1860.796	1	NRL
AS I	1860.870	5	NRL
GE I	1861.094	9	NRL
MN II	1861.440	1	NRL
CU II	1861.623		NRL
MN II	1861.667	100	NRL

Table III
Lines Below 2000A
(con't)

Element	Wavelength (Angstrom)	Plasma Intensity	Source
SI I	1861.795	200	NRL
CO II	1862.131	250	NRL
AL II	1862.318	800	NRL
FE I	1862.318	100	NRL
CO II	1862.360	10	NRL
V II	1862.370	300	NRL
MN II	1862.515	80	NRL
N II	1862.588	200	NRL
V II	1862.760	250	NRL
FE II	1862.774	8	NRL
MN II	1862.800	20	NRL
V II	1863.440	5	NRL
MN II	1863.480	20	NRL
FE I	1863.540	10	NRL
P I	1863.780	3	NRL
CO II	1863.860	40	NRL
ZN II	1864.117	100	NRL
ZN	1864.164	100	NRL
P II	1864.220	150	NRL
N II	1864.364		NRL
MN II	1864.400	80	NRL
NI II	1864.558	6	NRL
MN II	1864.615	40	NRL
P I	1864.640	3	NRL
FE II	1864.656	40	NRL
FE II	1864.743	400	NRL
P I	1864.780	110	NRL
CO I	1864.920	40	NRL
SI I	1865.028	200	NRL
GE I	1865.052	70	NRL
AS I	1865.100	8	NRL
MN II	1865.160	5	NRL
FE I	1865.300	300	NRL
MN II	1865.300	2	NRL
MN II	1865.520	10	NRL
NI II	1865.637	5	NRL
MG	1865.640	10	NRL
V II	1865.680	20	NRL
CR II	1865.800	20	NRL
MN II	1865.832	150	NRL

Table III
Lines Below 2000A
(con't)

Element	Wavelength (Angstrom)	Plasma Intensity	Source
V II	1865.990	30	NRL
ZN	1866.055	100	NRL
FE I	1866.070	80	NRL
ZN II	1866.077	100	NRL
V II	1866.080	1	NRL
AR II	1866.089	100	NRL
CR II	1866.220	300	NRL
MN II	1866.250	1	NRL
CO I	1866.270	60	NRL
ZN II	1866.366	10	NRL
CO I	1866.450	30	NRL
N II	1866.457		NRL
NI II	1866.499	5	NRL
V II	1866.680	50	NRL
FE I	1866.815	40	NRL
CO II	1867.028	50	NRL
V II	1867.470	200	NRL
MN II	1867.880	80	NRL
KR II	1867.889	40	NRL
ZN II	1867.994	40	NRL
N II	1868.240	10	NRL
MN II	1868.280		NRL
MN II	1868.580	20	NRL
AR II	1868.660	300	NRL
KR II	1868.662	120	NRL
SI II	1868.765	100	NRL
MN II	1868.780	40	NRL
MN II	1869.030	15	NRL
CO I	1869.160	50	NRL
SI II	1869.317	200	NRL
V II	1869.470	500	NRL
MN II	1869.550	1	NRL
AL	1869.560	20	NRL
FE II	1869.560	3	NRL
AS I	1869.940	15	NRL
SI II	1870.227	150	NRL
FE I	1870.360		NRL
CO I	1870.450	50	NRL
CR II	1870.460	20	NRL
NI II	1870.460	8	NRL

Table III
Lines Below 2000A
(con't)

Element	Wavelength (Angstrom)	Plasma Intensity	Source
KR II	1870.645	80	NRL
FE II	1870.720	1	NRL
AL	1870.750	10	NRL
SI II	1870.782	300	NRL
V II	1871.080	120	NRL
MN II	1871.390	20	NRL
KR II	1871.619	80	NRL
AS I	1871.680	30	NRL
AL	1871.740	60	NRL
ZN II	1872.125	100	NRL
FE I	1872.359	160	NRL
P II	1872.360	1	NRL
AR II	1872.589	100	NRL
FE II	1872.650	2	NRL
V I	1872.660	25	NRL
GE I	1872.974	6	NRL
AS I	1873.020	40	NRL
FE I	1873.052	160	NRL
MN II	1873.070	40	NRL
SI I	1873.104	250	NRL
AR II	1873.140	600	NRL
N I	1873.217	5	NRL
FE I	1873.259	100	NRL
V II	1873.390	100	NRL
KR II	1873.761	80	NRL
GE I	1874.256	200	NRL
V II	1874.450	120	NRL
FE I	1874.580	3	NRL
SI I	1874.842	180	NRL
FE II	1874.931	10	NRL
V II	1874.970	10	NRL
NI II	1875.069	3	NRL
FE I	1875.140		NRL
CR II	1875.220	60	NRL
KR II	1875.296	20	NRL
FE II	1875.430		NRL
FE II	1875.536	300	NRL
MN II	1875.590	30	NRL
AL	1875.630	20	NRL
MN I	1875.727	10	NRL

Table III
Lines Below 2000A
(con't)

Element	Wavelength (Angstrom)	Plasma Intensity	Source
SI I	1875.813	300	NRL
AL	1875.980	40	NRL
KR II	1875.999	200	NRL
CO I	1876.010	100	NRL
GE I	1876.010	10	NRL
V II	1876.060	200	NRL
MN II	1876.110	1	NRL
NI II	1876.180	1	NRL
FE II	1876.181	160	NRL
NI II	1876.418	2	NRL
FE I	1876.419	40	NRL
MN I	1876.445	15	NRL
V II	1876.470	100	NRL
CO I	1876.480	70	NRL
KR II	1876.491	20	NRL
P II	1876.790	100	NRL
FE II	1876.838	300	NRL
CO I	1876.880	80	NRL
AS I	1876.980	2	NRL
V II	1877.000	100	NRL
AL II	1877.040	100	NRL
MN II	1877.160	30	NRL
CO I	1877.400	150	NRL
FE II	1877.467	400	NRL
MN I	1877.518	20	NRL
AR II	1877.523	400	NRL
KR II	1877.613	40	NRL
NI II	1877.838	40	NRL
V II	1877.850		NRL
KR II	1878.041	240	NRL
FE I	1878.060	1	NRL
NI II	1878.103	2	NRL
V II	1878.190	5	NRL
FE I	1878.200		NRL
KR II	1878.256	40	NRL
CO I	1878.280	250	NRL
FE II	1878.390		NRL
AL II	1878.500	60	NRL
MN II	1878.520	3	NRL
N II	1878.624	200	NRL

Table III
Lines Below 2000A
(con't)

Element	Wavelength (Angstrom)	Plasma Intensity	Source
KR II	1878.830	160	NRL
FE I	1878.849	20	NRL
V II	1878.900	100	NRL
CR II	1879.050	200	NRL
MN II	1879.090	5	NRL
KR II	1879.141	40	NRL
AR II	1879.420	100	NRL
P II	1879.620	250	NRL
MN II	1879.660	1	NRL
AR II	1879.790	200	NRL
MN II	1879.850	40	NRL
FE I	1879.860	2	NRL
FE II	1880.046	40	NRL
FE I	1880.140	35	NRL
NE II	1880.210	200	NRL
CO I	1880.340	40	NRL
CR I	1880.390	50	NRL
V II	1880.811	1	NRL
CO I	1880.820	150	NRL
SI I	1880.968	500	NRL
FE II	1880.976	400	NRL
CR II	1881.060	120	NRL
NI II	1881.155	25	NRL
KR II	1881.418	20	NRL
GE I	1881.647	8	NRL
SI I	1881.854	300	NRL
CR I	1881.870	250	NRL
MN II	1881.900	30	NRL
AS I	1881.960	40	NRL
CU II	1882.208	1	NRL
MN I	1882.366	3	NRL
MN I	1882.900	1	NRL
FE I	1883.060		NRL
MN I	1883.085		NRL
CL II	1883.140	300	NRL
NI II	1883.170	4	NRL
CR II	1883.350	200	NRL
MN II	1883.780	10	NRL
NE II	1883.820	65	NRL
FE I	1883.910	40	NRL

Table III
Lines Below 2000A
(con't)

Element	Wavelength (Angstrom)	Plasma Intensity	Source
V II	1883.930	200	NRL
MN II	1883.980	1	NRL
CR II	1884.120	20	NRL
V II	1884.254	5	NRL
CO I	1884.450	100	NRL
CO I	1884.560	100	NRL
FE I	1884.730	2	NRL
V II	1885.520	3	NRL
NI II	1885.525	20	NRL
MN II	1885.700	20	NRL
V II	1885.900	100	NRL
NI II	1886.043	12	NRL
CR I	1886.340	500	NRL
AR II	1886.386	400	NRL
MN II	1887.310	2	NRL
N II	1887.404	350	NRL
CR I	1887.600	150	NRL
SI I	1887.693	450	NRL
FE I	1887.761	300	NRL
CR I	1887.850	50	NRL
MN II	1887.870	10	NRL
CO I	1887.890	120	NRL
CL II	1887.900	10	NRL
CR II	1887.960	120	NRL
NE II	1888.150	65	NRL
CR I	1888.170	150	NRL
FE I	1888.320	80	NRL
FE II	1888.733	400	NRL
AR II	1888.782	400	NRL
KR II	1888.783	160	NRL
AR II	1889.029	600	NRL
MN II	1889.050	20	NRL
N I	1889.056	10	NRL
CR I	1889.110	50	NRL
CR I	1889.200	50	NRL
CO I	1889.600	60	NRL
KR II	1889.679	80	NRL
NE II	1889.770	65	NRL
CO I	1889.870	100	NRL
MN II	1889.880	8	NRL

Table III
Lines Below 2000A
(con't)

Element	Wavelength (Angstrom)	Plasma Intensity	Source
AS I	1889.950	5	NRL
MN II	1890.170	5	NRL
# AS I	1890.420	1000	NRL
V II	1890.500	20	NRL
CR II	1890.550	600	NRL
CR I	1890.780	300	NRL
V I	1890.820	15	NRL
MN I	1890.962	1	NRL
MN II	1891.180	8	NRL
AS II	1891.278	50	NRL
MN I	1891.414	3	NRL
AS I	1891.470	5	NRL
MN II	1891.550	30	NRL
FE I	1891.740	200	NRL
CR I	1892.010	50	NRL
MN II	1892.020	50	NRL
MN II	1892.400	10	NRL
ZN	1892.866	75	NRL
SI I	1893.252	180	NRL
MN II	1893.288	60	NRL
CO I	1893.430	30	NRL
MN II	1893.490	30	NRL
SE	1893.500	120	NRL
CR I	1893.590	50	NRL
NI II	1893.600	1	NRL
MN II	1893.700	50	NRL
FE II	1894.006	200	NRL
CO I	1894.070	30	NRL
MN II	1894.190		NRL
ZN II	1894.259	75	NRL
V I	1894.470	40	NRL
AS II	1894.661	1	NRL
NI II	1895.082	6	NRL
GE I	1895.197	100	NRL
KR II	1895.408	20	NRL
V	1895.440	5	NRL
SI I	1895.461	100	NRL
CO II	1895.478	40	NRL
AL	1895.510	20	NRL
FE II	1895.675	200	NRL

Table III
Lines Below 2000A
(con't)

Element	Wavelength (Angstrom)	Plasma Intensity	Source
CR I	1895.780	50	NRL
ZN II	1896.056	2	NRL
NI II	1896.147	2	NRL
SI I	1896.339	100	NRL
MN II	1897.060	20	NRL
AR II	1897.352	51	NRL
AL II	1897.440	120	NRL
KR II	1897.457	160	NRL
CO I	1897.480	100	NRL
MN II	1897.480	100	NRL
V II	1897.700	10	NRL
CO I	1897.730	80	NRL
KR II	1898.048	200	NRL
MN II	1898.120	30	NRL
RE II	1898.360	8	VAF
V	1898.420	10	NRL
FE II	1898.538	200	NRL
SE	1898.555	800	NRL
KR II	1898.631	40	NRL
MN II	1898.640	30	NRL
V I	1898.780	25	NRL
CR II	1898.920	700	NRL
# OS	1899.000	25	VAF
AL II	1899.190	80	NRL
FE I	1899.210	20	NRL
AR II	1899.287	100	NRL
KR II	1899.501	40	NRL
MN II	1899.510	10	NRL
KR II	1899.629	160	NRL
AR II	1899.847	100	NRL
# SN II	1899.890	120	VAF
V I	1900.000	30	NRL
NI II	1900.025	4	NRL
MG	1900.040	100	NRL
S I	1900.270	670	NRL
AR II	1900.638	400	NRL
FE II	1900.667	10	NRL
NI II	1900.865	15	NRL
NI II	1900.921	2	NRL
GE I	1901.061	8	NRL

Table III
Lines Below 2000A
(con't)

Element	Wavelength (Angstrom)	Plasma Intensity	Source
SI I	1901.337	23	VAF
KR II	1901.490	280	NRL
ZN II	1901.523	60	NRL
AS I	1901.540	5	NRL
CO I	1901.750	200	NRL
MN II	1902.040	10	NRL
CO I	1902.150	100	NRL
AS I	1902.310	6	NRL
GE I	1902.396	2	NRL
# BI II	1902.410	10	VAF
CR I	1902.430	150	NRL
SI II	1902.459	100	NRL
KR II	1902.778	40	NRL
MN II	1902.948	500	NRL
KR II	1903.193	360	NRL
CR I	1903.300	50	NRL
FE I	1903.370	20	NRL
FE II	1903.370	20	NRL
GE I	1903.562	10	NRL
CR I	1903.570	50	NRL
AL II	1904.320	40	NRL
SI II	1904.326	500	NRL
V II	1904.540		NRL
SI I	1904.666	400	NRL
AL	1904.690	10	NRL
GE I	1904.702	400	NRL
CO I	1904.750	50	NRL
FE II	1904.784	300	NRL
CO I	1905.410	60	NRL
P I	1905.480	90	NRL
CO I	1905.870	200	NRL
SI II	1905.878	300	NRL
TI II	1906.300	300	NRL
MN II	1906.370	20	NRL
AL II	1906.400	60	NRL
P I	1906.400	60	NRL
V II	1906.451	1	NRL
KR II	1906.543	20	NRL
AL II	1906.600	40	NRL
V II	1906.618	1	NRL

Table III
Lines Below 2000A
(con't)

Element	Wavelength (Angstrom)	Plasma Intensity	Source
AL II	1906.670	40	NRL
CR I	1906.670	100	NRL
CO I	1906.720	50	NRL
BE I	1907.120	100	NRL
CR I	1907.280	100	NRL
KR II	1907.356	160	NRL
CR II	1907.360	60	NRL
V II	1907.361	1	NRL
MN II	1907.490	20	NRL
NE II	1907.490	350	NRL
NI II	1907.612	1	NRL
P I	1907.660	120	NRL
KR II	1907.760	120	NRL
V II	1907.800	300	NRL
MN II	1907.840	200	NRL
CO II	1907.951		NRL
AR II	1907.988	400	NRL
AS I	1908.130	1	NRL
TI II	1908.290	300	NRL
V II	1908.320	400	NRL
NI II	1908.326	1	NRL
GE I	1908.434	50	NRL
CR I	1908.460	200	NRL
MN II	1908.470	20	NRL
CO II	1908.517	30	NRL
MN II	1908.540	20	NRL
# TL II	1908.640	74	VAF
MN II	1908.910	100	NRL
TI II	1909.330	200	NRL
V II	1909.360	400	NRL
AR II	1909.569	50	NRL
CR I	1909.720	100	NRL
TI II	1909.740	200	NRL
KR II	1909.788	20	NRL
MN II	1909.830	60	NRL
KR II	1909.840	40	NRL
FE II	1909.950	1	NRL
KR II	1910.421	20	NRL
FE I	1910.530		NRL
KR II	1910.539	200	NRL

Table III
Lines Below 2000A
(con't)

Element	Wavelength (Angstrom)	Plasma Intensity	Source
SI II	1910.621	500	NRL
FE II	1910.669	160	NRL
CL II	1910.760	10	NRL
AL II	1910.830	40	NRL
AL II	1911.010	40	NRL
TI II	1911.010	50	NRL
AR II	1911.053	50	NRL
V II	1911.108	4	NRL
SI II	1911.265		NRL
CR I	1911.300	350	NRL
CR II	1911.360	140	NRL
MN II	1911.410	300	NRL
MN II	1911.760	2	NRL
KR II	1911.797	80	NRL
C	1911.830	10	NRL
V II	1911.880	400	NRL
NI II	1912.146	2	NRL
MN II	1912.150	1	NRL
GE II	1912.376	20	NRL
V II	1912.390	400	NRL
GE I	1912.409	100	NRL
BE I	1912.490	100	NRL
FE II	1912.582	2	NRL
C	1912.750	10	NRL
CR I	1912.790	200	NRL
AS II	1912.938	500	NRL
MN II	1912.960	30	NRL
V II	1913.100	200	NRL
V II	1913.700	350	NRL
MN II	1913.730	20	NRL
MN I	1913.752	15	NRL
SE	1913.788	700	NRL
TI II	1914.110	25	NRL
V II	1914.295	1	NRL
TI II	1914.320		NRL
KR II	1914.673	40	NRL
MN II	1914.680	200	NRL
S I	1914.680	500	NRL
ZN II	1914.806	60	NRL
V II	1914.910	150	NRL

Table III
Lines Below 2000A
(con't)

Element	Wavelength (Angstrom)	Plasma Intensity	Source
MN II	1915.100	6	VAF
MN II	1915.210	20	NRL
V II	1915.710		NRL
NE II	1916.090	550	NRL
CR I	1916.230	100	NRL
MN II	1916.340		NRL
V II	1916.404		NRL
MN II	1916.710	30	NRL
AS I	1917.210	20	NRL
FE II	1917.337	300	NRL
MN II	1917.380	3	NRL
KR II	1917.387	80	NRL
GE I	1917.592	200	NRL
MN II	1917.610	100	NRL
CO II	1917.621	60	NRL
V II	1917.790	150	NRL
FE II	1918.114	40	NRL
CR II	1918.300	80	NRL
MN I	1918.328	15	NRL
NI II	1918.370		NRL
KR II	1918.567	40	NRL
MN II	1918.643	200	NRL
MN II	1918.909	30	NRL
AU II	1918.930	35	VAF
ZN II	1918.962	50	NRL
SE	1919.190	600	NRL
AR II	1919.199	300	NRL
KR II	1919.199	160	NRL
V II	1919.350	200	NRL
KR II	1919.522	40	NRL
MN II	1919.639	300	NRL
BE I	1919.760	200	NRL
AR II	1920.007	200	NRL
MN II	1920.030	100	NRL
ZN II	1920.271	70	NRL
KR II	1920.467	40	NRL
NI II	1920.582	1	NRL
CU II	1920.672	5	NRL
CO II	1920.713	20	NRL
MN II	1920.950	2	NRL

Table III
Lines Below 2000A
(con't)

Element	Wavelength (Angstrom)	Plasma Intensity	Source
ZN	1920.951	10	NRL
MN II	1921.250	800	NRL
AL	1921.940	20	NRL
MN II	1922.031	100	NRL
CU II	1922.142	5	NRL
AL	1922.160	20	NRL
FE II	1922.234	10	NRL
AL	1922.260	10	NRL
MN I	1922.516	12	NRL
FE II	1922.797	400	NRL
CO II	1922.925	30	NRL
CR II	1923.020	160	NRL
MN II	1923.070	200	NRL
MN II	1923.340	200	NRL
CL II	1923.350	400	NRL
GE I	1923.467	100	NRL
CO I	1924.460	150	NRL
MG	1924.480	10	NRL
AL II	1924.830	200	NRL
V II	1924.870	300	NRL
MN II	1924.960	5	NRL
CO I	1925.050	120	NRL
MN II	1925.170	5	NRL
MN II	1925.520	300	NRL
MN II	1925.728	1	NRL
FE II	1925.983	400	NRL
AL II	1926.050	80	NRL
MN II	1926.167	40	NRL
MN II	1926.585	500	NRL
CO I	1926.900	100	NRL
MN II	1926.948	50	NRL
AL II	1926.960	60	NRL
AL	1927.090	40	NRL
AL	1927.130	10	NRL
MN II	1927.330	10	NRL
FE II	1927.481	20	NRL
NI II	1927.707	2	NRL
MN II	1928.150	40	NRL
C II	1928.300	10	NRL
P II	1928.300	5	NRL

Table III
Lines Below 2000A
(con't)

	Element	Wavelength (Angstrom)	Plasma Intensity	Source
	MG	1928.430	10	NRL
	CU II	1928.450	1	NRL
	NI II	1929.063	1	NRL
	AS I	1929.140	3	NRL
	FE II	1929.194	20	NRL
	CO I	1929.340	150	NRL
	CU II	1929.608	25	NRL
	V II	1929.610	600	NRL
	BE I	1929.670	300	NRL
	ZN II	1929.670	3	NRL
	CU II	1929.751	25	NRL
	GE I	1929.826	400	NRL
	CR II	1929.960	240	NRL
	AL II	1929.990	240	NRL
	NE II	1930.020	450	NRL
	MN II	1930.060	3	NRL
	GE I	1930.271	8	NRL
	MN II	1930.450	30	NRL
	FE II	1930.888	20	NRL
	CO I	1930.900	60	NRL
#	C I	1930.905	67	VAF
	CO I	1931.000	100	NRL
	AL II	1931.060	200	NRL
	ZN II	1931.073	40	NRL
	C	1931.150	100	NRL
	KR II	1931.277	40	NRL
	MN II	1931.404	300	NRL
	AR II	1931.419	100	NRL
	KR II	1931.565	240	NRL
	CO I	1931.890	80	NRL
	AR II	1932.230	200	NRL
	MN II	1932.270	10	NRL
	AL II	1932.390	240	NRL
	FE II	1932.477	300	NRL
	V II	1932.550	5	NRL
	MN II	1932.600	20	NRL
	CR II	1932.640	100	NRL
	V II	1932.990	30	NRL
	CO I	1933.030	30	NRL
	V II	1933.280	100	NRL

Table III
Lines Below 2000A
(con't)

Element	Wavelength (Angstrom)	Plasma Intensity	Source
MN II	1933.320	15	NRL
FE II	1933.451	10	NRL
AS II	1933.530		NRL
AR II	1933.694	200	NRL
KR II	1933.784	120	NRL
KR II	1933.852	80	NRL
MN II	1933.930	80	NRL
MN II	1933.930	80	NRL
GE I	1934.048	100	NRL
CO I	1934.340	120	NRL
FE II	1934.475	1	NRL
AL II	1934.520	400	NRL
FE I	1934.528	500	NRL
AL II	1934.730	300	NRL
MN II	1934.780	20	NRL
MN II	1935.068	100	NRL
FE II	1935.296	300	NRL
MN II	1935.384	20	NRL
FE II	1935.450		NRL
CO I	1935.460		NRL
V II	1935.531	1	NRL
CR II	1935.580	500	NRL
MN II	1936.089	1	NRL
CO I	1936.580	300	NRL
MN II	1936.740	10	NRL
FE II	1936.799	400	NRL
S II	1936.840	300	NRL
AL II	1936.930	300	NRL
# AS I	1936.960	56	VAF
PT	1937.000	22	VAF
AR II	1937.041	100	NRL
CO II	1937.070	30	NRL
GE I	1937.146	8	NRL
FE I	1937.274	500	NRL
MN II	1937.280	2	NRL
V II	1937.440	600	NRL
MN II	1937.460	15	NRL
GE I	1937.482	300	NRL
CR II	1937.560	400	NRL
AS I	1937.590	900	NRL

Table III
Lines Below 2000A
(con't)

Element	Wavelength (Angstrom)	Plasma Intensity	Source
NI II	1937.661	1	NRL
V II	1937.680	70	NRL
AS II	1937.869		NRL
GE II	1938.008	100	NRL
MN II	1938.200	20	NRL
MN II	1938.270	10	NRL
GE I	1938.300	300	NRL
CR II	1938.420	60	NRL
KR II	1938.427	20	NRL
V II	1938.500	100	NRL
NI II	1938.579	2	NRL
V II	1938.700	80	NRL
ZN	1938.775	12	NRL
NE II	1938.820	350	NRL
GE II	1938.891	100	NRL
FE II	1938.899	160	NRL
MN II	1938.920	1	NRL
CO I	1938.940	100	NRL
KR II	1939.037	40	NRL
CR II	1939.150	120	NRL
AL II	1939.270	300	NRL
V II	1939.320	80	NRL
NI II	1939.560		NRL
CO I	1939.750	30	NRL
MN II	1939.770	1	NRL
CR II	1939.900	100	NRL
NI II	1939.901	10	NRL
MN II	1939.960	20	NRL
KR II	1940.112	20	NRL
CO I	1940.160	150	NRL
MN II	1940.190	8	NRL
ZN II	1940.413	40	NRL
FE I	1940.649	500	NRL
V II	1940.860	400	NRL
AR II	1941.072	300	NRL
V II	1941.270	300	NRL
CO II	1941.281	800	NRL
AS I	1941.360	5	NRL
V II	1941.400	300	NRL
SI II	1941.667	500	NRL

Table III
Lines Below 2000A
(con't)

Element	Wavelength (Angstrom)	Plasma Intensity	Source
AS II	1941.876	1	NRL
MN II	1941.890	20	NRL
KR II	1941.944	40	NRL
FE II	1941.990	2	NRL
MN II	1942.080	20	NRL
# HG II	1942.270	120	VAF
MN II	1942.350	5	NRL
V II	1942.350	400	NRL
MN II	1942.660	40	NRL
NI II	1943.060	2	NRL
AS II	1943.305		NRL
AS II	1943.317	40	NRL
SI	1943.420		NRL
MN II	1943.541	100	NRL
CO I	1943.640	120	NRL
BE I	1943.680	500	NRL
NI II	1943.744	2	NRL
KR II	1943.765	40	NRL
V II	1943.990		NRL
GE I	1944.116	150	NRL
MN II	1944.160	100	NRL
CO II	1944.181	20	NRL
MN II	1944.360	20	NRL
SI II	1944.586	150	NRL
CU II	1944.597	40	NRL
GE I	1944.731	300	NRL
MN II	1944.770	40	NRL
FE I	1945.070	200	NRL
CO I	1945.090	120	NRL
MN II	1945.150	200	NRL
FE I	1945.294	400	NRL
V II	1945.350	300	NRL
SI II	1945.504	300	NRL
ZN II	1945.538	60	NRL
P II	1945.540	50	NRL
V II	1945.640	300	NRL
MN II	1945.830	5	NRL
CO I	1945.860		NRL
CR II	1945.980	200	NRL
AS II	1946.137	1	NRL

Table III
Lines Below 2000A
(con't)

Element	Wavelength (Angstrom)	Plasma Intensity	Source
FE I	1946.216	40	NRL
MN II	1946.340	30	NRL
FE II	1946.429	2	NRL
CU II	1946.493	10	NRL
KR II	1946.677	20	NRL
CO I	1946.790	250	NRL
AR II	1946.795	200	NRL
FE II	1946.850	2	NRL
MN II	1946.940	20	NRL
FE I	1946.983	600	NRL
GE II	1947.550	2	NRL
AS II	1947.572		NRL
CO I	1947.580	50	NRL
SI II	1947.769	100	NRL
MN II	1947.932	200	NRL
CO I	1948.090	100	NRL
FE II	1948.372	200	NRL
ZN II	1948.458	30	NRL
CR II	1948.510	200	NRL
MN II	1948.720	10	NRL
MN II	1948.720	80	NRL
KR II	1948.752	240	NRL
CO I	1949.000	150	NRL
CR II	1949.000	800	NRL
MN I	1949.100	3	NRL
SI	1949.170		NRL
CR II	1949.220	700	NRL
SI II	1949.331	100	NRL
MN II	1949.560	10	NRL
SI II	1949.564	100	NRL
CR II	1950.060	1000	NRL
CO II	1950.097	500	NRL
MN II	1950.143	200	NRL
FE I	1950.223	500	NRL
MN II	1950.310	30	NRL
SB I	1950.390	18	VAF
S II	1950.530	300	NRL
MN II	1950.690	2	NRL
ZN	1950.769	10	NRL
V II	1950.770	5	NRL

Table III
Lines Below 2000A
(con't)

Element	Wavelength (Angstrom)	Plasma Intensity	Source
MN II	1950.922	40	NRL
AS II	1951.393		NRL
CO I	1951.440	120	NRL
FE I	1951.556	500	NRL
CO I	1951.900	250	NRL
ZN II	1951.911	60	NRL
NI II	1951.920		NRL
AU I	1951.930	18	VAF
FE I	1952.262	500	NRL
GE II	1952.290	3	NRL
AS II	1952.541		NRL
CU II	1952.576	5	NRL
FE I	1952.579	500	NRL
MN II	1952.860	2	NRL
SI	1952.960		NRL
ZN II	1952.999	80	NRL
FE I	1953.001	500	NRL
MN II	1953.230	300	NRL
MN II	1953.350	3	NRL
NI II	1953.407	40	NRL
MN II	1953.510	10	NRL
CO I	1953.710	80	NRL
GE I	1953.802	100	NRL
BI I	1953.890	14	VAF
MN II	1953.900	10	NRL
MN II	1953.980	5	NRL
P I	1954.010	12	NRL
V II	1954.061	2	NRL
MN II	1954.100	5	NRL
P I	1954.200	21	NRL
CO I	1954.220	300	NRL
MN II	1954.320	30	NRL
NI II	1954.709	1	NRL
MN II	1954.810	200	NRL
ZN II	1954.872	75	NRL
SI I	1954.966	500	NRL
MN II	1955.060	1	NRL
GE I	1955.115	200	NRL
P I	1955.160	21	NRL
CO I	1955.170	300	NRL

Table III
Lines Below 2000A
(con't)

Element	Wavelength (Angstrom)	Plasma Intensity	Source
FE II	1955.639	20	NRL
FE I	1955.690	400	NRL
P I	1955.790	21	NRL
CR II	1955.930	300	NRL
FE I	1956.026	500	NRL
FE II	1956.085	10	NRL
CO I	1956.220	150	NRL
KR II	1956.412	40	NRL
BE I	1956.630	100	NRL
SI	1956.750		NRL
AS II	1956.835		NRL
CO II	1957.418	150	NRL
O II	1957.420	10	NRL
CU II	1957.518	20	NRL
KR II	1957.542	20	NRL
CO I	1957.690	120	NRL
FE I	1957.838	600	NRL
V I	1957.900	250	NRL
SI I	1957.965	100	NRL
MN II	1957.980	40	NRL
FE II	1958.086	5	NRL
CO I	1958.100	80	NRL
FE II	1958.121	100	NRL
V I	1958.180		NRL
AL II	1958.250	40	NRL
S II	1958.380	300	NRL
KR II	1958.427	40	NRL
V II	1958.446	1	NRL
CO I	1958.550	250	NRL
MN II	1958.580	5	NRL
FE I	1958.598	600	NRL
V I	1958.600		NRL
FE I	1958.739	300	NRL
AL II	1958.770	20	NRL
AS I	1958.820	20	NRL
FE I	1958.840		NRL
AS I	1958.910	40	NRL
MN II	1958.932	80	NRL
CO I	1958.940	150	NRL
AL II	1959.000	10	NRL

Table III
Lines Below 2000A
(con't)

Element	Wavelength (Angstrom)	Plasma Intensity	Source
AL	1959.100	10	NRL
V I	1959.120	150	NRL
MN II	1959.246	300	NRL
V I	1959.360	150	NRL
MN II	1959.690	5	NRL
GE II	1959.781	30	NRL
V I	1959.970	200	NRL
V II	1960.137	3	NRL
FE I	1960.139	600	NRL
SI	1960.170		NRL
# SE I	1960.260	40	VAF
MN II	1960.270	10	NRL
AL II	1960.320	40	NRL
O II	1960.340	100	NRL
MN II	1960.370	10	NRL
V II	1960.445		NRL
AL II	1960.640	40	NRL
MN II	1960.670		NRL
AL II	1960.840	40	NRL
AS I	1960.890	2	NRL
MN II	1960.900	5	NRL
SE	1960.901	1000	NRL
V II	1960.980	20	NRL
CO I	1961.000	150	NRL
FE I	1961.236	500	NRL
CO I	1961.260	80	NRL
AR II	1961.361	400	NRL
MN II	1961.390	20	NRL
CO I	1961.590	250	NRL
V I	1961.690	300	NRL
KR II	1961.863	40	NRL
GE I	1962.013	500	NRL
FE I	1962.031	500	NRL
FE II	1962.071	1	NRL
FE I	1962.107	600	NRL
AR II	1962.161	300	NRL
O II	1962.240	300	NRL
AL II	1962.590	100	NRL
AL II	1962.720	120	NRL
FE I	1962.746	10	NRL

Table III
Lines Below 2000A
(con't)

Element	Wavelength (Angstrom)	Plasma Intensity	Source
MN II	1962.800	30	NRL
FE I	1962.871	400	NRL
CR II	1963.000	300	NRL
FE II	1963.110		NRL
FE I	1963.110	500	NRL
KR II	1963.361	240	NRL
GE I	1963.373	100	NRL
CO I	1963.380	120	NRL
V I	1963.470	350	NRL
CO I	1963.550	200	NRL
FE I	1963.629	200	NRL
HF II	1963.820	150	VAF
O II	1963.840	200	NRL
NI I	1963.850	50	NRL
CO I	1963.920	80	NRL
CO I	1964.030	200	NRL
FE I	1964.043	400	NRL
V II	1964.091	1	NRL
MN II	1964.190	10	NRL
V I	1964.270	300	NRL
FE II	1964.342	240	NRL
O II	1964.350	10	NRL
MN II	1964.450	80	NRL
ZN II	1964.538	80	NRL
BE I	1964.590	900	NRL
MN II	1964.700	20	NRL
AL	1964.990	60	NRL
AS II	1965.036	3	NRL
V I	1965.070	300	NRL
V I	1965.260	300	NRL
AL II	1965.310	40	NRL
NI II	1965.357	8	NRL
GE I	1965.383	200	NRL
MN II	1965.680	60	NRL
MN II	1966.040	20	NRL
MN II	1966.240	2	NRL
GE II	1966.317	40	NRL
AL	1966.380	10	NRL
V I	1966.520	300	NRL
CO I	1966.680	90	NRL

Table III
Lines Below 2000A
(con't)

Element	Wavelength (Angstrom)	Plasma Intensity	Source
V II	1966.752	3	NRL
V I	1966.760	300	NRL
AR II	1966.953	100	NRL
V II	1967.185	4	NRL
MN II	1967.590	30	NRL
FE II	1967.670	1	NRL
CO I	1967.780	100	NRL
V I	1967.980	400	NRL
CU II	1968.012	2	NRL
FE II	1968.060	20	NRL
FE II	1968.210	1	NRL
AS II	1968.660	20	NRL
CO I	1968.690	250	NRL
AS II	1968.855	20	NRL
FE II	1968.872	20	NRL
NI I	1968.900	50	NRL
CO II	1968.918	30	NRL
CO I	1968.930	250	NRL
MN II	1969.240	200	NRL
KR II	1969.278	40	NRL
ZN II	1969.404	100	NRL
V I	1969.570	75	NRL
CO I	1969.680	30	NRL
MN II	1969.870	50	NRL
MN II	1969.950	100	NRL
KR II	1970.125	80	NRL
MN II	1970.160	5	NRL
FE II	1970.329	2	NRL
CU II	1970.495	15	NRL
FE II	1970.680	20	NRL
CO I	1970.710	500	NRL
CO I	1970.770		NRL
FE I	1970.771	10	NRL
V II	1970.877		NRL
GE I	1970.880	500	NRL
MN II	1970.947	5	NRL
AS II	1970.953		NRL
FE II	1970.989	1	NRL
S II	1970.990	400	NRL
CO I	1971.160	300	NRL

Table III
Lines Below 2000A
(con't)

Element	Wavelength (Angstrom)	Plasma Intensity	Source
MN II	1971.220	100	NRL
AS II	1971.580		NRL
MN II	1971.710	20	NRL
CO I	1971.750	150	NRL
# AS I	1971.970	39	VAF
AR II	1972.274	200	NRL
V II	1972.278	1	NRL
V I	1972.480	75	NRL
CO I	1972.520	300	NRL
AS I	1972.620	800	NRL
V II	1972.620	200	NRL
CO I	1972.820		NRL
# RE II	1973.130	49	VAF
AR II	1973.484	200	NRL
CO I	1973.620	50	NRL
AS I	1973.650	3	NRL
CO I	1973.850	250	NRL
FE I	1973.911	20	NRL
KR II	1974.015	20	NRL
FE I	1974.059	20	NRL
MN II	1974.100	100	NRL
V II	1974.278	1	NRL
CO I	1974.390	150	NRL
AR II	1974.462	300	NRL
ZN II	1974.467	12	NRL
CU II	1974.468		NRL
FE II	1974.489	1	NRL
MG	1974.750	10	NRL
MN II	1974.820	80	NRL
CO II	1975.002	20	NRL
V II	1975.021	1	NRL
MN II	1975.240	30	NRL
KR II	1975.251	280	NRL
AS II	1975.270	1	NRL
CO I	1975.360	60	NRL
V I	1975.420	100	NRL
V II	1975.489	1	NRL
FE II	1975.542	20	NRL
CO I	1975.670	200	NRL
CO I	1975.940	60	NRL

Table III
Lines Below 2000A
(con't)

Element	Wavelength (Angstrom)	Plasma Intensity	Source
MN II	1976.002	10	NRL
MN II	1976.180	2	NRL
KR II	1976.252	40	NRL
MN II	1976.270	2	NRL
V II	1976.620	600	NRL
AR II	1976.765	300	NRL
MN II	1976.870	40	NRL
NI I	1976.870	150	NRL
CO I	1976.970	300	NRL
CU II	1977.027	15	NRL
ZN II	1977.159	25	NRL
AR II	1977.200	20	NRL
IN II	1977.450	2	VAF
ZN II	1977.494	5	NRL
SI I	1977.597	400	NRL
V II	1977.600	5	NRL
MN II	1977.720	40	NRL
# AU I	1978.190	77	VAF
CO I	1978.360	100	NRL
CO I	1978.530	120	NRL
MN II	1978.750	20	NRL
KR II	1978.904	80	NRL
V II	1978.960	200	NRL
V II	1979.087	5	NRL
SI I	1979.206	400	NRL
GE II	1979.274	30	NRL
V II	1979.618	1	NRL
AS II	1979.755	10	NRL
CU II	1979.956	200	NRL
NI II	1980.010	15	NRL
V II	1980.040	400	NRL
CO I	1980.590	150	NRL
V II	1980.590	250	NRL
SI I	1980.617	300	NRL
NI II	1980.699	3	NRL
AL	1980.820	20	NRL
MN II	1980.840	2	NRL
CO I	1980.890	400	NRL
AS II	1980.992	2	NRL
MN II	1981.010	10	NRL

Table III
Lines Below 2000A
(con't)

Element	Wavelength (Angstrom)	Plasma Intensity	Source
MN II	1981.160	20	NRL
V II	1981.245		NRL
KR II	1981.264	40	NRL
AS II	1981.466	10	NRL
V II	1981.530	80	NRL
NI I	1981.610	100	NRL
KR II	1981.653	40	NRL
MN II	1981.670	20	NRL
AR II	1981.740	100	NRL
S II	1981.800	200	NRL
V I	1981.850	5	NRL
CO I	1981.970	200	NRL
V I	1982.060	200	NRL
ZN II	1982.111	100	NRL
V II	1982.210	5	NRL
V II	1982.410	80	NRL
ZN II	1982.429	10	NRL
V I	1982.450	200	NRL
CO I	1982.520	200	NRL
MN II	1982.600	10	NRL
CO I	1982.810	80	NRL
KR II	1982.866	120	NRL
FE II	1983.050		NRL
CO II	1983.162	100	NRL
MN II	1983.170	60	NRL
SI I	1983.232	300	NRL
AR II	1983.299	100	NRL
V I	1983.370	250	NRL
AS II	1983.581	100	NRL
KR II	1983.634	40	NRL
AL II	1983.650	60	NRL
AS II	1983.787	3	NRL
AR II	1983.820	100	NRL
CO II	1983.947	20	NRL
V II	1984.050	100	NRL
SI I	1984.069	300	NRL
MN II	1984.430	20	NRL
V I	1984.430	30	NRL
SI I	1984.439	200	NRL
CU II	1984.764	1	NRL

Table III
Lines Below 2000A
(con't)

Element	Wavelength (Angstrom)	Plasma Intensity	Source
CO II	1984.827	10	NRL
V I	1984.910	250	NRL
FE II	1984.940		NRL
BE I	1985.130	300	NRL
S II	1985.170	200	NRL
CO I	1985.250	100	NRL
MN II	1985.250	2	NRL
CO I	1985.360	100	NRL
CR II	1985.420	440	NRL
ZN II	1985.608	70	NRL
CR II	1985.670	240	NRL
CO I	1985.880	40	NRL
CO I	1986.310	60	NRL
SI I	1986.363	500	NRL
MN II	1986.400	20	NRL
FE II	1986.423	40	NRL
GE II	1986.600	3	NRL
ZN II	1986.988	100	NRL
CO I	1987.030	150	NRL
CO I	1987.150	120	NRL
CO I	1987.240	100	NRL
V I	1987.250	100	NRL
V II	1987.314	10	NRL
C II	1987.330	40	NRL
CR II	1987.430	100	NRL
V II	1987.547	4	NRL
CO I	1987.650	200	NRL
C II	1987.760	120	NRL
V II	1987.820	3	NRL
GE I	1987.849	300	NRL
AS II	1988.018	20	NRL
V II	1988.031	8	NRL
MN II	1988.060	30	NRL
C II	1988.090	80	NRL
GE I	1988.267	600	NRL
C II	1988.510	40	NRL
AR II	1988.620	300	NRL
V II	1988.674		NRL
AL II	1988.710	40	NRL
AS II	1988.955	100	NRL

Table III
Lines Below 2000A
(con't)

Element	Wavelength (Angstrom)	Plasma Intensity	Source
SI I	1988.993	50	VAF
CR I	1989.000	200	NRL
GE I	1989.117	100	NRL
CO I	1989.280	100	NRL
MN II	1989.410	50	NRL
AU II	1989.630	20	VAF
AS I	1989.700	22	VAF
CO I	1989.800	250	NRL
CU II	1989.855	90	NRL
CR I	1989.920	750	NRL
CU II	1990.180	1	NRL
NI I	1990.250	200	NRL
CR I	1990.270	400	NRL
CO I	1990.340	300	NRL
KR II	1990.341	20	NRL
AS I	1990.350	200	NRL
AS I	1990.480	6	VAF
AL II	1990.530	800	NRL
C	1990.530	10	NRL
CR II	1990.790	200	NRL
FE II	1990.805	1	NRL
AL II	1991.050	20	NRL
AS I	1991.130	100	NRL
C	1991.140	5	NRL
CR I	1991.220	750	NRL
N II	1991.301	10	NRL
V II	1991.310		NRL
CO II	1991.574		NRL
CO I	1991.800	30	NRL
SI I	1991.852	500	NRL
CR I	1992.120	300	NRL
MN II	1992.300	20	NRL
KR II	1992.464	40	NRL
CR I	1992.650	250	NRL
AS II	1992.682	5	NRL
CO I	1992.790	200	NRL
CO I	1993.250	100	NRL
FE II	1993.289	160	NRL
ZN II	1993.367	50	NRL
CR II	1993.370	300	NRL

Table III
Lines Below 2000A
(con't)

Element	Wavelength (Angstrom)	Plasma Intensity	Source
AR II	1993.555	20	NRL
NI II	1993.570	2	NRL
MN II	1993.600	10	NRL
C I	1993.620	3	VAF
CR II	1993.630	500	NRL
KR II	1993.763	120	NRL
CU II	1993.810	1	NRL
NI II	1993.906	1	NRL
MN II	1993.990	100	NRL
CR I	1994.100	400	NRL
TE I	1994.180	6	VAF
MN II	1994.230	300	NRL
NI I	1994.290	100	NRL
V II	1994.340	100	NRL
CR I	1994.550	750	NRL
FE II	1994.857	400	NRL
AS I	1994.890	20	NRL
CO I	1994.980	150	NRL
V II	1995.020	5	NRL
SE I	1995.112	1	VAF
MN II	1995.275	1	NRL
AS I	1995.430	100	NRL
CR I	1995.690	250	NRL
NI II	1995.723	15	NRL
MN I	1996.056	50	NRL
AL	1996.300	20	NRL
CR II	1996.620	60	NRL
ZN II	1996.922	50	NRL
FE II	1997.034	4	NRL
CR I	1997.090	250	NRL
CO II	1997.095	10	NRL
CR I	1997.300	500	NRL
GE I	1997.806	150	NRL
CR I	1997.900	600	NRL
MN II	1997.940	20	NRL
BE I	1998.010	100	NRL
CR II	1998.140	40	NRL
ZN II	1998.238	20	NRL
# GE I	1998.240	34	VAF
AS II	1998.340	10	NRL

Table III
Lines Below 2000A
(con't)

Element	Wavelength (Angstrom)	Plasma Intensity	Source
CO I	1998.490	250	NRL
GE I	1998.887	1000	NRL
S II	1998.920	200	NRL
ZN II	1998.977	25	NRL
FE II	1999.430	200	NRL
FE II	1999.462	150	NRL
MN I	1999.511	100	NRL
NI I	1999.530		NRL
FE II	1999.727	20	NRL
MN II	1999.920	80	NRL

Explanation of Table IV

This table includes all argon lines found in the expanded MIT Tables. The data are taken from one of four sources. These are the original MIT Tables[1] (no reference in column four), the NRL compilations[2] (NRL), Fassel[3] (VAF), and the NBS compilations[4] (NBS). For the data arising from the original MIT Tables, the intensity data are from the discharge tube column.

References

1. G.R. Harrison, *Massachusetts Institute of Technology Wavelength Tables* (John Wiley & Sons, New York, 1969).
2. R.L. Kelley and L.J. Palumbo, *Atomic and Ionic Emission Lines below 2000 Angstroms—Hydrogen through Krypton,* NRL Report 7599 (Naval Research Laboratory, Washington, D.C., 1973).
3. R.K. Winge, V.J. Peterson, and V.A. Fassel, *Applied Spectroscopy,* **33,** 3, pp. 206–219 (1979).
4. National Bureau of Standards, *Atomic Transition Probabilities,* NSRDS–NBS, **22,** II, pp. 187–210 (1969).

Table IV
Argon Lines

Element	Wavelength (Angstrom)	Plasma Intensity	Source
AR II	1854.986	50	NRL
AR II	1866.089	100	NRL
AR II	1868.660	300	NRL
AR II	1872.589	100	NRL
AR II	1873.140	600	NRL
AR II	1877.523	400	NRL
AR II	1879.420	100	NRL
AR II	1879.790	200	NRL
AR II	1886.386	400	NRL
AR II	1888.782	400	NRL
AR II	1889.029	600	NRL
AR II	1897.352	51	NRL
AR II	1899.287	100	NRL
AR II	1899.847	100	NRL
AR II	1900.638	400	NRL
AR II	1907.988	400	NRL
AR II	1909.569	50	NRL
AR II	1911.053	50	NRL
AR II	1919.199	300	NRL
AR II	1920.007	200	NRL
AR II	1931.419	100	NRL
AR II	1932.230	200	NRL
AR II	1933.694	200	NRL
AR II	1937.041	100	NRL
AR II	1941.072	300	NRL
AR II	1946.795	200	NRL
AR II	1961.361	400	NRL
AR II	1962.161	300	NRL
AR II	1966.953	100	NRL
AR II	1972.274	200	NRL
AR II	1973.484	200	NRL
AR II	1974.462	300	NRL
AR II	1976.765	300	NRL
AR II	1977.200	20	NRL
AR II	1981.740	100	NRL
AR II	1983.299	100	NRL
AR II	1983.820	100	NRL
AR II	1988.620	300	NRL
AR II	1993.555	20	NRL
AR	2122.600	5	

Table IV
Argon Lines
(con't)

Element	Wavelength (Angstrom)	Plasma Intensity	Source
AR	2131.000	2	
AR	2137.200	40	
AR	2139.200	5	
AR	2140.010	10	
AR	2148.150	5	
AR II	2152.600	63	
AR II	2159.000	63	
AR	2162.740	63	
AR	2164.170	100	
AR II	2170.800	2	
AR II	2171.410	63	
AR	2177.040	20	
AR II	2179.250	40	
AR II	2182.740	5	
AR II	2183.300	40	
AR	2183.700	20	
AR	2187.340	20	
AR	2197.560	10	
AR II	2202.770	2	
AR	2203.220	2	
AR	2208.900	10	
AR	2212.220	20	
AR	2213.000	63	
AR	2215.200	2	
AR	2218.870	63	
AR	2221.940	40	
AR	2222.930	63	
AR	2224.760	40	
AR	2227.700	63	
AR II	2229.500	2	
AR II	2235.770	40	
AR II	2238.200	63	
AR	2244.800	2	
AR	2245.970	5	
AR	2248.000	20	
AR II	2250.300	20	
AR II	2252.260	5	
AR	2252.870	2	
AR II	2261.800	10	
AR	2272.600	2	

Table IV
Argon Lines
(con't)

Element	Wavelength (Angstrom)	Plasma Intensity	Source
AR	2280.600	63	
AR	2281.200	20	
AR	2282.050	40	
AR II	2282.640	20	
AR	2284.020	10	
AR	2284.910	5	
AR	2287.190	20	
AR	2287.800	63	
AR	2288.400	2	
AR	2289.780	20	
AR	2293.270	20	
AR	2298.170	40	
AR	2300.070	2	
AR II	2300.190	10	
AR	2304.000	2	
AR	2304.500	2	
AR	2309.160	40	
AR	2313.740	63	
AR	2314.990	40	
AR	2316.310	100	
AR	2317.370	10	
AR II	2317.770	2	
AR	2322.110	5	
AR	2324.400	5	
AR	2331.450	63	
AR	2337.790	63	
AR	2339.800	10	
AR	2344.220	4	
AR	2350.500	20	
AR	2352.730	5	
AR	2353.420	10	
AR	2354.140	20	
AR II	2354.790	5	
AR	2357.600	63	
AR II	2360.070	40	
AR	2364.140	40	
AR	2366.780	10	
AR II	2369.280	2	
AR	2369.970	2	
AR	2371.740	20	

Table IV
Argon Lines
(con't)

Element	Wavelength (Angstrom)	Plasma Intensity	Source
AR	2376.300	2	
AR	2377.200	2	
AR	2379.880	2	
AR	2381.180	2	
AR II	2383.500	40	
AR	2385.000	40	
AR II	2387.960	40	
AR	2389.110	5	
AR	2390.940	10	
AR	2398.390	2	
AR	2399.260	2	
AR II	2404.400	10	
AR	2405.270	2	
AR II	2410.970	5	
AR	2412.500	2	
AR II	2414.260	5	
AR	2415.800	2	
AR II	2420.490	10	
AR	2422.730	2	
AR	2424.000	2	
AR II	2424.700	2	
AR II	2430.060	5	
AR	2438.780	2	
AR II	2440.070	2	
AR	2454.270	10	
AR II	2459.970	5	
AR	2474.020	5	
AR II	2475.480	5	
AR II	2479.080	5	
AR	2480.870	10	
AR	2481.500	5	
AR II	2482.170	5	
AR	2491.060	5	
AR II	2499.550	2	
AR	2500.420	5	
AR II	2515.600	10	
AR II	2516.810	20	
AR II	2522.530	5	
AR	2525.510	5	
AR	2528.330	10	

Table IV
Argon Lines
(con't)

Element	Wavelength (Angstrom)	Plasma Intensity	Source
AR II	2528.710	5	
AR	2534.740	40	
AR II	2535.280	2	
AR II	2536.040	20	
AR	2540.080	2	
AR	2544.720	40	
AR II	2549.840	2	
AR	2556.630	5	
AR	2559.310	5	
AR II	2562.120	20	
AR II	2564.450	10	
AR	2569.250	2	
AR II	2570.010	2	
AR	2570.460	5	
AR	2617.640	2	
AR	2621.030	10	
AR	2624.630	5	
AR II	2627.410	2	
AR	2647.290	10	
AR	2651.950	2	
AR	2683.140	2	
AR II	2692.620	20	
AR	2708.280	40	
AR	2724.840	2	
AR	2732.530	20	
AR	2733.060	5	
AR	2741.090	2	
AR	2744.820	20	
AR	2753.920	10	
AR II	2754.910	2	
AR	2757.260	5	
AR	2762.180	5	
AR II	2764.660	10	
AR	2767.990	2	
AR	2769.740	20	
AR	2795.330	2	
AR	2795.450	2	
AR	2796.650	2	
AR	2806.160	40	
AR	2843.370	2	

Table IV
Argon Lines
(con't)

Element	Wavelength (Angstrom)	Plasma Intensity	Source
AR II	2844.120	5	
AR II	2847.810	5	
AR	2855.330	5	
AR	2860.730	5	
AR II	2865.850	20	
AR	2874.550	5	
AR	2878.760	2	
AR	2879.310	5	
AR	2884.210	5	
AR II	2891.610	40	
AR	2893.980	2	
AR	2896.750	20	
AR	2897.300	10	
AR	2915.620	5	
AR	2924.660	20	
AR	2931.490	20	
AR II	2932.600	20	
AR II	2935.570	5	
AR II	2942.900	100	
AR II	2955.390	40	
AR	2956.550	10	
AR	2957.560	5	
AR	2960.270	20	
AR II	2979.050	40	
AR	3000.140	5	
AR II	3000.450	10	
AR	3003.000	3	
AR II	3014.490	10	
AR II	3026.750	5	
AR II	3028.930	16	
AR II	3033.520	8	
AR	3046.100	5	
AR	3053.200	3	
AR I	3056.280	3	
AR	3060.940	10	
AR I	3062.060	3	
AR I	3063.440	5	
AR	3066.890	3	
AR II	3082.990	5	
AR	3085.050	5	

Table IV
Argon Lines
(con't)

Element	Wavelength (Angstrom)	Plasma Intensity	Source
AR II	3088.240	10	
AR II	3093.410	50	
AR	3094.980	3	
AR I	3100.090	5	
AR	3102.630	3	
AR II	3104.380	10	
AR	3109.750	3	
AR I	3110.660	3	
AR	3116.630	3	
AR	3117.850	3	
AR I	3120.060	3	
AR I	3130.800	20	
AR I	3132.870	3	
AR	3136.550	5	
AR II	3137.660	5	
AR	3139.020	25	
AR I	3142.600	3	
AR I	3151.520	3	
AR I	3152.290	3	
AR II	3153.800	5	
AR I	3160.060	5	
AR II	3161.380	5	
AR	3161.450	3	
AR	3165.310	5	
AR	3169.680	50	
AR I	3172.180	5	
AR I	3172.960	1	VAF
AR	3181.050	25	
AR II	3186.190	3	
AR I	3186.630	5	
AR	3194.250	10	
AR I	3195.120	5	
AR I	3200.390	1	VAF
AR I	3201.120	3	
AR I	3202.850	5	
AR I	3203.660	10	
AR	3204.340	10	
AR II	3205.030	5	
AR I	3207.500	10	
AR II	3207.610	10	

Table IV
Argon Lines
(con't)

Element	Wavelength (Angstrom)	Plasma Intensity	Source
AR	3212.540	10	
AR	3216.750	10	
AR	3217.700	3	
AR II	3221.640	3	
AR II	3222.420	3	
AR I	3225.580	20	
AR II	3225.970		NBS
AR	3226.000	3	
AR I	3229.910	3	
AR I	3234.490	1	VAF
AR	3236.820	5	
AR	3243.710	20	
AR	3247.550	3	
AR	3249.820	25	
AR I	3257.580	1	VAF
AR	3259.710	5	
AR	3263.600	10	
AR I	3263.780	3	
AR I	3264.290	3	
AR II	3269.050	5	
AR I	3271.160	10	
AR	3273.360	5	
AR I	3278.930	3	
AR I	3279.250	3	
AR	3281.720	16	
AR	3285.870	10	
AR I	3289.390	3	
AR I	3289.950	3	
AR	3291.470	5	
AR	3293.660	25	
AR	3293.950	5	
AR I	3300.390	20	
AR	3301.870	10	
AR	3306.500	3	
AR	3307.240	16	
AR	3309.390	3	
AR	3310.470	3	
AR	3311.260	5	
AR	3312.970	3	
AR I	3319.345	3	VAF

Table IV
Argon Lines
(con't)

Element	Wavelength (Angstrom)	Plasma Intensity	Source
AR I	3320.060	3	
AR I	3321.580	5	
AR I	3322.440	5	
AR I	3323.820	32	
AR I	3325.500	2	VAF
AR	3336.200	5	
AR	3344.790	3	
AR II	3350.940	5	
AR	3358.560	3	
AR I	3359.480	10	
AR II	3361.730	5	
AR I	3363.470	20	
AR II	3365.540	5	
AR	3366.590	5	
AR	3370.970	10	
AR I	3372.880	3	
AR I	3373.481	2	VAF
AR II	3376.460	25	
AR II	3379.480	3	
AR	3379.580	10	
AR I	3381.490	20	
AR I	3387.580	20	
AR I	3388.350	20	
AR	3388.540	25	
AR I	3389.850	20	
AR I	3390.290	3	
AR I	3392.310	3	
AR I	3392.780	2	VAF
AR I	3392.810	100	
AR I	3393.752	4	VAF
AR II	3397.900	20	
AR I	3397.900	20	
AR I	3406.180	2	VAF
AR II	3414.460	5	
AR I	3416.800	5	
AR I	3417.680	3	
AR I	3418.510	3	
AR	3421.640	10	
AR II	3429.640	3	
AR	3430.440	5	

Table IV
Argon Lines
(con't)

Element	Wavelength (Angstrom)	Plasma Intensity	Source
AR II	3432.590		NBS
AR I	3442.580	10	
AR	3452.320	3	
AR	3454.100	10	
AR I	3454.940	20	
AR I	3457.810	3	
AR I	3461.078	4	VAF
AR II	3464.140	16	
AR II	3465.800	3	
AR	3466.340	10	
AR	3470.270	3	
AR	3476.740	20	
AR II	3478.240	5	
AR	3480.520	5	
AR I	3483.170	5	
AR II	3487.330	3	
AR I	3490.500	3	
AR II	3490.890	5	
AR	3491.240	16	
AR	3491.540	50	
AR I	3493.270	1	VAF
AR	3499.490	10	
AR I	3506.490	2	VAF
AR	3509.780	16	
AR	3511.150	3	
AR	3514.390	130	
AR	3517.900	3	
AR	3520.000	16	
AR	3521.270	10	
AR II	3521.980	5	
AR II	3531.220	3	
AR	3535.330	16	
AR	3543.160	10	
AR	3545.580	320	
AR II	3545.840	130	
AR	3548.510	25	
AR	3550.030	5	
AR I	3553.580	16	
AR I	3554.306	18	VAF
AR I	3555.970	100	

Table IV
Argon Lines
(con't)

Element	Wavelength (Angstrom)	Plasma Intensity	Source
AR I	3556.010	2	VAF
AR II	3556.910	10	
AR	3559.510	16	
AR II	3561.040	16	
AR II	3562.190	5	
AR I	3563.290	3	VAF
AR I	3564.270	100	
AR II	3564.340	2	VAF
AR	3565.020	10	
AR I	3567.657	9	VAF
AR I	3572.290	2	VAF
AR	3576.620	320	
AR	3581.620	16	
AR	3582.350	50	
AR I	3582.700	32	
AR I	3588.110	3	
AR	3588.440	320	
AR I	3599.670	20	
AR II	3600.220	3	
AR II	3601.510	5	
AR	3603.460	3	
AR	3603.910	3	
AR II	3605.890	16	
AR I	3606.522	9	VAF
AR II	3611.840	5	
AR	3622.150	16	
AR I	3632.684	4	VAF
AR I	3634.461	4	VAF
AR II	3634.830	5	
AR	3635.670	3	
AR	3637.050	10	
AR	3637.890	3	
AR II	3639.850	25	
AR I	3643.090	100	
AR I	3643.120	2	VAF
AR I	3649.832	7	VAF
AR	3650.900	5	
AR II	3655.290	16	
AR	3656.050	10	
AR I	3659.530	1	VAF

Table IV
Argon Lines
(con't)

Element	Wavelength (Angstrom)	Plasma Intensity	Source
AR II	3660.440	16	
AR I	3663.760	5	
AR	3669.620	10	
AR I	3670.670	2	VAF
AR II	3671.010	3	
AR	3673.260	5	
AR I	3675.230	1	VAF
AR	3678.270	10	
AR II	3680.060	10	
AR II	3682.560	5	
AR I	3690.896	2	VAF
AR I	3696.510	20	
AR II	3706.940	5	
AR	3709.900	5	
AR II	3713.030	3	
AR	3714.740	3	
AR	3717.170	10	
AR II	3718.210	16	
AR	3720.430	10	
AR II	3724.510	5	
AR	3729.290	200	
AR	3735.490	5	
AR II	3737.890	16	
AR I	3743.760	100	
AR	3746.460	5	
AR	3746.920	5	
AR	3750.500	5	
AR II	3753.530	5	
AR II	3754.060	3	
AR	3756.680	3	
AR	3763.520	10	
AR	3765.270	16	
AR II	3766.130	10	
AR I	3770.369	2	VAF
AR	3770.540	16	
AR	3774.540	3	
AR I	3775.450	10	
AR II	3777.550	3	
AR	3780.840	50	
AR I	3781.361	1	VAF

Table IV
Argon Lines
(con't)

Element	Wavelength (Angstrom)	Plasma Intensity	Source
AR	3786.400	16	
AR II	3796.600	5	
AR	3799.390	16	
AR II	3803.190	16	
AR	3808.610	10	
AR	3809.490	25	
AR II	3819.040	5	
AR	3823.290	3	
AR II	3825.700	10	
AR	3826.830	16	
AR II	3830.430	10	
AR I	3834.679	6	VAF
AR	3841.540	5	
AR II	3844.569		NBS
AR	3844.750	5	
AR	3845.420	10	
AR	3850.570	400	
AR II	3855.159		NBS
AR I	3864.260	10	
AR I	3866.280	5	
AR	3868.530		
AR	3872.150	10	
AR	3875.260	25	
AR I	3876.070	10	
AR	3880.340	5	
AR	3891.400	16	
AR	3891.970	25	
AR I	3894.660	1	VAF
AR II	3895.260	3	
AR I	3899.860	1	VAF
AR	3900.630	10	
AR	3911.580	10	
AR	3914.760	25	
AR	3917.770	3	
AR II	3922.360		NBS
AR II	3925.710	3	
AR	3928.620	130	
AR	3931.240	16	
AR	3932.550	25	
AR	3944.270	50	

Table IV
Argon Lines
(con't)

Element	Wavelength (Angstrom)	Plasma Intensity	Source
AR II	3946.100	25	
AR I	3947.504	8	VAF
AR I	3948.979	27	VAF
AR	3952.740	16	
AR	3958.390	10	
AR	3968.360	200	
AR	3974.480	10	
AR	3974.760	16	
AR	3979.360	25	
AR I	3979.710	10	
AR	3988.180	5	
AR	3992.060	25	
AR II	3994.810	10	
AR	4011.230	5	
AR	4013.870	200	
AR	4031.410	2	
AR I	4032.970	20	
AR	4033.830	32	
AR II	4035.470	32	
AR	4038.820	40	
AR II	4042.910	79	
AR I	4044.418	31	VAF
AR I	4045.966	3	VAF
AR II	4047.510	2	
AR II	4052.940	20	
AR I	4054.525	2	VAF
AR II	4057.669		NBS
AR II	4060.599		NBS
AR II	4065.109		NBS
AR II	4072.010	160	
AR	4072.400	40	
AR	4076.640	20	
AR	4076.960	10	
AR II	4079.600	20	
AR	4080.670	10	
AR	4082.400	32	
AR II	4097.150	5	
AR II	4099.470	5	
AR	4103.910	200	
AR	4112.830	20	

Table IV
Argon Lines
(con't)

Element	Wavelength (Angstrom)	Plasma Intensity	Source
AR II	4114.520	2	
AR II	4116.390	10	
AR	4128.650	20	
AR	4129.700	10	
AR II	4131.730	79	
AR II	4147.379		NBS
AR	4152.540	20	
AR	4154.500	79	
AR	4156.110	20	
AR I	4158.590	50	VAF
AR I	4164.179	17	VAF
AR	4168.410	2	
AR	4168.700	2	
AR	4168.980	5	
AR	4175.400	10	
AR	4176.330	20	
AR	4178.390	20	
AR	4179.310	20	
AR I	4181.883	28	VAF
AR	4189.670	10	
AR I	4190.712	32	VAF
AR I	4191.028	32	VAF
AR I	4198.317	50	VAF
AR II	4199.930	5	
AR I	4200.675	50	VAF
AR II	4201.580	2	
AR	4201.990	20	
AR	4203.430	20	
AR	4217.450	10	
AR	4218.690	20	
AR	4222.670	20	
AR II	4226.650	2	
AR II	4226.989		NBS
AR II	4227.020	10	
AR	4228.140	40	
AR	4229.890	10	
AR II	4237.230	40	
AR	4243.570	20	
AR II	4243.710	2	
AR	4249.370	20	

Table IV
Argon Lines
(con't)

Element	Wavelength (Angstrom)	Plasma Intensity	Source
AR	4250.410	2	
AR I	4251.185	8	VAF
AR	4254.950	10	
AR	4255.620	5	
AR	4258.590	5	
AR I	4259.361	50	VAF
AR	4265.520	2	
AR I	4266.286	32	VAF
AR	4266.530	200	
AR	4271.240	2	
AR I	4272.168	43	VAF
AR	4275.190	10	
AR II	4277.550	79	
AR	4282.900	40	
AR	4289.090	5	
AR	4294.970	20	
AR II	4297.990	20	
AR	4299.240	5	
AR I	4300.100	40	VAF
AR II	4300.660	32	
AR II	4309.110	2	
AR II	4309.250	5	
AR	4310.470	20	
AR	4331.250	200	
AR	4332.060	79	
AR I	4333.560	38	VAF
AR I	4335.337	11	VAF
AR	4336.510	5	
AR II	4337.100	32	
AR I	4345.167	9	VAF
AR	4348.110	500	
AR	4352.230	32	
AR II	4359.670	2	
AR II	4362.070	20	
AR I	4363.794	1	VAF
AR II	4367.870	10	
AR	4368.360	3	VAF
AR II	4370.760	32	
AR	4371.360	79	
AR	4374.870	5	

Table IV
Argon Lines
(con't)

Element	Wavelength (Angstrom)	Plasma Intensity	Source
AR	4375.960	20	
AR	4379.740	79	
AR	4383.790	10	
AR I	4385.080	5	
AR II	4394.650	2	
AR	4400.090	32	
AR	4401.020	40	
AR	4404.910	2	
AR	4420.900	40	
AR I	4423.994	1	VAF
AR	4426.010	320	
AR	4430.180	100	
AR	4431.020	79	
AR II	4433.830	20	
AR II	4439.480	2	
AR II	4440.090	5	
AR I	4445.840	5	
AR II	4448.880	16	
AR	4456.610	2	
AR	4460.530	20	
AR	4461.460	5	
AR II	4474.770	20	
AR I	4479.310	5	
AR II	4481.830	79	
AR II	4490.990	20	
AR II	4498.550	10	
AR	4502.950	20	
AR	4505.160	2	
AR I	4507.450	2	
AR	4509.870	2	
AR I	4510.733	21	VAF
AR I	4522.323	6	VAF
AR II	4530.570	10	
AR	4534.780	20	
AR	4535.510	10	
AR II	4537.670	10	
AR I	4541.600	20	
AR	4544.746	32	
AR	4545.080	200	
AR	4547.780	20	

Table IV
Argon Lines
(con't)

Element	Wavelength (Angstrom)	Plasma Intensity	Source
AR I	4554.319	16	
AR II	4561.030	10	
AR	4563.780	20	
AR	4564.430	20	
AR	4564.820	2	
AR I	4569.690	2	
AR	4579.390	79	
AR I	4584.958	10	
AR I	4586.610	10	
AR I	4587.210	5	
AR	4587.900	2	
AR I	4589.288	79	
AR II	4589.930	160	
AR II	4593.440	2	
AR I	4596.097	6	VAF
AR II	4598.770	20	
AR II	4609.600	320	
AR II	4611.250	5	
AR I	4625.460	10	
AR I	4626.780	32	
AR I	4628.441	4	VAF
AR II	4637.250	32	
AR I	4642.148	79	
AR I	4647.493	40	
AR I	4651.388	20	
AR	4657.940	160	
AR II	4666.280	2	
AR II	4682.290	10	
AR I	4702.316	7	VAF
AR II	4703.360	10	
AR I	4708.460	2	
AR I	4709.080	10	
AR I	4709.500	32	
AR I	4718.100	2	
AR I	4719.220	2	
AR I	4719.940	20	
AR	4721.620	10	
AR	4724.100	5	
AR	4726.910	200	
AR	4727.480	5	

Table IV
Argon Lines
(con't)

Element	Wavelength (Angstrom)	Plasma Intensity	Source
AR I	4730.660	5	
AR II	4730.660	5	
AR II	4732.080	5	
AR	4735.930	400	
AR I	4746.823	79	
AR I	4748.230	5	
AR I	4752.938	160	
AR	4764.890	160	
AR I	4768.672	160	
AR I	4770.340	2	
AR	4786.190	2	
AR I	4791.150	2	
AR II	4792.120	20	
AR I	4798.742	32	
AR I	4804.330	5	
AR	4806.070	500	
AR I	4825.970	2	
AR I	4832.380	5	
AR I	4832.790	5	
AR I	4834.100	32	
AR I	4835.970	32	
AR I	4836.691	160	
AR I	4846.730	5	
AR	4847.900	79	
AR	4859.440	5	
AR I	4865.910	10	
AR II	4867.590	5	
AR I	4867.840	10	
AR I	4872.730	10	
AR I	4876.260	200	
AR	4878.370	2	
AR	4879.900	320	
AR	4882.250	10	
AR I	4883.270	32	
AR I	4883.860	5	
AR I	4886.290	32	
AR I	4887.947	200	
AR	4889.060	3	
AR I	4894.692	160	
AR	4901.260	2	

Table IV
Argon Lines
(con't)

Element	Wavelength (Angstrom)	Plasma Intensity	Source
AR II	4904.750	32	
AR I	4908.520	10	
AR II	4914.320	2	
AR I	4917.850	5	
AR I	4921.042	79	
AR I	4929.160	2	
AR	4933.240	32	
AR I	4937.718	32	
AR II	4942.960	5	
AR I	4944.800	5	
AR II	4949.450	2	
AR I	4951.750	10	
AR II	4952.919		NBS
AR I	4955.210	2	
AR I	4956.753	100	
AR	4965.120	40	
AR	4972.160	20	
AR	4973.530	2	
AR I	4974.180	10	
AR I	4975.660	2	
AR I	4985.090	10	
AR I	4989.945	79	
AR	5004.318	20	
AR	5006.840	2	
AR I	5007.090	2	
AR	5009.350	200	
AR II	5017.160	63	
AR I	5017.250	5	
AR II	5017.630	5	
AR	5029.640	5	
AR I	5032.025	63	
AR I	5034.250	10	
AR I	5035.880	5	
AR I	5040.510	10	
AR I	5041.230	10	
AR I	5044.150	2	
AR I	5047.300	2	
AR I	5048.813	500	
AR I	5054.180	320	
AR I	5056.530		

Table IV
Argon Lines
(con't)

Element	Wavelength (Angstrom)	Plasma Intensity	Source
AR I	5060.080	500	
AR	5062.070	200	
AR I	5063.990	5	
AR I	5065.480	5	
AR I	5068.390	5	
AR I	5069.660	5	
AR I	5070.990	40	
AR I	5071.300	5	
AR I	5073.080	200	
AR I	5078.030	40	
AR I	5081.440	10	
AR I	5082.740	20	
AR I	5087.087	63	
AR II	5090.550	5	
AR I	5093.320	10	
AR	5098.970	20	
AR I	5099.640	5	
AR I	5104.740	20	
AR I	5118.200	63	
AR I	5121.880	5	
AR I	5127.780	63	
AR I	5134.170	2	
AR I	5141.810	20	
AR II	5141.810	20	
AR	5145.360	200	
AR I	5151.395	200	
AR I	5152.300		NBS
AR I	5153.110	20	
AR I	5159.690	10	
AR I	5162.284	500	
AR II	5162.800	2	
AR II	5165.820	20	
AR II	5176.280	10	
AR I	5177.535	40	
AR I	5187.746	790	
AR I	5192.720	63	
AR I	5194.020	5	
AR I	5194.770	20	
AR I	5198.960	2	
AR I	5205.790	10	

Table IV
Argon Lines
(con't)

Element	Wavelength (Angstrom)	Plasma Intensity	Source
AR I	5207.170	10	
AR I	5208.040	10	
AR I	5210.488	200	
AR I	5214.768	200	
AR II	5215.109		NBS
AR I	5216.280	63	
AR II	5216.840	10	
AR I	5219.300	40	
AR I	5221.270	500	
AR I	5222.900	20	
AR I	5229.860	40	
AR I	5234.740	5	
AR I	5236.210	20	
AR I	5239.710	2	
AR I	5241.096	63	
AR I	5242.130	2	
AR I	5246.240	40	
AR I	5246.760	5	
AR I	5249.200	40	
AR I	5252.786	320	
AR I	5254.476	63	
AR	5263.020	2	
AR I	5267.480	2	
AR I	5279.050	20	
AR	5280.400	63	
AR	5281.660	2	
AR I	5283.430	20	
AR I	5286.080	63	
AR	5286.920	20	
AR I	5290.000	20	
AR I	5296.320	5	
AR II	5305.770	10	
AR I	5309.517	200	
AR I	5317.726	63	
AR I	5324.800	5	
AR I	5328.020	20	
AR I	5341.780	10	
AR I	5344.280	5	
AR I	5345.810	20	
AR I	5347.412	200	

Table IV
Argon Lines
(con't)

Element	Wavelength (Angstrom)	Plasma Intensity	Source
AR I	5350.580	20	
AR	5353.460	20	
AR	5356.490	10	
AR I	5369.970	5	
AR I	5373.493	500	
AR I	5387.370	40	
AR I	5389.100	40	
AR I	5390.720	40	
AR I	5393.269		NBS
AR I	5393.971	200	
AR	5397.600	10	
AR I	5399.010	20	
AR	5402.690	10	
AR	5407.440	20	
AR I	5410.470	500	
AR I	5413.320	10	
AR I	5417.220	10	
AR I	5421.346	500	
AR I	5422.550	2	
AR I	5429.690	20	
AR I	5430.270	10	
AR I	5439.970	500	
AR I	5442.220	500	
AR I	5443.210	100	
AR I	5443.880	20	
AR I	5448.610	10	
AR I	5449.310	5	
AR I	5451.650	500	
AR	5454.410	10	
AR I	5456.010	5	
AR I	5457.370	200	
AR I	5457.419		NBS
AR I	5457.750	10	
AR I	5459.610	20	
AR I	5467.130	6	
AR	5469.200	2	
AR	5469.650	20	
AR I	5473.440	500	
AR I	5483.320	10	
AR I	5486.470	20	

Table IV
Argon Lines
(con't)

Element	Wavelength (Angstrom)	Plasma Intensity	Source
AR I	5488.460	2	
AR I	5490.130	63	
AR I	5492.060	40	
AR I	5493.490	20	
AR I	5495.872	1000	
AR	5498.240	5	
AR I	5499.000	10	
AR	5500.380	2	
AR I	5505.180	10	
AR I	5506.112	500	
AR I	5507.630	10	
AR	5514.450	2	
AR I	5518.200	5	
AR I	5523.700	5	
AR I	5524.930	320	
AR I	5528.930	40	
AR I	5534.450	63	
AR	5537.390	2	
AR I	5540.900	40	
AR I	5541.460	2	
AR I	5542.730	2	
AR	5545.080	2	
AR I	5552.760	10	
AR I	5553.400	2	
AR	5554.070	5	
AR I	5558.702	500	
AR I	5559.620	200	
AR I	5560.220	10	
AR I	5565.960	5	
AR I	5572.548	500	
AR I	5574.200	5	
AR II	5577.700	5	
AR	5578.560	5	
AR	5580.950	2	
AR I	5581.830	63	
AR I	5588.690	500	
AR	5591.750	5	
AR I	5597.460	500	
AR I	5598.500	20	
AR I	5600.430	40	

Table IV
Argon Lines
(con't)

Element	Wavelength (Angstrom)	Plasma Intensity	Source
AR I	5601.080	63	
AR I	5601.850	2	
AR I	5604.360	20	
AR I	5605.250	5	
AR I	5606.732	500	
AR I	5608.900	20	
AR I	5611.350	20	
AR I	5617.970	63	
AR I	5618.009		NBS
AR	5619.000	5	
AR I	5620.660	2	
AR I	5620.890	63	
AR I	5623.760	63	
AR I	5630.440	10	
AR I	5635.540	63	
AR	5635.910	2	
AR I	5637.290	20	
AR I	5639.110	100	
AR I	5641.340	63	
AR I	5648.660	200	
AR I	5649.259		NBS
AR I	5650.703	1600	
AR	5654.480	5	
AR I	5659.130	500	
AR I	5662.000	5	
AR I	5665.820	5	
AR	5673.010	2	
AR I	5681.900	500	
AR I	5683.730	40	
AR I	5687.400	20	
AR I	5689.640	200	
AR I	5689.910	200	
AR II	5691.710	2	
AR I	5700.860	63	
AR I	5712.509		NBS
AR II	5724.329		NBS
AR I	5737.960	5	
AR I	5738.400	20	
AR I	5739.517	500	
AR I	5747.180	2	

Table IV
Argon Lines
(con't)

Element	Wavelength (Angstrom)	Plasma Intensity	Source
AR I	5758.840	5	
AR II	5772.116	100	
AR I	5774.000	40	
AR I	5781.550	2	
AR I	5783.520	40	
AR I	5789.480	20	
AR I	5790.390	5	
AR I	5802.082	40	
AR I	5834.263	63	
AR I	5843.740	2	
AR I	5860.315	63	
AR I	5870.260	2	
AR I	5882.625	100	
AR I	5888.592	320	
AR I	5912.084	500	
AR I	5916.580	5	
AR I	5927.130	10	
AR I	5928.805	200	
AR I	5940.860	2	
AR I	5942.668	40	
AR I	5943.890	2	
AR I	5949.260	10	
AR II	5950.909		NBS
AR	5964.460	2	
AR I	5968.310	2	
AR I	5971.590	5	
AR I	5981.900	5	
AR I	5987.289	40	
AR I	5988.110	2	
AR I	5994.660	2	
AR I	5999.000	20	
AR I	6005.740	4	
AR I	6013.680	6	
AR I	6025.140	10	
AR I	6032.124	63	
AR I	6043.230	100	
AR I	6052.721	32	
AR I	6059.373	100	
AR I	6064.750	6	
AR II	6077.429		NBS

Table IV
Argon Lines
(con't)

Element	Wavelength (Angstrom)	Plasma Intensity	Source
AR I	6081.230	4	
AR I	6085.860	2	
AR I	6090.760	10	
AR I	6098.807	63	
AR I	6101.160	6	
AR II	6103.560	8	
AR I	6104.600	6	
AR I	6105.645	63	
AR I	6113.470	8	
AR II	6114.920	100	
AR I	6119.670	2	
AR I	6121.859		NBS
AR II	6123.380	6	
AR I	6127.380	16	
AR I	6128.710	8	
AR II	6138.670	6	
AR I	6145.430	100	
AR I	6155.230	63	
AR I	6165.110	8	
AR I	6170.183	100	
AR	6172.280	20	
AR I	6173.106	100	
AR I	6179.410	4	
AR I	6212.507	100	
AR I	6215.945	63	
AR I	6222.610	4	
AR I	6230.910	4	
AR II	6239.730	4	
AR II	6243.130	16	
AR I	6243.390	6	
AR I	6244.729		NBS
AR I	6248.400	16	
AR I	6278.640	6	
AR I	6296.876	20	
AR I	6307.662	32	
AR I	6309.140	8	
AR	6324.450	6	
AR	6332.510	4	
AR	6333.210	4	
AR	6348.270	2	

Table IV
Argon Lines
(con't)

Element	Wavelength (Angstrom)	Plasma Intensity	Source
AR I	6349.200	2	
AR I	6364.890	20	
AR I	6369.577	32	
AR I	6384.719	100	
AR	6396.630	2	
AR II	6399.230	8	
AR	6403.100	2	
AR I	6416.315	100	
AR	6418.430	4	
AR	6422.940	5	
AR I	6431.570	16	
AR II	6437.630	4	
AR	6441.950	6	
AR	6443.890	4	
AR I	6466.560	20	
AR	6468.080	4	
AR	6472.470	4	
AR I	6481.150	8	
AR II	6483.100	16	
AR I	6493.970	16	
AR I	6499.100	6	
AR II	6500.250	6	
AR I	6513.840	8	
AR I	6538.115	32	
AR	6545.110	2	
AR I	6571.370	2	
AR I	6581.600	2	
AR I	6594.660	2	
AR I	6596.100	8	
AR I	6598.660	6	
AR I	6604.020	2	
AR I	6604.853	32	
AR	6621.010	4	
AR I	6623.780	4	
AR I	6632.040	8	
AR II	6638.240	32	
AR II	6639.720	20	
AR II	6643.790	100	
AR I	6656.880	6	
AR I	6660.640	100	

Table IV
Argon Lines
(con't)

Element	Wavelength (Angstrom)	Plasma Intensity	Source
AR I	6664.020	100	
AR II	6666.360	10	
AR I	6668.270	8	
AR I	6672.100	2	
AR I	6677.282	32	
AR II	6684.360	32	
AR I	6684.730	6	
AR I	6689.910	2	
AR I	6698.450	6	
AR I	6698.850	100	
AR I	6719.200	100	
AR I	6722.900	4	
AR I	6752.832	200	
AR I	6754.300	8	
AR I	6756.100	100	
AR II	6756.610	10	
AR I	6766.560	100	
AR I	6779.850	4	
AR II	6799.320	4	
AR II	6808.550	6	
AR I	6818.260	4	
AR II	6818.390	8	
AR I	6827.240	32	
AR I	6851.860	4	
AR II	6861.300	8	
AR II	6863.520	16	
AR I	6871.290	160	
AR I	6879.590	40	
AR II	6886.570	16	
AR I	6887.100	20	
AR I	6888.170	100	
AR I	6925.009		NBS
AR I	6937.666	100	
AR I	6951.460	20	
AR I	6960.230	20	
AR I	6965.430	400	
AR II	6990.160	4	
AR I	6992.170	4	
AR I	7030.262	100	
AR II	7055.010	4	

Table IV
Argon Lines
(con't)

Element	Wavelength (Angstrom)	Plasma Intensity	Source
AR I	7067.217	400	
AR I	7068.730	32	
AR II	7077.030	2	
AR I	7086.700	16	
AR I	7107.496	200	
AR I	7125.800	32	
AR I	7147.041	32	
AR I	7158.830	32	
AR I	7162.570	8	
AR I	7176.340	4	
AR I	7202.550	2	
AR I	7206.986	100	
AR I	7229.930	4	
AR II	7233.580	8	
AR I	7265.230	3	
AR I	7267.200	2	
AR I	7270.660	10	
AR I	7272.936	100	
AR I	7285.440	6	
AR I	7311.710	100	
AR I	7316.000	32	
AR II	7348.110	4	
AR I	7350.780	6	
AR I	7353.179		NBS
AR I	7353.316	100	
AR I	7372.118	100	
AR II	7380.450	8	
AR I	7383.980	400	
AR I	7392.970	16	
AR I	7412.310	16	
AR I	7422.260	6	
AR I	7425.240	13	
AR I	7435.330	32	
AR I	7436.250	10	
AR II	7440.460	100	
AR I	7443.400	25	
AR I	7471.180	4	
AR I	7484.240	16	
AR I	7491.880	2	
AR I	7503.867	630	

Table IV
Argon Lines
(con't)

Element	Wavelength (Angstrom)	Plasma Intensity	Source
AR II	7505.130	100	
AR I	7510.420	10	
AR I	7514.651	200	
AR II	7589.330	250	
AR II	7618.030	79	
AR I	7618.330	32	
AR I	7628.860	50	
AR I	7635.105	500	
AR II	7654.060	6	
AR I	7662.300	2	
AR I	7667.030	4	
AR I	7670.040	50	
AR II	7683.490	2	
AR	7689.360	10	
AR I	7690.100	2	
AR I	7704.810	20	
AR I	7723.760	200	
AR I	7724.206	200	
AR I	7798.550	32	
AR I	7814.330	10	
AR II	7849.420	16	
AR I	7855.730	8	
AR I	7860.440	2	
AR I	7861.910	16	
AR I	7868.200	40	
AR I	7891.075	100	
AR I	7910.230	4	
AR I	7916.450	20	
AR I	7948.175	400	
AR I	7956.990	10	
AR I	7960.840	2	
AR I	7965.080	3	
AR I	8006.156	630	
AR I	8014.786	790	
AR I	8016.739		NBS
AR II	8017.550	63	
AR I	8021.900	2	
AR I	8037.230	20	
AR I	8046.130	50	
AR I	8053.307	100	

Table IV
Argon Lines
(con't)

Element	Wavelength (Angstrom)	Plasma Intensity	Source
AR I	8066.600	20	
AR I	8079.680	20	
AR I	8082.930	5	
AR II	8086.140	2	
AR I	8094.060	20	
AR I	8103.692	2000	
AR I	8115.311	5000	
AR II	8119.180	50	
AR	8127.300	2	
AR I	8143.500	10	
AR I	8151.860	3	
AR	8166.210	2	
AR I	8171.950	10	
AR	8178.840	40	
AR I	8178.960	20	
AR I	8203.419		NBS
AR II	8217.850	3	
AR I	8224.720	6	
AR I	8249.580	4	
AR I	8255.070	50	
AR I	8264.521	1000	
AR I	8291.880	8	
AR I	8332.210	20	
AR I	8353.500	4	
AR I	8367.030	3	
AR I	8384.730	63	
AR I	8392.280	79	
AR I	8399.350	20	
AR I	8408.208	2000	
AR I	8424.647	2000	
AR I	8437.710	6	
AR I	8443.440	20	
AR I	8490.300	40	
AR I	8496.640	2	
AR I	8521.441	2000	
AR I	8561.380	3	
AR I	8578.060	5	
AR I	8579.490	4	
AR II	8604.040	8	
AR I	8605.780	160	

Table IV
Argon Lines
(con't)

Element	Wavelength (Angstrom)	Plasma Intensity	Source
AR I	8620.470	100	
AR I	8667.943	400	
AR I	8678.430	63	
AR I	8690.120	2	
AR I	8700.950	3	
AR I	8713.790	5	
AR I	8736.190	2	
AR I	8736.630	20	
AR I	8739.510	3	
AR I	8761.720	200	
AR II	8771.880	100	
AR I	8784.590	32	
AR I	8799.130	100	
AR	8805.160	3	
AR I	8840.390	3	
AR I	8840.820	20	
AR I	8849.970	160	
AR I	8874.840	4	
AR II	8904.530	2	
AR II	8926.070	3	
AR I	8962.190	40	
AR I	8964.480	10	
AR I	8967.390	2	
AR I	8970.980	2	
AR I	8988.200	3	
AR I	8994.090	10	
AR II	9017.590	50	
AR I	9057.230	4	
AR I	9057.510	2	
AR I	9066.770	40	
AR I	9073.340	50	
AR I	9075.420	63	
AR I	9122.966	500	
AR II	9138.450	2	
AR II	9150.770	2	
AR I	9180.170	6	
AR I	9194.680	160	
AR	9198.610	50	
AR II	9210.370	2	
AR I	9221.080	5	

Table IV
Argon Lines
(con't)

Element	Wavelength (Angstrom)	Plasma Intensity	Source
AR I	9224.498	1000	
AR II	9279.720	20	
AR I	9291.580	100	
AR	9322.840	2	
AR I	9328.080	2	
AR I	9334.800	8	
AR I	9340.590	3	
AR I	9354.218	200	
AR I	9362.500	4	
AR II	9374.150	10	
AR I	9377.630	5	
AR I	9402.690	20	
AR I	9408.660	3	
AR II	9418.570	2	
AR I	9446.570	2	
AR I	9459.090	100	
AR II	9475.200	32	
AR I	9478.390	50	
AR I	9486.020	2	
AR I	9547.730	2	
AR I	9555.200	4	
AR I	9561.600	5	
AR I	9595.090	4	
AR I	9657.784	1600	
AR I	9666.860	50	
AR I	9673.390	6	
AR I	9677.800	8	
AR I	9784.501	1000	
AR I	9800.920	4	
AR II	9829.860	4	
AR I	9882.180	6	
AR II	9906.120	100	
AR I	9951.880	20	